Advanced Structured Materials

Volume 186

Series Editors

Common engineering materials are reaching their limits in many applications, and new developments are required to meet the increasing demands on engineering materials. The performance of materials can be improved by combining different materials to achieve better properties than with a single constituent, or by shaping the material or constituents into a specific structure. The interaction between material and structure can occur at different length scales, such as the micro, meso, or macro scale, and offers potential applications in very different fields.

This book series addresses the fundamental relationships between materials and their structure on overall properties (e.g., mechanical, thermal, chemical, electrical, or magnetic properties, etc.). Experimental data and procedures are presented, as well as methods for modeling structures and materials using numerical and analytical approaches. In addition, the series shows how these materials engineering and design processes are implemented and how new technologies can be used to optimize materials and processes.

Advanced Structured Materials is indexed in Google Scholar and Scopus.

Holm Altenbach · Victor A. Eremeyev ·
Leonid A. Igumnov · Anatoly Bragov
Editors

Deformation and Destruction of Materials and Structures Under Quasi-static and Impulse Loading

Springer

Editors
Holm Altenbach
Fakultät für Maschinenbau
Institut für Mechanik
Otto-von-Guericke-Universität
Magdeburg, Germany

Victor A. Eremeyev
Dipartimento di Ingegneria civile,
ambientale e architettura
Università degli Studi di Cagliari
Cagliari, Italy

Leonid A. Igumnov
National Research Lobachevsky State
University
Nizhny Novgorod, Russia

Anatoly Bragov
National Research Lobachevsky State
University
Nizhny Novgorod, Russia

ISSN 1869-8433 ISSN 1869-8441 (electronic)
Advanced Structured Materials
ISBN 978-3-031-22092-0 ISBN 978-3-031-22093-7 (eBook)
https://doi.org/10.1007/978-3-031-22093-7

This Springer imprint is published by the registered company Springer Nature Switzerland AG
The registered company address is: Gewerbestrasse 11, 6330 Cham, Switzerland

Preface

This book is dedicated to the 100th birthday of Andrey Grigorievich Ugodchikov, who was one of the leading specialists in mechanics of the Soviet Union and Russia, an expert in the field of apparatus and machine strength. A talented scientist and engineer, teacher and organizer A. G. Ugodchikov had many contributions to both the development of mechanics in general and applied problems of strength in particular as well as to the education of scientific and academic staff. He created in Nizhny Novgorod a well-known in Russia and abroad multidisciplinary scientific school of mechanics on theoretical and applied problems of strength of machine-building structures under multifactor effects of high parameters.

A. G. Ugodchikov performed a great scientific and organizational activity being the rector of Gorky (now Nizhny Novgorod) University from 1969 to 1988. He laid foundations for the successful development of the university, its material and scientific base, for the growth of high-qualified scientific and academic staff, and for broadening the horizons of research work thus allowing the university to confidently occupy its place among the key national research universities in Russia.

Research activities of A. G. Ugodchikov began with the methods of the theory of functions of a complex variable for solving a fairly wide range of problems in the plane theory of elasticity—Muskhelishvili's method of complex potentials. This elegant method allows one to find a solution to the formulated problem for the given boundary shape and boundary conditions when there is a function that conformally maps a circle onto a simply connected domain occupied by an elastic medium. A. G. Ugodchikov managed to solve complex applied problems on the stress-strain state of diesel engine components. He kept lifespan interest to the development of the methods of complex potentials and generalized analytical functions for solving applied problems. His first investigations had already necessitated computer modeling of problems in the mechanics of deformable solid. This approach became for him and his students a specific feature of the emerging school of specialists in strength mechanics in Nizhny Novgorod.

The research area of A. G. Ugodchikov expanded during his long creative life and the results of his research work are noticeable in the following scientific directions

- Theoretical and experimental foundations of quasi-static deformation of elastic-plastic viscoplastic materials and structural elements made of them,
- Experimental studies of material deformation and fracture under quasi-static and impulsive loading conditions,
- Theoretical and numerical methods for studying the stress-strain state of elasto-plastic plates and shells under dynamic, pulsed loading and their interaction with other media,
- Theoretical foundations of methods for optimizing deformable systems and means of their numerical implementation,
- Theoretical foundations of deformation and fracture of solid deformable bodies under pulsed thermal and force loads of high parameters,
- Theoretical foundations and numerical methods for studying the kinetics of stress-strain state of deformable bodies and structural units under the effects of highly parametric interdependent fields of various nature.

As organizer, A. G. Ugodchikov managed to found in Nizhny Novgorod a specialized university department, a research institute and prepare numerous followers and students who formed the basis of university teams.

Below are given a list of selected publications of Andrey G. Ugochikov

1. Ugodchikov A. G. Stress determination when pressing several round washers into a plate bounded by a curve of a particular type, Dokl. Akad. Nauk SSSR. 1951. Vol. 77, No. 2. pp. 213–216 (in Russian)
2. Ugodchikov A. G. On calculating mounting stresses in machine-building parts, Applied Mechanics. 1957. Vol. 3, No. 3. pp. 202–207 (in Russian)
3. Ugodchikov A. G. To the solution of the first basic problem of the plane theory of elasticity for doubly-connected domains, Dokl. Akad. Nauk of the Ukrainian SSR. 1960. No. 11. pp. 1480–1484 (in Russian)
4. Ugodchikov A. G. To the solution of the generalized biharmonic problem of the plane theory of elasticity for double-connected domains. Dokl. Akad. Nauk of the Ukrainian SSR. 1961. No. 11. pp. 1440–1443 (in Russian)
5. Ugodchikov A. G. Construction of conformally mapping functions. Kiev: Naukova dumka. 1966 (in Russian)
6. Ugodchikov A. G., Dlugach M. I., Stepanov A. E. Solution of boundary value problems of the plane theory of elasticity on digital and analog machines. Moscow: Vysshaya shkola, 1970 (in Russian)
7. Ugodnikov A. G., Khutoryansky N. M. The boundary element method in mechanics of deformable solids. Kazan: KSU Publishing House, 1986, 296 p. (in Russian)
8. Ugodchikov A. G. On equations of the dynamics of a deformable solid body, Dokl. Akad. Nauk SSSR. 1991. Vol. 317, No. 4 (in Russian)
9. Ugodchikov A. G. Moment dynamics of a linear elastic body. Dokl. RAS, 1995, Vol. 340, No. 1 (in Russian)
10. Malkov V. P., Ugodchikov A. G. Optimization of elastic systems. Moscow: Nauka, 1981 (in Russian)

11. Ugodchikov A. G., Korotkikh Yu. G. Some methods for solving physically nonlinear problems in the theory of plates and shells on EDC. Kiev: Naukova Dumka, 1971 (in Russian)
12. Bazhenov V. G., Kochetkov A. V., Mikhailov G. S., Ugodchikov A. G. The Interaction of elastoplastic thin-walled structural elements with shock waves in ideal compressible media, Izv. USSR Academy of Sciences. MTT. 1979. No. 2. pp. 141–149 (in Russian)
13. Bazhenov V. G., Kochetkov A. V., Krylov, S. V., Ugodchikov A. G. High-speed impact of elasto-plastic thin-walled structures on compressible fluid surface, Izv. USSR Academy of Sciences. MTT. 1984. Vol. 5, pp. 161–169 (in Russian)
14. Ugodchikov A. G., Short, J. G. Some solution methods for EDC physically nonlinear problems of the plates and shells theory. Kiev: Naukova dumka, 1971 (in Russian)
15. Ugodchikov A. G., Korotkikh Yu. G. Equations of state under low-cycle loading (Chap. 6: Equations of the thermoviscoplasticity theory with combined hardening). Moscow: Nauka, 1981. pp. 129–167 (in Russian)
16. Ugodchikov A. G., Khutoryansky N. M. The boundary element method in the mechanics of a deformable solid. Kazan: KSU Publishing House, 1986 (in Russian)

Magdeburg, Germany — Holm Altenbach
Cagliari, Italy — Victor A. Eremeyev
Nizhny Novgorod, Russia — Leonid A. Igumnov
Nizhny Novgorod, Russia — Anatoly Bragov
October 2022

Contents

Contributors

Dmitry R. Abashev Department of Technical Mechanics, Moscow Polytechnic University, Moscow, Russian Federation

Sergei M. Aizikovich Don State Technical University, Rostov-on-Don, Russia

Amin Alibakhshi Department of Mechanical Engineering, Science and Research Branch, Islamic Azad University, Tehran, Iran

Svetlana A. Atroshenko Institute for Problems in Mechanical Engineering, Russian Academy of Sciences, Saint Petersburg, Russia;
Saint Petersburg State University, Saint Petersburg, Russia

Olga M. Babeshko Kuban State University, Krasnodar, Russia

Vladimir A. Babeshko Scientific Center, Russian Academy of Sciences, Rostov-On-Don, Russia;
Kuban State University, Krasnodar, Russia

Dmitry V. Balandin Research and Education Mathematical Center, National Research Lobachevsky State University of Nizhny Novgorod, Nizhny Novgorod, Russian Federation

Regina A. Bardakova Don State Technical University, Rostov-on-Don, Russia

Artem V. Basalin National Research Lobachevsky State University of Nizhny Novgorod, Nizhni Novgorod, Russia

Emilia Bazhlekova Institute of Mathematics and Informatics, Bulgarian Academy of Sciences, Acad. G. Bonchev St, Sofia, Bulgaria

Valentin S. Bondar Department of Technical Mechanics, Moscow Polytechnic University, Moscow, Russian Federation

Anatoly M. Bragov National Research Lobachevsky State University of Nizhny Novgorod, Nizhny Novgorod, Russia

Shahriar Dastjerdi Division of Mechanics, Civil Engineering Department, Akdeniz University, Antalya, Turkey

Maria D. Datcheva Institute of Mechanics, Bulgarian Academy of Sciences, Acad. G. Bonchev St, Sofia, Bulgaria

Sergey A. Davydov Moscow Aviation Institute, Moscow, Russian Federation

Nadezhda S. Dyukina National Research Lobachevsky State University of Nizhny Novgorod, Nizhny Novgorod, Russia

Victor A. Eremeyev Department of Mechanics of Materials and Structures, Faculty of Civil and Environmental Engineering, Gdansk University of Technology, Gdansk, Poland;
DICAAR, Università degli Studi di Cagliari, Cagliari, Italy

Olga V. Evdokimova Scientific Center, Russian Academy of Sciences, Rostov-On-Don, Russia

Grigory. V. Fedotenkov Institute of Mechanics, Moscow Aviation Institute, Lomonosov Moscow State University, Moscow, USSR

Elena. S. Feoktistova Moscow Aviation Institute, Moscow, Russia

Vyacheslav A. Firsov Tupolev Kazan National Research Technical University, Kazan, Russia

Ruslan K. Gazizullin Tupolev Kazan National Research Technical University, Kazan, Russia;
N.I. Lobachevsky Institute of Mathematics and Mechanics, Kazan Federal University, Kazan, Russia

Mikhail E. Gonov National Research Lobachevsky State University of Nizhny Novgorod, Nizhny Novgorod, Russia

Leonid A. Igumnov Research Institute for Mechanics, National Research Lobachevsky State University of Nizhny Novgorod, Nizhny Novgorod, Russian Federation

Ivan O. Kharchevnikov Don State Technical University, Rostov-on-Don, Russia

Aleksandr Yu. Konstantinov National Research Lobachevsky State University of Nizhny Novgorod, Nizhny Novgorod, Russia

Ekaterina A. Korovaytseva Institute of Mechanics, Lomonosov Moscow State University, Moscow, Russia

Anzhelika A. Kotova Don State Technical University, Rostov-on-Don, Russia

Andrey K. Lomunov National Research Lobachevsky State University of Nizhny Novgorod, Nizhny Novgorod, Russia

Quoc Chien Mai Moscow Aviation Institute, Moscow, Russian Federation

Mohammad Malikan Department of Mechanics of Materials and Structures, Faculty of Civil and Environmental Engineering, Gdansk University of Technology, Gdansk, Poland

Andrey L. Nikolaev Don State Technical University, Rostov-on-Don, Russia

Anatoly. S. Okonechnikov Moscow Aviation Institute, Moscow, Russia

Vitaly N. Paimushin Tupolev Kazan National Research Technical University, Kazan, Russia;
N.I. Lobachevsky Institute of Mathematics and Mechanics, Kazan Federal University, Kazan, Russia

Egor V. Petrakov National Research Lobachevsky State University of Nizhny Novgorod, Nizhny Novgorod, Russian Federation

Andrey N. Petrov Research Institute for Mechanics, National Research Lobachevsky State University of Nizhny Novgorod, Nizhny Novgorod, Russian Federation

Sergey G. Pshenichnov Institute of Mechanics, Lomonosov Moscow State University, Moscow, Russia

Marina Yu. Ryazantseva Research Institute of Mechanics, Lomonosov Moscow State University, Moscow, Russian Federation

Evgeniy V. Sadyrin Don State Technical University, Rostov-on-Don, Russia

Georgii G. Savenkov Saint Petersburg State Technological Institute (Technical University), Saint Petersburg, Russia;
Physico-Technical Institute named after A.F. Ioffe RAS, Saint Petersburg, Russia

Victor M. Shishkin Vyatka State University, Kirov, Russia

Dmitry V. Tarlakovskii Moscow Aviation Institute, Moscow, Russian Federation; Research Institute of Mechanics, Lomonosov Moscow State University, Moscow, Russian Federation;
N.I. Lobachevsky Institute of Mathematics and Mechanics, Kazan Federal University, Kazan, Russian Federation

Anatoliy V. Vestyak Moscow Aviation Institute, Moscow, Russian Federation

Tatiana N. Yuzhina National Research Lobachevsky State University of Nizhny Novgorod, Nizhny Novgorod, Russia

Igor Yu. Zabiyaka Don State Technical University, Rostov-on-Don, Russia

Andrei V. Zemskov Moscow Aviation Institute, Research Institute of Mechanics Lomonosov Moscow State University, Moscow, Russian Federation

Chapter 1
A Review of Hyperelastic Constitutive Models for Dielectric Elastomers

Amin Alibakhshi, Shahriar Dastjerdi, Mohammad Malikan, and Victor A. Eremeyev

Abstract Dielectric elastomers are smart materials that are essential components in soft systems and structures. The core element of a dielectric elastomer is soft matter, which is mainly rubber-like and elastomeric. These soft materials show a nonlinear behaviour and have a nonlinear strain–stress curve. The best candidates for modelling the nonlinear behaviour of such materials are hyperelastic strain energy functions. Hyperelastic functions have been extensively used for modelling dielectric elastomer smart structures. This review paper introduces hyperelastic constitutive laws for modelling dielectric elastomers. For this purpose, first, a general scheme of hyperelastic models is expressed. Then, some well-known hyperelastic models are introduced. Finally, we review in detail the utilized hyperelastic models for different configurations of dielectric elastomers. Possible future works in this field are outlined eventually.

Keywords Dielectric elastomers · Hyperelastic models · Nonlinear elasticity · Smart polymers

---

A. Alibakhshi
Department of Mechanical Engineering, Science and Research Branch, Islamic Azad University, 1477893855 Tehran, Iran
e-mail: alibakhshiamin@yahoo.com

S. Dastjerdi
Division of Mechanics, Civil Engineering Department, Akdeniz University, Antalya 07058, Turkey
e-mail: dastjerdi_shahriar@yahoo.com

M. Malikan · V. A. Eremeyev (✉)
Department of Mechanics of Materials and Structures, Faculty of Civil and Environmental Engineering, Gdansk University of Technology, 80-233, Gdansk, Poland
e-mail: victor.eremeev@unica.it; victor.eremeev@pg.edu.pl

M. Malikan
e-mail: mohammad.malikan@pg.edu.pl

V. A. Eremeyev
DICAAR, Università degli Studi di Cagliari, Via Marengo, 2, 09123 Cagliari, Italy

H. Altenbach et al. (eds.), *Deformation and Destruction of Materials and Structures Under Quasi-static and Impulse Loading*, Advanced Structured Materials 186,
https://doi.org/10.1007/978-3-031-22093-7_1

1.1 Introduction

For decades, research on various mechanical structures has been a major topic among scientists [1–5]. The literature in this field has mainly focused on conventional materials possessing a physically linear behaviour. However, many other materials in the world have a nonlinear behaviour. Many human tissues and bodies of animals and plants show a nonlinear behaviour and are modelled as soft structures. These soft structures have been the inspiration of new fields, specifically soft robotics [6, 7]. The main components in this field are soft materials [8, 9]. Using soft materials, we can fabricate soft systems and structures. Dielectric elastomers (DEs) [10, 11] have emerged as powerful smart materials that have been widely employed as soft actuators and soft sensors for developing soft materials and structures. The simplest form of a DE includes a soft membrane that is covered with compliant electrodes [12]. The soft membrane is a major part of a DE. Rubber-like and elastomeric materials with material nonlinearity are used as soft membranes in DEs. To induce deformation in DEs, it is necessary to apply electromechanical loading to them [13, 14]. Generally, a potential difference (voltage) and tensile mechanical load are utilized to induce deformation in DEs. For this reason, DEs are considered smart electromechanical systems.

In response to electromechanical loading, DEs deform nonlinearly and encounter nonlinear oscillation and vibration [15]. From mechanical and vibrational points of view, their potential energy should be calculated for modelling the deformation of DEs. Because DEs consist of nonlinear materials as the core element, linear elasticity cannot be employed for calculating their potential energy. The best alternative to linear elasticity is nonlinear elasticity. Usually, the nonlinear elasticity of DE is captured using hyperelastic constitutive laws [16, 17]. Hyperelastic models have been extensively employed for modelling static and dynamic responses of DEs. There is a large volume of published studies describing the nonlinear response of different types of DEs based on hyperelastic models [18–21]. As observed from prior studies, the knowledge of hyperelastic models is essential for the accurate modelling of DEs. This paper attempts to provide a more detailed investigation of hyperelastic models for DEs.

This review paper is organized as follows. First, in Sect. 1.2, the theory of DEs is described in brief. Then, in Sect. 1.3, the general formulation of hyperelasticity is expressed. Next, in Sect. 1.4, the available hyperelastic models for diverse geometries of DEs are reviewed. Finally, in Sect. 1.5, the main conclusions and perspectives for hyperelastic models and nonlinear elasticity in DEs are expressed.

1.2 Theory of Dielectric Elastomers

The basic theory of dielectric elastomers (DEs) originates from the theory of *electro-elasticity* [22]. Many researchers have tried to develop and review the *electro-elasticity* theory. For instance, in [23, 24], authors developed and reviewed some boundary-value problems of *electro-elasticity* by using the continuum mechanics notation and finite strain theory. A full discussion of *electro-elasticity* lies beyond the scope of this study, but the above-stated literature was introduced for more information.

In line with electro-elasticity, Suo [25] introduced the *theory* of dielectric elastomers. The basic equations for electromechanical large deformations of DEs have been formulated and extended. In another paper, Zhao and Suo [26] considered the electro-elasticity of DEs and discussed the electrical and mechanical equations.

Based on the studies mentioned above, in general, a DE consists of a membrane sandwiched between two compliant electrodes. The membrane may take different geometries such as square, rectangular, spherical, tubular, cylindrical, plate-like, and beam-like [27–33]. Tensile mechanical loads are applied to the membrane, and a potential difference (voltage) is applied to the electrodes. The electrodes are located on the top and bottom surfaces of the soft membrane. When the electrodes are connected to the potential difference, one surface gains the positive electric charge, and another surface gains the negative electric charge. These opposite electric charges attract each other and thereby induce large deformation in the DE such that in the thickness direction, the membrane shrinks, and in the in-plane direction, it stretches and expands. During this process, depending on the type of loading, DEs experience different responses and behaviours. The potential difference and tensile load can be time-varying or static. When electrical or mechanical loadings depend upon time, the response of DE becomes complicated, and, in this sense, they undergo nonlinear vibrations (Fig. 1.1).

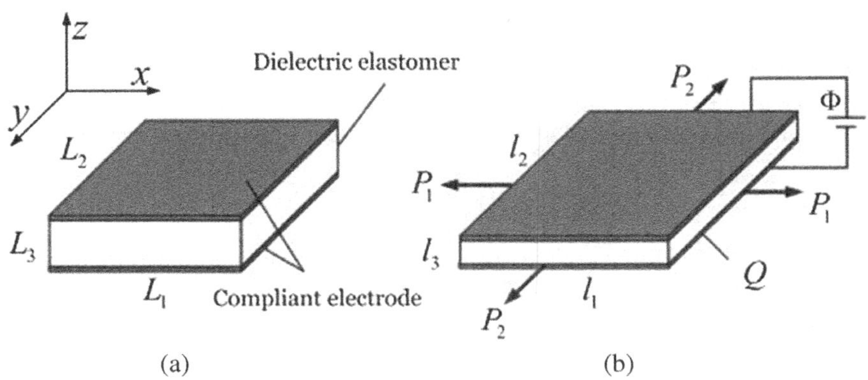

Fig. 1.1 The schematic view of a dielectric elastomer membrane with voltage **Φ** and pressures $\boldsymbol{P}_1$, $\boldsymbol{P}_2$. **a** reference configuration, **b** current configuration [34]

1.3 Hyperelasticity

Dielectric elastomers are soft structures whose structural materials are mainly rubbery and elastomeric. For instance, polydimethylsiloxane, silicone, rubber, and VHB-based elastomers have been extensively used for structural materials of DEs [35]. However, these materials possess an inherent nonlinear trend in their stress–strain curves. Thus, they do not follow linear elasticity and Hooke's law. Therefore, the best candidates to describe the nonlinear elasticity of DE structures are hyperelastic material models. Up to now, diverse hyperelastic models have been employed for DE systems [36, 37], e.g., the neo-Hookean model, Mooney–Rivlin model, Gent model and modified versions of the Gent model, and Ogden model.

1.3.1 General Form of Hyperelasticity

Hyperelastic models are expressed in a functional form called the strain energy function and are mainly denoted by W. The hyperelasticity function may take numerous mathematical forms, such as polynomial, exponential, and logarithmic. The strain energy functions are mainly formulated in terms of material parameters and principal invariants of right or left Cauchy–Green deformation tensors. The material parameters are obtained from empirical tests, and principal invariants are expressed based on the field of deformation and the corresponding deformation gradient. One important difference between hyperelasticity and linear elasticity is that the former requires two configurations to describe the deformation. They are reference configuration (nominal quantity) and current configuration (true quantity). Generally, hyperelasticity is a subfield of finite strain theory, and in this field, tensors play a crucial role in the description of hyperelastic models. Therefore, it seems that a good knowledge of tensors may help researchers in understanding hyperelasticity. Before mathematically speaking of hyperelasticity, we introduce some very important tensors.

The first one and maybe the most important tensor is the deformation gradient tensor. When the solids deform, we should explore how two elements dx and dX change and find their relation. We formulate the deformation gradient tensor as [38]

$$d\vec{x} = \mathbf{F} d\vec{X}, \tag{1.1}$$

where $\mathbf{F}$ is the deformation gradient tensor (material deformation gradient). Depending upon the coordinate system, $\mathbf{F}$ may take different forms.

Based on the deformation gradient tensor, two other important tensors are introduced, and the hyperelastic models are formulated according to these tensors. They are the right and left Cauchy–Green deformation tensors. We formulate these tensors as [39]

$$\mathbf{C} = \mathbf{F}^T\mathbf{F}, \tag{1.2}$$

$$\mathbf{b} = \mathbf{F}\mathbf{F}^T. \tag{1.3}$$

In the above equation, $\mathbf{C}$ is the right Cauchy–Green deformation tensor, and $\mathbf{b}$ stands for the left Cauchy–Green deformation tensor; the letter T is the symbol of the *transpose operation.*

Using the *eigenvalue problem*, we can derive the principal invariants of the right and left Cauchy–Green deformation tensor, which are the main elements of hyperelastic models. The invariants of the right Cauchy–Green deformation tensor are written as

$$\begin{aligned} I_1 &= \mathrm{tr}(\mathbf{C}), \\ I_2 &= \frac{1}{2}\left(\mathrm{tr}\mathbf{C}^2 - \mathrm{tr}\left(\mathbf{C}^2\right)\right), \\ I_3 &= \det(\mathbf{C}) = J^2. \end{aligned} \tag{1.4}$$

In the above equation, I_1 and I_2 are respectively the first and second invariants; "tr" stands for *trace*; "det" shows the *determinant*; J represents the determinant of $\mathbf{F}$, i.e., $J = \det(\mathbf{F})$.

The invariants of the left Cauchy–Green deformation tensor are expressed as

$$\begin{aligned} I_1 &= \mathrm{tr}\ (\mathbf{b}), \\ I_2 &= \frac{1}{2}\left(\mathrm{tr}\ \mathbf{b}^2 - \mathrm{tr}\left(\mathbf{b}^2\right)\right), \\ I_3 &= \det(\mathbf{b}) = J^2. \end{aligned} \tag{1.5}$$

The general form of a strain energy function takes the following form:

$$W = W_{\mathrm{iso}} + W_{\mathrm{aniso}}, \tag{1.6}$$

in which W_{iso} is the isotropic part; W_{aniso} stands for the anisotropic part. It is noted that the majority of materials for DEs are considered to be incompressible. The isotropic part itself is formulated as $W_{\mathrm{iso}} = W_{\mathrm{dev}} + W_{\mathrm{vol}}$ in which W_{dev} stands for the isochoric deformation and W_{vol} shows volume change. Due to the incompressibility, the volumetric part (W_{vol}) becomes zero. In the next section, the common hyperelastic models for DEs are introduced and formulated.

1.3.2 Neo-Hookean Model

Neo-Hookean constitutive law is the simplest hyperelastic model that has been used for DEs. This model has been developed for both compressibility and incompressibility states. The neo-Hookean materials model for the compressibility state is expressed as

$$W_{\text{Neo-Hookean}} = c_1(I_1 - 3 - 2\ln J) + D_1(J - 1)^2, \quad (1.7)$$

where c_1 and D_1 are material constants. Considering the incompressibility, the neo-Hookean model is formulated as

$$\psi_{\text{Neo-Hookean}} = c_1(I_1 - 3). \quad (1.8)$$

1.3.3 Gent Model

Some constituent materials for DEs, such as the VHB-based elastomers, have revealed the strain-stiffening effect in response to external loading. The strain-stiffening effect defines a specified value of the stretch in elastomers. The Gent model for a compressible nonlinear elastic material is written as [40]

$$W_{\text{Gent}} = -\frac{\mu J_{lim}}{2}\ln\left(1 - \frac{I_1 - 3}{J_{lim}}\right) + \frac{\kappa}{2}\left(\frac{J^2 - 1}{2} - \ln J\right)^4, \quad (1.9)$$

in which μ and κ stand for material parameters; J_{lim} is a dimensionless parameter, the so-called stiffening parameter (Gent parameters) that measures the strength of the strain-stiffening; as J_{lim} is decreased, the strain-stiffening effect increases. The Gent model for the incompressibility condition takes the following form [41]:

$$\psi_{\text{Gent}} = -\frac{\mu J_{lim}}{2}\ln\left(1 - \frac{I_1 - 3}{J_{lim}}\right). \quad (1.10)$$

1.3.4 Mooney–Rivlin Model

Another hyperelastic model that has been utilized for DEs is the Mooney–Rivlin model. This model is a good candidate for deformations with large strains. The absence of the strain-stiffening effect is, however, the model's most significant flaw. The Mooney–Rivlin model for the compressibility condition is expressed as

$$W_{\text{Mooney-Rivlin}} = C_{10}\left(\overline{I}_1 - 3\right) + C_{01}\left(\overline{I}_2 - 3\right) + \frac{1}{D_1}(J-1)^2, \qquad (1.11)$$

where C_{10} and C_{01} represent material constants. The Mooney–Rivlin constitutive law for an incompressible material is expressed as

$$\psi_{Mooney-Rivlin} = C_1\left(\overline{I}_1 - 3\right) + C_2\left(\overline{I}_2 - 3\right), \qquad (1.12)$$

where C_1 and C_2 stand for material constants. In Eqs. (1.11) and (1.12), $\overline{I}_1 = J^{-2/3} I_1$ and $\overline{I}_1 = J^{-4/3} I_2$.

1.3.5 Gent–Gent Model

The Gent–Gent model is a modified version of the Gent model. The existence of the second principal invariant differentiates between the original Gent model and the Gent–Gent model. The Gent model with the incompressibility condition is formulated as [42, 43]

$$W_{\text{Gent-Gent}} = -c_1 J_{lim} \ln\left(1 - \frac{I_1 - 3}{J_{lim}}\right) + c_2 \ln\left(\frac{I_2}{3}\right), \qquad (1.13)$$

where c_1 and c_2 are material parameters.

1.4 Studies on Dielectric Elastomers Based on the Type of Hyperelastic Model

This section reviews the published literature on dielectric elastomers (DEs) based on hyperelastic models. First, the literature based on the Gent strain energy function is reviewed. After that, the studies based on the neo-Hookean model, Mooney–Rivlin model, and Ogden model are reviewed. Finally, we discuss modified versions of the Gent model for DEs.

1.4.1 Studies Based on Gent Model

A large amount of literature on DEs based on the Gent model is available. For instance, electromechanical instability in dynamical modes in a DE balloon was identified by Chen et al. [44], who utilized the Gent hyperelastic model. Furthermore, by plotting the voltage-stretch curve and pressure-stretch curve, the electromechanical instabilities of an interconnected DE spherical shell were studied by Sun et al. [45], who utilized the Gent hyperelastic material model.

By depicting the time-stretch response and voltage-stretch curve, the oscillations and instability of a balloon made of DE were analysed by Chen and Wang [46] with the aid of the incompressible Gent constitutive model. Furthermore, nucleation and Propagation of Wrinkles were investigated in [47] by using the Gent model to consider the strain-stiffening.

The response of interconnected DE balloons was studied and simulated by Chen and Wang [48], who utilized the Gent model. Mao et al. [49] addressed the wrinkling phenomenon experimentally in DE balloons by using the Gent model. They used the stretch limit in the Gent model as $J_m = 220$. Snap-through instability and electrical breakdown were also explored in their study.

In a deep analysis conducted by Lv et al. [50], dynamic characteristics and instabilities of an electromechanically actuated hyperelastic balloon were assessed, where a Gent model was used to capture the strain-stiffening, and the damping effect was also considered.

In [51], by implementing the Gent model with stretch limits as $J_m = 270$ and 97.2, new electromechanical instabilities in DE spherical shells were identified and discussed. Wang et al. [52] by employing the Gent model analysed the anomalous bulging behaviour in a DE spherical balloon.

In [53], using a visco-hyperelastic Gent model, the delayed electromechanical characteristic of a spherical balloon was explored. In that study, a rheological model with two springs and one dashpot was employed. In addition, the Gent energy function in conjunction with the viscoelastic effect was employed for investigating the wrinkling behaviour of a DE balloon (see [54]).

Zhang et al. [55] harnessed the dielectric breakdown and instabilities in a DE with the Gent hyperelastic material model. Zhou et al. [56] considered the Gent model to study the nonlinear behaviour of a DE membrane. They utilized Equi-biaxial loading and incorporated the viscoelasticity for the system.

Zhu et al. [57] investigated the response of DE encountering the wrinkling phenomenon, employed the Gent constitutive law, and plotted the voltage-stretch curves to provide a profound analysis. Finally, with the aid of the finite strain theory and the Gent model, Garnell et al. [58] explored the sound radiation and vibrations of a DE membrane.

A systematic investigation was conducted by Wang et al. [59] to assess the vibrational behaviour of circular DEs. They reported the chaos and quasi-periodic motion in DEs when the Gent model is employed. The influence of geometrical sizes on the nonlinear vibration of a DE membrane was assessed in [60], who employed the Gent model and assumed that the elastomer is incompressible. Alibakhshi and Heidari [61] investigated nonlinear vibrations of a microbeam made of DEs employing the Gent model, and different phenomena were identified in that paper. In a series of papers [62–67], the thermoelasticity of DEs has been investigated, where the Gent material law was considered to calculate the elastic energy part.

1.4.2 Studies Based on Neo-Hookean, Mooney–Rivlin, and Ogden Models

In this subsection, we concentrate on the neo-Hookean model, Mooney–Rivlin, and Ogden models. Zhu et al. [68] analysed the nonlinear vibrations of a DE spherical shell by modelling the elastomer based on the neo-Hookean strain energy function. They analytically and numerically solved the problem and discussed the equilibrium stretches of the elastomers.

The performance of a DE in electromechanical situations was addressed by Zhang and Chen [69], who adopted the neo-Hookean model in an incompressible condition. The bifurcation phenomenon in a spherical balloon was analysed by Liang and Cai [70], who employed the Ogden hyperelastic model and applied the pressure and voltage to the balloon.

In [71], random vibrations of a spherical balloon were analysed with the help of the neo-Hookean strain energy function. The author of that paper used the method of stochastic averaging to solve the random problem analytically. In another paper [72], the random response of a spherical balloon made of DE was investigated by using the Mooney–Rivlin strain energy function. In that paper, stochastic averaging and Monte Carlo simulation were implemented to help the author to identify different aspects of stochastic problems.

With the application of inflation pressure and potential difference, the bifurcation of a DE balloon was analysed by Xie et al. [73]. They utilized the Ogden, neo-Hookean, and Mooney–Rivlin strain energy functions and compared the results with the Gent model. In another paper published [74], the free vibration of a DE spherical balloon was analytically and numerically solved using the neo-Hookean model, Runge–Kutta method, and Newton-harmonic balance.

Static pull-in and snap-through instabilities and DC static instability in a DE balloon with the aid of the neo-Hookean, Mooney–Rivlin, and Ogden models were analysed by Sharma et al. [75]. In [76, 77], the nonlinear vibration and resonance of DE balloons were explored numerically and analytically using multiple timescales and incremental harmonic balance methods. In those papers, the neo-Hookean strain energy function was applied. The parametric excitation of a DE was analysed using the neo-Hookean model (see [78]), where the equilibrium points and primary and secondary resonances were addressed. In [79, 80], by utilizing the Ogden and Mooney–Rivlin strain energy functions, the dynamic response of DEs was experimentally and analytically captured.

Kim et al. analyzed vibration frequencies of a DE membrane by implementing the neo-Hookean, Ogden, and Mooney–Rivlin models [81]. Dai and Wang [82] carried out a dynamic analysis of the in-plane oscillations of neo-Hookean DEs. They applied the incompressibility conditions for DE and depicted time-stretch response and phase portraits in nonlinear vibration analysis.

The nonlinear response of a DE-based smart system was studied by Srivastava and Basu [83], who utilized the single-term Ogden strain energy function in their work. A neo-Hookean-based viscoelastic model was implemented in [84] to assess the performance of a circular DE. Random oscillation of a DE balloon was reported in [85], in which the neo-Hookean model was adopted. Some researchers have tried to develop DE-based systems incorporating neo-Hookean and Mooney–Rivlin models in a thermoelectricity theory [86–89].

1.4.3 Studies Based on Gent–Gent Model

Researchers decided to discard this limitation because the Gent model cannot accurately capture the deformation of hyperelastic materials at large strains. Some papers extended the Gent model and by which its new versions were introduced. However, the literature on modelling DEs based on the modified version of DEs is very limited.

In recent years, some researchers utilized the Gent–Gent model for DEs, which is a modified version of the Gent model. For instance, Alibakhshi and Heidari investigated the nonlinear vibration and chaos in DE balloons based on the Gent–Gent model [90] and concluded that this hyperelastic model is influential in controlling instabilities and chaos in such systems. Alibakhshi et al. [91] analysed the nonlinear resonance of a DE membrane employing the Gent–Gent model. They also considered a new version of the Gent model in that paper introduced by Bien-aimé et al. [92]. Chen et al. using a compressible Gent–Gent model, researched elastic waves of a DE laminate [93].

1.5 Future Works on Nonlinear Elasticity of Dielectric Elastomers

Researchers have been working on developing the responsiveness of DEs based on anisotropy in recent years. They are considering fibre-reinforced hyperelastic materials for such smart structures. The hyperelastic models incorporating fibre-reinforcement are utilized for this kind of system [94–99]. The Holzapfel–Gasser–Ogden hyperelastic model has been used in the majority of investigations in this field [100, 101]. Thus, future works developing fibre-reinforcement of DEs seem to be of interest.

Another emerging topic in this field of study is employing hyperelastic models with the inclusion of humidity. It has been reported that humidity might affect the performance of DEs in real-world applications [102–104]. Thus, this topic may also be interesting to researchers and can be a major topic of study in future works.

1.6 Conclusion

This paper reviewed the hyperelastic strain energy functions utilized for dielectric elastomers. First, the mathematical formulations for nonlinear electro-elasticity and finite strain theory were explained. Then, the studies for DEs based on the type of hyperelastic model were reviewed. The following are some of the findings of this research:

- The most widely used hyperelastic model for dielectric elastomers is the Gent model.
- After the Gent model, the neo-Hookean model has been a candidate for capturing the nonlinear elasticity of dielectric elastomers. This model has been used for dielectric elastomers but not as much as the Gent model.
- The modified versions of the Gent model, such as the Gent–Gent model, are new hyperelastic materials for analysing dielectric elastomers.
- The role of the second invariant of the Cauchy–Green deformation tensor is prominent for dielectric hyperelastic smart structures.
- In the majority of the literature on DEs, incompressibility condition has been assumed.
- The type of hyperelastic model defines the range of chaos, nonlinear vibrations, and electromechanical instabilities.

For further reading on nonlinear elasticity, we refer to classic books [105–107] as well as corresponding chapters [108–111]. In particular, in [110, 111] other useful models of elastomers could be found.

References

1. Dastjerdi S, Malikan M, Eremeyev VA, Akgöz B, Civalek Ö (2021) On the generalized model of shell structures with functional cross-sections. Compos Struct 272:114192. https://doi.org/10.1016/j.compstruct.2021.114192
2. Dastjerdi S, Akgöz B, Civalek Ö (2020) On the effect of viscoelasticity on behavior of gyroscopes. Int J Eng Sci 149:103236. https://doi.org/10.1016/j.ijengsci.2020.103236
3. Dastjerdi S, Malikan M, Dimitri R, Tornabene F (2021) Nonlocal elasticity analysis of moderately thick porous functionally graded plates in a hygro-thermal environment. Compos Struct 255:112925. https://doi.org/10.1016/j.compstruct.2020.112925
4. Dastjerdi S, Akgöz B, Civalek Ö, Malikan M, Eremeyev VA (2020) On the non-linear dynamics of torus-shaped and cylindrical shell structures. Int J Eng Sci 156:103371. https://doi.org/10.1016/j.ijengsci.2020.103371
5. Malikan M, Jabbarzadeh M, Dastjerdi S (2017) Non-linear static stability of bi-layer carbon nanosheets resting on an elastic matrix under various types of in-plane shearing loads in thermo-elasticity using nonlocal continuum. Microsyst Technol 23(7):2973–2991. https://doi.org/10.1007/s00542-016-3079-9
6. Majidi C (2019) Soft-matter engineering for soft robotics. Adv Mater Technol 4(2):1800477. https://doi.org/10.1002/admt.201800477
7. Boyraz P, Runge G, Raatz A (2018) An overview of novel actuators for soft robotics. High-Throughput 7(3). https://doi.org/10.3390/act7030048

8. Miriyev A, Stack K, Lipson H (2017) Soft material for soft actuators. Nat Commun 8(1):596. https://doi.org/10.1038/s41467-017-00685-3
9. Liu J, Gao Y, Lee YJ, Yang S (2020) Responsive and foldable soft materials. Trends Chem 2(2):107–122. https://doi.org/10.1016/j.trechm.2019.11.003
10. Lu T, Ma C, Wang T (2020) Mechanics of dielectric elastomer structures: a review. Extrem Mech Lett 38:100752. https://doi.org/10.1016/j.eml.2020.100752
11. Moretti G, Rosset S, Vertechy R, Anderson I, Fontana M (2020) A review of dielectric elastomer generator systems. Adv Intell Syst 2(10):2000125. https://doi.org/10.1002/aisy.202070103
12. Zhang J, Chen H, Sheng J, Liu L, Wang Y, Jia S (2014) Dynamic performance of dissipative dielectric elastomers under alternating mechanical load. Appl Phys A Mater Sci Process 115(1):59–67. https://doi.org/10.1007/s00339-013-8092-6
13. Heidari H, Alibakhshi A, Azarboni HR (2020) Chaotic motion of a parametrically excited dielectric elastomer. Int J Appl Mech 12(3):2050033. https://doi.org/10.1142/S1758825120500337
14. Zhao X, Wang Q (2014) Harnessing large deformation and instabilities of soft dielectrics: theory, experiment, and application. Appl Phys Rev 1(2):021304. https://doi.org/10.1063/1.4871696
15. Zhang J, Chen H, Li B, McCoul D, Pei Q (2015) Coupled nonlinear oscillation and stability evolution of viscoelastic dielectric elastomers. Soft Matter 11(38):7483–7493. https://doi.org/10.1039/c5sm01436k
16. Marckmann G, Verron E (2006) Comparison of hyperelastic models for rubber-like materials. Rubber Chem Technol 79(5):835–858. https://doi.org/10.5254/1.3547969
17. Mihai LA, Chin LK, Janmey PA, Goriely A (2015) A comparison of hyperelastic constitutive models applicable to brain and fat tissues. J R Soc Interface 12(110):20150486. https://doi.org/10.1098/rsif.2015.0486
18. Ahmadi S, Gooyers M, Soleimani M, Menon C (2013) Fabrication and electromechanical examination of a spherical dielectric elastomer actuator. Smart Mater Struct 22(11):115004. https://doi.org/10.1088/0964-1726/22/11/115004
19. Li B, Zhang J, Chen H, Li D (2016) Voltage-induced pinnacle response in the dynamics of dielectric elastomers. Phys Rev E 93(5):052506. https://doi.org/10.1103/PhysRevE.93.052506
20. Mao R, Wu B, Carrera E, Chen W (2019) Electrostatically tunable small-amplitude free vibrations of pressurized electro-active spherical balloons. Int J Non Linear Mech 117:103237. https://doi.org/10.1016/j.ijnonlinmec.2019.103237
21. Gei M, Colonnelli S, Springhetti R (2014) The role of electrostriction on the stability of dielectric elastomer actuators. Int J Solids Struct 51(3–4):848–860. https://doi.org/10.1016/j.ijsolstr.2013.11.011
22. Vu DK, Steinmann P, Possart G (2007) Numerical modelling of non-linear electroelasticity. Int J Numer Methods Eng 70(6):685–704. https://doi.org/10.1002/nme.1902
23. Dorfmann A, Ogden RW (2005) Nonlinear electroelasticity. Acta Mech 174(3–4):167–183. https://doi.org/10.1007/s00707-004-0202-2
24. Dorfmann L, Ogden RW (2017) Nonlinear electroelasticity: material properties, continuum theory and applications. Proc R Soc A Math Phys Eng Sci 473(2204):20170311. https://doi.org/10.1098/rspa.2017.0311
25. Suo Z (2010) Theory of dielectric elastomers. Acta Mech Solida Sin 23(6):549–578. https://doi.org/10.1016/S0894-9166(11)60004-9
26. Zhao X, Suo Z (2010) Theory of dielectric elastomers capable of giant deformation of actuation. Phys Rev Lett 104(17):178302. https://doi.org/10.1103/PhysRevLett.104.178302
27. Zeng C, Gao X (2020) Stability of an anisotropic dielectric elastomer plate. Int J Non Linear Mech 124:103510. https://doi.org/10.1016/j.ijnonlinmec.2020.103510
28. Hansy-Staudigl E, Krommer M, Humer A (2019) A complete direct approach to nonlinear modeling of dielectric elastomer plates. Acta Mech 230(11):3923–3943. https://doi.org/10.1007/s00707-019-02529-1

29. Staudigl E, Krommer M, Vetyukov Y (2018) Finite deformations of thin plates made of dielectric elastomers: modeling, numerics, and stability. J Intell Mater Syst Struct 29(17):3495–3513. https://doi.org/10.1177/1045389X17733052
30. Ariana A, Mohammadi AK (2020) Nonlinear dynamics and bifurcation behavior of a sandwiched micro-beam resonator consist of hyper-elastic dielectric film. Sens Actuators, A Phys 312:112113. https://doi.org/10.1016/j.sna.2020.112113
31. Feng C, Jiang L, Lau WM (2011) Dynamic characteristics of a dielectric elastomer-based microbeam resonator with small vibration amplitude. J Micromechanics Microengineering 21(9):095002. https://doi.org/10.1088/0960-1317/21/9/095002
32. Feng C, Yu L, Zhang W (2014) Dynamic analysis of a dielectric elastomer-based microbeam resonator with large vibration amplitude. Int J Non Linear Mech 65:63–68. https://doi.org/10.1016/j.ijnonlinmec.2014.05.004
33. Garcia LA, Trindade MA (2019) Finite element modeling and parametric analysis of a dielectric elastomer thin-walled cylindrical actuator. J Brazilian Soc Mech Sci Eng 41(1):18. https://doi.org/10.1007/s40430-018-1527-4
34. Sheng J, Chen H, Li B, Wang Y (2014) Nonlinear dynamic characteristics of a dielectric elastomer membrane undergoing in-plane deformation. Smart Mater Struct 23(4):045010. https://doi.org/10.1088/0964-1726/23/4/045010
35. Zhang J, Chen H, Li D (2016) Method to control dynamic snap-through instability of dielectric elastomers. Phys Rev Appl 6(6):064012. https://doi.org/10.1103/PhysRevApplied.6.064012
36. Hackett RM (2017) Hyperelasticity primer. Springer, Cham
37. Mihai LA, Goriely A (2017) How to characterize a nonlinear elastic material? A review on nonlinear constitutive parameters in isotropic finite elasticity. Proc R Soc A: Math, Phys Eng Sci 473:20170607. https://doi.org/10.1098/rspa.2017.0607
38. Salençon J (2001) Handbook of continuum mechanics: general concepts thermoelasticity. Springer, Berlin
39. Chaves EWV (2013) Notes on continuum mechanics. Springer, Barcelona
40. Horgan CO, Saccomandi G (2004) Constitutive models for compressible nonlinearly elastic materials with limiting chain extensibility. J Elast 77(2):123–138. https://doi.org/10.1007/s10659-005-4408-x
41. Gent AN (1996) A new constitutive relation for rubber. Rubber Chem Technol 69(1):59–61. https://doi.org/10.5254/1.3538357
42. Ogden RW, Saccomandi G, Sgura I (2004) Fitting hyperelastic models to experimental data. Comput Mech 34(6):484–502. https://doi.org/10.1007/s00466-004-0593-y
43. Zhou L, Wang S, Li L, Fu Y (2018) An evaluation of the Gent and Gent–Gent material models using inflation of a plane membrane. Int J Mech Sci 146–147:39–48. https://doi.org/10.1016/j.ijmecsci.2018.07.035
44. Chen F, Zhu J, Wang MY (2015) Dynamic electromechanical instability of a dielectric elastomer balloon. EPL 112(4):47003. https://doi.org/10.1209/0295-5075/112/47003
45. Sun W, Wang H, Zhou J (2015) Actuation and instability of interconnected dielectric elastomer balloons. Appl Phys A Mater Sci Process 119(2):443–449. https://doi.org/10.1007/s00339-015-9001-y
46. Chen F, Wang MY (2015) Dynamic performance of a dielectric elastomer balloon actuator. Meccanica 50(11):2731–2739. https://doi.org/10.1007/s11012-015-0206-0
47. Mao G, Huang X, Diab M, Li T, Qu S, Yang W (2015) Nucleation and propagation of voltage-driven wrinkles in an inflated dielectric elastomer balloon. Soft Matter 11(33):6569–6575. https://doi.org/10.1039/c5sm01102g
48. Chen F, Wang MY (2016) Simulation of networked dielectric elastomer balloon actuators. IEEE Robot Autom Lett 1(1):221–226. https://doi.org/10.1109/LRA.2016.2514350
49. Mao G, Huang X, Diab M, Liu J, Qu S (2016) Controlling wrinkles on the surface of a dielectric elastomer balloon. Extrem Mech Lett 9:139–146. https://doi.org/10.1016/j.eml.2016.06.001
50. Lv X, Liu L, Liu Y, Leng J (2018) Dynamic performance of dielectric elastomer balloon incorporating stiffening and damping effect. Smart Mater Struct 27(10):105036. https://doi.org/10.1088/1361-665X/aab9db

51. Liang X, Cai S (2018) New electromechanical instability modes in dielectric elastomer balloons. Int J Solids Struct 132–133:96–104. https://doi.org/10.1016/j.ijsolstr.2017.09.021
52. Wang F, Yuan C, Lu T, Wang TJ (2017) Anomalous bulging behaviors of a dielectric elastomer balloon under internal pressure and electric actuation. J Mech Phys Solids 102:1–16. https://doi.org/10.1016/j.jmps.2017.01.021
53. Lv X, Liu L, Leng J, Liu Y, Cai S (2019) Delayed electromechanical instability of a viscoelastic dielectric elastomer balloon. Proc R Soc A Math Phys Eng Sci 475(2229):20190316. https://doi.org/10.1098/rspa.2019.0316
54. Mao G, Xiang Y, Huang X, Hong W, Lu T, Qu S (2018) Viscoelastic effect on the wrinkling of an inflated dielectric-elastomer balloon. J Appl Mech Trans ASME 85(7):071003. https://doi.org/10.1115/1.4039672
55. Zhang H, Wang Y, Godaba H, Khoo BC, Zhang Z, Zhu J (2017) Harnessing dielectric breakdown of dielectric elastomer to achieve large actuation. J Appl Mech Trans ASME 84(12):121011. https://doi.org/10.1115/1.4038174
56. Zhou J, Jiang L, Khayat RE (2015) Investigation on the performance of a viscoelastic dielectric elastomer membrane generator. Soft Matter 11(15):2983–2992. https://doi.org/10.1039/c5sm00036j
57. Zhu J, Kollosche M, Lu T, Kofod G, Suo Z (2012) Two types of transitions to wrinkles in dielectric elastomers. Soft Matter 8(34):8840–8846. https://doi.org/10.1039/c2sm26034d
58. Garnell E, Rouby C, Doaré O (2019) Dynamics and sound radiation of a dielectric elastomer membrane. J Sound Vib 459:114836. https://doi.org/10.1016/j.jsv.2019.07.002
59. Wang F, Lu T, Wang TJ (2016) Nonlinear vibration of dielectric elastomer incorporating strain stiffening. Int J Solids Struct 87:70–80. https://doi.org/10.1016/j.ijsolstr.2016.02.030
60. Zhang J, Chen H (2020) Voltage-induced beating vibration of a dielectric elastomer membrane. Nonlinear Dyn 100(3):2225–2239. https://doi.org/10.1007/s11071-020-05678-4
61. Alibakhshi A, Heidari H (2022) Nonlinear dynamic responses of electrically actuated dielectric elastomer-based microbeam resonators. J Intell Mater Syst Struct 33(4):558–571. https://doi.org/10.1177/1045389x211023584
62. Deng L, He Z, Li E, Chen S (2018) Maximum actuation strain for dissipative dielectric elastomers with simultaneous effect of prestretch and temperature. J Appl Polym Sci 135(8):45850. https://doi.org/10.1002/app.45850
63. Liu L, Chen H, Li B, Wang Y, Li D (2015) Thermal and strain-stiffening effects on the electromechanical breakdown strength of dielectric elastomers. Appl Phys Lett 107(6):062906. https://doi.org/10.1063/1.4928712
64. Sheng J, Chen H, Liu L, Zhang J, Wang Y, Jia S (2013) Dynamic electromechanical performance of viscoelastic dielectric elastomers. J Appl Phys 114(13):134101. https://doi.org/10.1063/1.4823861
65. Sheng J, Chen H, Li B (2011) Effect of temperature on the stability of dielectric elastomers. J Phys D Appl Phys 44(36):365406. https://doi.org/10.1088/0022-3727/44/36/365406
66. Sheng J, Chen H, Li B, Wang Y (2013) Influence of the temperature and deformation-dependent dielectric constant on the stability of dielectric elastomers. J Appl Polym Sci 128(4):2402–2407. https://doi.org/10.1002/app.38361
67. Alibakhshi A, Heidari H (2020) Nonlinear resonance analysis of dielectric elastomer actuators under thermal and isothermal conditions. Int J Appl Mech 12(9):2050100. https://doi.org/10.1142/S1758825120501008
68. Zhu J, Cai S, Suo Z (2010) Nonlinear oscillation of a dielectric elastomer balloon. Polym Int 59(3):378–383. https://doi.org/10.1002/PI.2767
69. Zhang J, Chen H (2014) Electromechanical performance of a viscoelastic dielectric elastomer balloon. Int J Smart Nano Mater 5(2):893930. https://doi.org/10.1080/19475411.2014.893930
70. Liang X, Cai S (2015) Shape bifurcation of a spherical dielectric elastomer balloon under the actions of internal pressure and electric voltage. J Appl Mech Trans ASME 82(10):101002. https://doi.org/10.1115/1.4030881
71. Jin X, Huang Z (2017) Random response of dielectric elastomer balloon to electrical or mechanical perturbation. J Intell Mater Syst Struct 28(2):195–203. https://doi.org/10.1177/1045389X16649446

72. Jin X, Wang Y, Huang Z (2017) On the ratio of expectation crossings of random-excited dielectric elastomer balloon. Theor Appl Mech Lett 7(2):100–104. https://doi.org/10.1016/j.taml.2017.03.005
73. Xie YX, Liu JC, Fu YB (2016) Bifurcation of a dielectric elastomer balloon under pressurized inflation and electric actuation. Int J Solids Struct 78–79:182–188. https://doi.org/10.1016/j.ijsolstr.2015.08.027
74. Tang D, Lim CW, Hong L, Jiang J, Lai SK (2017) Analytical asymptotic approximations for large amplitude nonlinear free vibration of a dielectric elastomer balloon. Nonlinear Dyn 88(3):2255–2264. https://doi.org/10.1007/s11071-017-3374-8
75. Sharma AK, Arora N, Joglekar MM (2018) DC dynamic pull-in instability of a dielectric elastomer balloon: an energy-based approach. Proc R Soc A: Math, Phys Eng Sci 474:20170900. https://doi.org/10.1098/rspa.2017.0900
76. Wang Y, Zhang L, Zhou J (2020) Incremental harmonic balance method for periodic forced oscillation of a dielectric elastomer balloon. Appl Math Mech (English Ed) 41(3):459–470. https://doi.org/10.1007/s10483-020-2590-7
77. Alibakhshi A, Heidari H (2019) Analytical approximation solutions of a dielectric elastomer balloon using the multiple scales method. Eur J Mech A/Solids 74:485–496. https://doi.org/10.1016/j.euromechsol.2019.01.009
78. Zhu J, Cai S, Suo Z (2010) Resonant behavior of a membrane of a dielectric elastomer. Int J Solids Struct 47(24):3254–3262. https://doi.org/10.1016/j.ijsolstr.2010.08.008
79. Fox JW, Goulbourne NC (2009) Electric field-induced surface transformations and experimental dynamic characteristics of dielectric elastomer membranes. J Mech Phys Solids 57(8):1417–1435. https://doi.org/10.1016/j.jmps.2009.03.008
80. Fox JW, Goulbourne NC (2008) On the dynamic electromechanical loading of dielectric elastomer membranes. J Mech Phys Solids 56(8):2669–2686. https://doi.org/10.1016/j.jmps.2008.03.007
81. Kim TJ, Liu Y, Leng J (2018) Cauchy stresses and vibration frequencies for the instability parameters of dielectric elastomer actuators. J Appl Polym Sci 135(21):46215. https://doi.org/10.1002/app.46215
82. Dai HL, Wang L (2015) Nonlinear oscillations of a dielectric elastomer membrane subjected to in-plane stretching. Nonlinear Dyn 82(4):1709–1719. https://doi.org/10.1007/s11071-015-2271-2
83. Kumar Srivastava A, Basu S (2019) Modelling the performance of devices based on thin dielectric elastomer membranes. Mech Mater 137:103136. https://doi.org/10.1016/j.mechmat.2019.103136
84. Wang Z, He T (2018) Electro-viscoelastic behaviors of circular dielectric elastomer membrane actuator containing concentric rigid inclusion. Appl Math Mech 39(4):547–560. https://doi.org/10.1007/s10483-018-2318-8
85. Jin X, Wang Y, Chen MZQ, Huang Z (2017) Response analysis of dielectric elastomer spherical membrane to harmonic voltage and random pressure. Smart Mater Struct 26(3):035063. https://doi.org/10.1088/1361-665X/aa5e44
86. Li P, Zhang H, Wang Q, Shao B, Fan H (2020) Effect of temperature on the performance of laterally constrained dielectric elastomer actuators with failure modes. J Appl Polym Sci 137(35):49037. https://doi.org/10.1002/app.49037
87. Liu L, Liu Y, Yu K, Leng J (2014) Thermoelectromechanical stability of dielectric elastomers undergoing temperature variation. Mech Mater 72:33–45. https://doi.org/10.1016/j.mechmat.2013.05.013
88. Liu L, Liu Y, Li B, Yang K, Li T, Leng J (2011) Thermo-electro-mechanical instability of dielectric elastomers. Smart Mater Struct 20(7):075004. https://doi.org/10.1088/0964-1726/20/7/075004
89. Liu L, Liu Y, Luo X, Li B, Leng J (2012) Electromechanical instability and snap-through instability of dielectric elastomers undergoing polarization saturation. Mech Mater 55:60–72. https://doi.org/10.1016/j.mechmat.2012.07.009

90. Alibakhshi A, Heidari H (2020) Nonlinear dynamics of dielectric elastomer balloons based on the Gent–Gent hyperelastic model. Eur J Mech A/Solids 82:103986. https://doi.org/10.1016/j.euromechsol.2020.103986
91. Alibakhshi A, Imam A, Haghighi SE (2021) Effect of the second invariant of the Cauchy–Green deformation tensor on the local dynamics of dielectric elastomers. Int J Non Linear Mech. 137:103807. https://doi.org/10.1016/j.ijnonlinmec.2021.103807
92. Bien-aimé LKM, Blaise BB, Beda T (2020) Characterization of hyperelastic deformation behavior of rubber-like materials. SN Appl Sci 2(4):648. https://doi.org/10.1007/s42452-020-2355-6
93. Chen Y, Wu B, Su Y, Chen W (2020) Effects of strain stiffening and electrostriction on tunable elastic waves in compressible dielectric elastomer laminates. Int J Mech Sci 176:105572. https://doi.org/10.1016/j.ijmecsci.2020.105572
94. Allahyari E, Asgari M (2021) Nonlinear dynamic analysis of anisotropic fiber-reinforced dielectric elastomers: a mathematical approach. J Intell Mater Syst Struct 32(18–19):2300–2324. https://doi.org/10.1177/1045389X21995879
95. Li C, Xie Y, Li G, Yang X, Jin Y, Li T (2015) Electromechanical behavior of fiber-reinforced dielectric elastomer membrane. Int J Smart Nano Mater 6(2). https://doi.org/10.1080/19475411.2015.1061234
96. Xiao R, Gou X, Chen W (2016) Suppression of electromechanical instability in fiber-reinforced dielectric elastomers. AIP Adv 6(3):035321. https://doi.org/10.1063/1.4945399
97. Ahmadi A, Asgari M (2020) Nonlinear coupled electro-mechanical behavior of a novel anisotropic fiber-reinforced dielectric elastomer. Int J Non-Linear Mech 119:103364. https://doi.org/10.1016/j.ijnonlinmec.2019.103364
98. Allahyari E, Asgari M (2021) Fiber reinforcement characteristics of anisotropic dielectric elastomers: a constitutive modeling development. Mech Adv Mater Struct (in press). https://doi.org/10.1080/15376494.2021.1958275
99. Allahyari E, Asgari M (2020) Effect of fibers configuration on nonlinear vibration of anisotropic dielectric elastomer membrane. Int J Appl Mech 12(10):2050114. https://doi.org/10.1142/S1758825120501148
100. Nolan DR, Gower AL, Destrade M, Ogden RW, McGarry JP (2014) A robust anisotropic hyperelastic formulation for the modelling of soft tissue. J Mech Behav Biomed Mater 39:48–60. https://doi.org/10.1016/j.jmbbm.2014.06.016
101. Holzapfel GA, Gasser TC, Ogden RW (2000) A new constitutive framework for arterial wall mechanics and a comparative study of material models. J Elast 61(1–3):1–48. https://doi.org/10.1023/A:1010835316564
102. Zhang J, Liu X, Liu L, Yang Z, Li P, Chen H (2020) Modeling and experimental study on dielectric elastomers incorporating humidity effect. EPL 129(5):57002. https://doi.org/10.1209/0295-5075/129/57002
103. Zuo Y, Ding Y, Zhang J, Zhu M, Liu L, Zhao J (2021) Humidity effect on dynamic electromechanical properties of polyacrylic dielectric elastomer: an experimental study. Polymers (Basel) 13(5):784. https://doi.org/10.3390/polym13050784
104. Zhang J, Tang L, Liu L, Zhao J, Yang Z, Li P (2021) Modeling of humidity effect on electromechanical properties of viscoelastic dielectric elastomer. Int J Mech Sci 193:106177. https://doi.org/10.1016/j.ijmecsci.2020.106177
105. Ogden RW (1997) Non-linear elastic deformations. Dover, New York
106. Gurtin ME (1983) Topics in finite elasticity. SIAM, Philadelphia
107. Lurie AI (1990) Non-linear theory of elasticity, series in applied mathematics and mechanics, vol 36. North-Holland, Amsterdam
108. Ogden RW (2001) Elements of the theory of finite elasticity. In: Fu YB, Ogden RW (eds) Nonlinear elasticity: theory and applications. Cambridge University Press, Cambridge, pp 1–58
109. Ogden RW, Holzapfel GA (eds) (2006) Mechanics of biological tissue. Springer, Berlin

110. Eremeyev VA, Cloud MJ, Lebedev LP (2018) Applications of tensor analysis in continuum mechanics. World Scientific, New Jersey
111. Altenbach H, Eremeyev VA (2014) Basic equations of continuum mechanics. In: Altenbach H, Öchsner A (eds) Plasticity of pressure-sensitive materials. Engineering materials. Springer, Berlin, Heidelberg, pp 1–47. https://doi.org/10.1007/978-3-642-40945-5_1

Chapter 2
Evolution of the Microstructure of Obstacles from FCC Alloys Under High-Velocity Impact Conditions

Svetlana A. Atroshenko and Georgii G. Savenkov

Abstract The article is devoted to the analysis of the behavior of various FCC metals under shock loading with irregularly shaped projectiles at speeds of 1.5–2.0 km/s.

Keywords Aluminum alloy · Stainless steel · Aluminum bronze · Shock loading

2.1 Introduction

Protection of various modern technical objects experiencing high-energy loads from the impact of irregularly shaped strikers (for example, ammunition fragments and micrometeorites) with impact velocities over 1.5–2.0 km/s is of significant scientific and practical interest. Computer simulation of the processes of high-speed interaction of strikers with various objects in order to create optimal structures requires deep knowledge of the physical and mechanical properties and processes occurring in the thickness, at least of the material of the target (object). However, the existing technical measuring instruments make it possible to register only the kinematic parameters of deformation and destruction of the obstacle and the striker, while the development of internal processes remains inaccessible for visualization. In addition, the physical processes of high-speed deformation and fracture occurring in obstacles are highly dependent on the set of contact boundaries that are inherent in irregularly shaped strikers, and in computer modeling and experimental studies, strikers, as a rule, have

S. A. Atroshenko (✉)
Institute for Problems in Mechanical Engineering, Russian Academy of Sciences, Saint Petersburg, Russia
e-mail: satroshe@mail.ru

Saint Petersburg State University, Saint Petersburg, Russia

G. G. Savenkov
Saint Petersburg State Technological Institute (Technical University), Saint Petersburg, Russia
e-mail: sav-georgij@yandex.ru

Physico-Technical Institute named after A.F. Ioffe RAS, Saint Petersburg, Russia

H. Altenbach et al. (eds.), *Deformation and Destruction of Materials and Structures Under Quasi-static and Impulse Loading*, Advanced Structured Materials 186,
https://doi.org/10.1007/978-3-031-22093-7_2

the correct geometric shape (cylinder, sphere). As a result, there is a significant loss in the accuracy of calculations.

Therefore, the study of the evolution of the microstructure of the target material under the conditions of a high-speed impact with an irregularly shaped striker comes to the fore, despite the fact that such studies are of a post-factor nature and the state of the microstructure at the moment of dynamic action may differ from that after dynamic loading. However, it can be noted that it is the processes of restructuring the internal structure of the material during its dynamic deformation and destruction that determine the physical and mechanical properties of the medium.

In this work, we investigated the microstructure of samples cut from targets made of an aluminum alloy of the AMg6 type (but additionally alloyed with scandium), aluminum bronze and stainless steel 18Cr–10Ni–Ti, pierced by compact irregularly shaped impactors (otherwise called impact "nuclei" [1, 2]). The initial velocity of the impact of the striker with the obstacle was ~1.8–2.0 km/s.

2.2 Samples, Materials and Experimental Technique

Samples for metallographic studies were cut from targets with a diameter of 90 mm and a thickness of 11 mm. The view of an aluminum alloy target after penetration is shown in Fig. 2.1a. Samples prepared for metallographic studies are shown in Fig. 2.1b, c.

The standard mechanical characteristics of the investigated alloys were as follows: $\sigma_{0,2} = 325$ MPa, $\sigma_B = 430$ MPa, $\delta_5 = 25\%$, $\psi = 30\%$—for aluminum alloy; $\sigma_{0,2} = 275$ MPa, $\sigma_B = 620$ MPa, $\delta_5 = 59\%$, $\psi = 70\%$—for stainless steel; $\sigma_{0,2} = 385$ MPa, $\sigma_B = 725$ MPa, $\delta_5 = 35\%$, $\psi = 36\%$—for aluminum bronze.

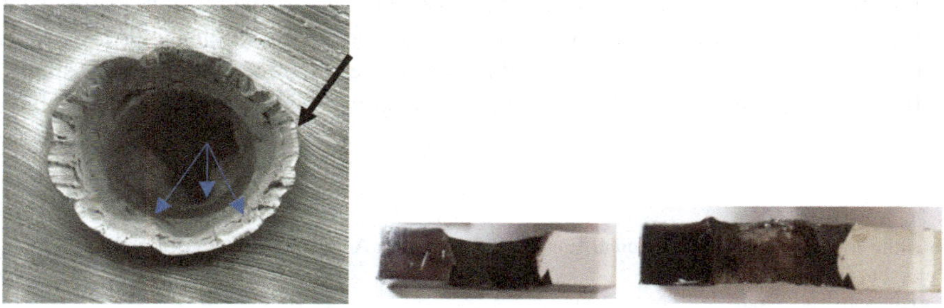

a) b) c)

Fig. 2.1 General view of an aluminum target after penetration (**a**) (the "crest" of the cavity is indicated by a thick arrow) and samples for metallographic studies (**b**, **c**): **b**—stainless steel; **c**—aluminum bronze

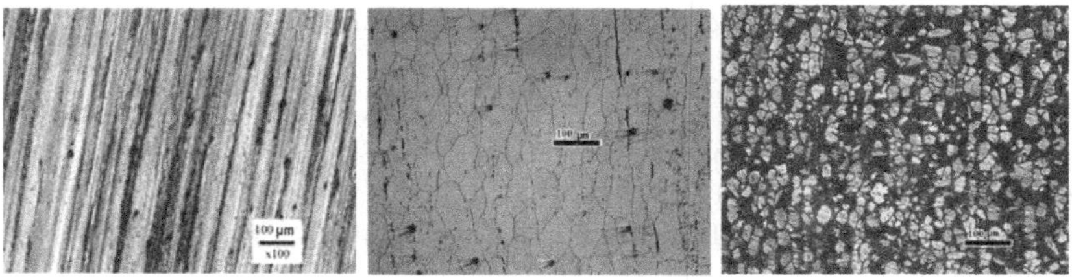

a) b) c)

Fig. 2.2 Initial structures of the investigated alloys: **a**—aluminum alloy; **b**—steel; **c**—bronze

The structure was studied using an Axio Observer Z1 M optical microscope in a bright field and C-DIC contrast. The initial structures of the studied alloys are shown in Fig. 2.2.

2.3 Results of Microstructural Studies and Their Analysis

2.3.1 General Remarks and Results

In our case, all targets are classified as obstacles of finite thickness. In this case, the mutual influence of the cavity formation zone and the rear free surface becomes a significant factor [3, 4]. The compression wave caused by the impact is reflected from the free surface in the form of a tensile wave, which has time to return to the cavity even before the plastic flow of the material stops. If the duration and amplitude of tensile stresses exceed the critical values, then the destruction of the barrier material near the rear surface occurs—it is spalling. If a spall does not form, the unloading changes the stress–strain state in the cavity formation zone (compression along one axis is transformed into a volumetric oppositely deformed state), which contributes to the facilitated penetration of the obstacle by the striker. In our studies, for all three samples on the rear surface, zones of destruction were revealed, which are obviously associated with spalling phenomena (Fig. 2.1).

In all targets, irrespective of the alloy grade, nodal points (indicated by arrows in Fig. 2.1) between the so-called terrace ledges were found in the macro-relief of the surfaces of the holes formed [4].

Also three (in stainless steel and aluminum bronze) or four zones of penetration (in aluminum alloy) stand out (Fig. 2.1) in the punched targets. The first zone, due to the non-stationary stage of penetration [5], corresponds to the depth of ~0.3–0.5 of the penetration thickness. The second (and the third for the barrier made of aluminum alloy) is ~1/3 of the thickness (stationary stage of penetration), and the third (the fourth, for the aluminum alloy) (associated with spalling phenomena) is ~1/6–1/3 of the thickness.

2.3.2 Evolution of Microstructure in Aluminum Alloy

The original stripe structure (Fig. 2.2a) at the very top of the ridge changes to a fragmented one (Fig. 2.3a, b), which indicates large plastic deformations and high temperatures in this layer. Further, as the projectile velocity decreases, the strip structure remains unchanged (Fig. 2.3c), however, periodic protrusions are formed along the edge of the cavity (Fig. 2.3d).

The formation of such protrusions may be due to the fact that as the striker moves, the temperature of the material in front of it increases; in combination with high pressures, this can lead to the appearance of local areas of melting and, accordingly, to a decrease in the friction force. This leads to stress release and a change in the friction mechanism from dry to plastic. The transition to a section with a lower temperature creates some stoppers for the movement of the striker, which again leads to a change in the friction mechanism from plastic to dry. Such a sequence of changes in mechanisms and the appearance of stoppers can lead to the appearance of periodic protrusions.

Finally, when the speed of the striker becomes much less than the initial one, rotational (vortex) modes of plastic deformation begin to appear intensively (Fig. 2.3e, f). One of the probable causes of the vorticity of the target material may be that as the projectile velocity decreases, the material in front of it may already be in a

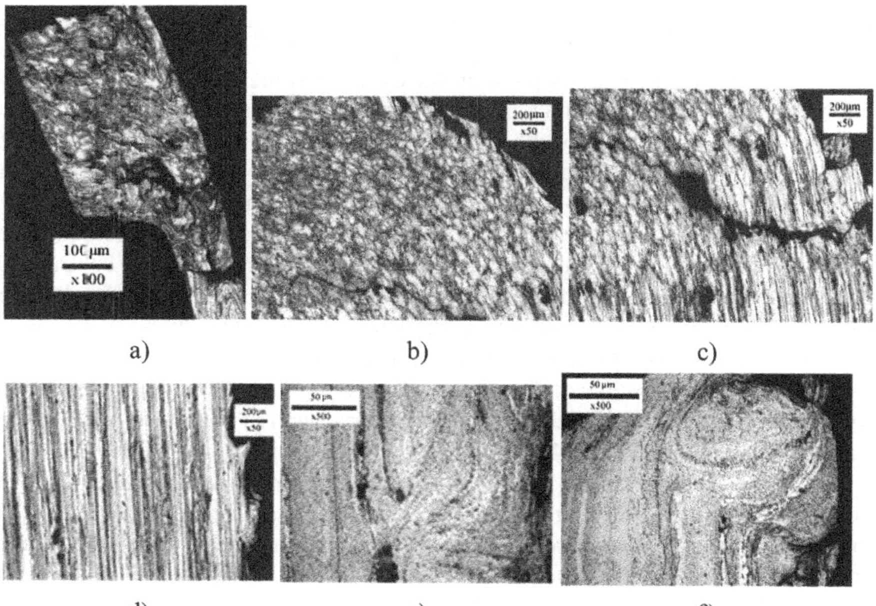

Fig. 2.3 Consecutive change in aluminum alloy of the profile of the relief as the target is penetrated: **a**, **b**—fragmented structure; **c**—strip structure; **d**—protrusions; **e**, **f**—rotational modes

friable state due to the appearance of tensile stresses arising from the interaction of the forward and backward unloading waves. The friability of the material increases the number of degrees of freedom, which can lead to the formation of structures that experience pure rotation due to shear stresses.

2.3.3 *Microstructure Evolution in Stainless Steel*

As noted above, there are three zones of penetration in the stainless steel barrier (Fig. 2.1b), corresponding to three stages of penetration.

In the first zone, at a distance of 1.5–3.8 mm from the frontal surface and 0.15–0.75 mm from the edge of the cavity, an area of rotational plastic deformation was found, consisting of disks with a practically ideal circular shape. The radius of the disks ranges from 35 to 325 μm (Fig. 2.4). Inside the large circles are smaller circular elements and friable elements (pores). The microhardness inside the circular formations is higher than the initial hardness of the material outside these regions (HV = 2.62 GPa) by ~25% and on average is HV ≈ 3.14 GPa. Near the circular formations, closer to the edge of the cavity, adiabatic shear bands (ASB) were found, which are apparently associated with a high local temperature near the contact surface between the impactor and the target [6] and with the instability of plastic flow resulting from the effect of thermal softening at adiabatic or almost adiabatic plastic deformation. It can be assumed that at the first stage of penetration, when the velocity parameters of the striker are still sufficiently high, the target material does not have time to dissipate the mechanical energy of the striker due to the collective motion and multiplication of dislocations. Therefore, to maintain the dissipation rate at the required level, additional rotational modes of plastic deformation are switched on [7].

The second stage of penetration is characterized by the formation of a wavy relief of the cavern edges on the contact surface between the striker and the target material. Also for the first zone, signs of rotational (wave-like) plastic deformation inside the

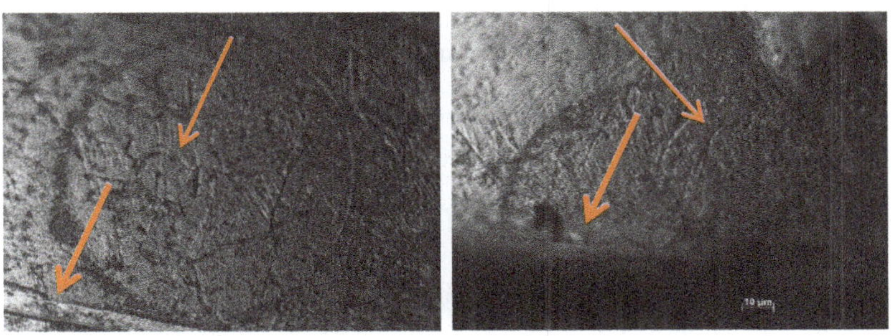

a) b)

Fig. 2.4 Region of rotational plasticity in stainless steel ×1000 (**a**, **b**) (thin arrows indicate small circular formations, thick arrows indicate adiabatic shear bands)

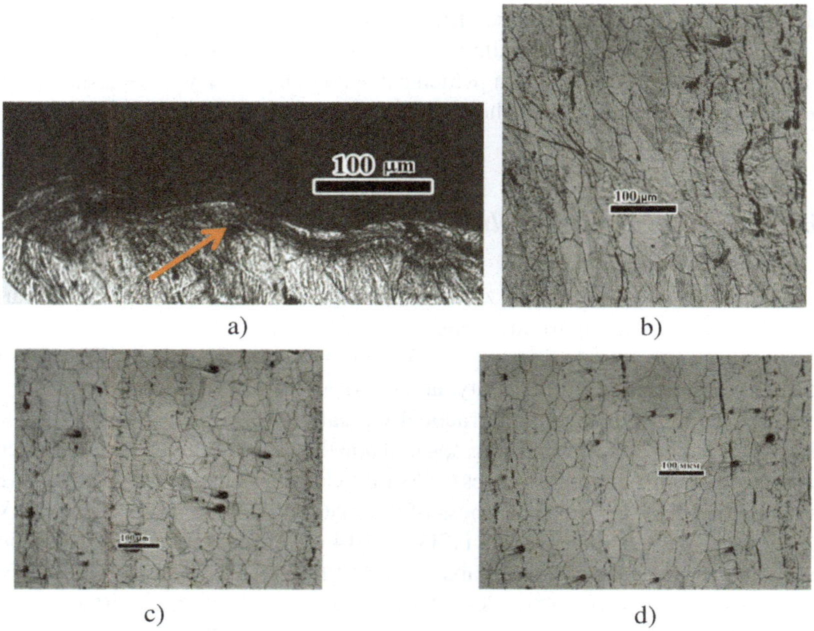

Fig. 2.5 Steel microstructure in the second target zone: ASB (indicated by an arrow) and wavy traces of plastic deformation—(**a**, **b**), crushing of δ-ferrite colonies—(**c**) and traces of grain coarsening—(**d**)

grain structure were found (Fig. 2.5b). In addition, this zone is characterized by recrystallization processes, which manifest themselves in the form of crumbling of ferrite colonies (Fig. 2.5c) or coarsening of grains (Fig. 2.5d).

In the third target zone, associated with the interference of loading and unloading waves, two types of cracks were observed: parallel to the rear surface and shear cracks perpendicular to them (Fig. 2.6). If the first type of cracks is caused by tensile stresses during the interference of unloading waves, then the second ones can be due to their appearance to shear stresses in the direct load wave, when they reach a critical value [8].

2.3.4 Evolution of Microstructure in Aluminum Bronze

In the first zone of the aluminum bronze target, adiabatic shear bands were revealed as characteristic features (Fig. 2.7a). In the second zone, along with PAS, there are also zones of dynamic recrystallization (Fig. 2.7b, c).

Fig. 2.6 Spall and shear cracks in steel (×100)

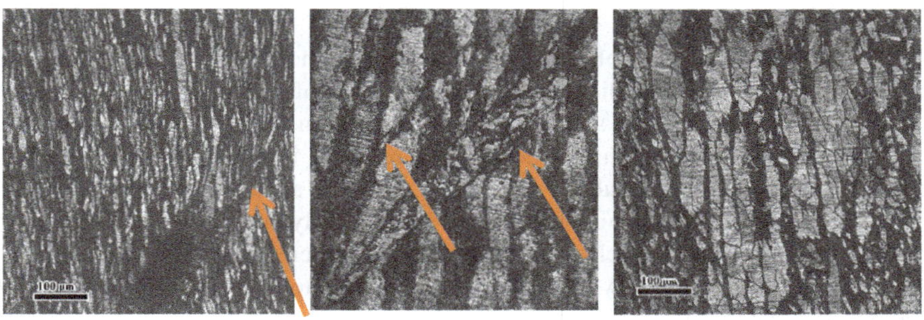

a) b) c)

Fig. 2.7 Adiabatic shear bands—(**a**, **b**) (indicated by arrows), dynamic recrystallization zones—(**b**) and recrystallized grains—(**c**)

In the third zone, on the sides of the cavity, there are cracks up to 3 mm long, oriented relative to the direction of the striker's action in different directions at an angle of 45° (Fig. 2.8).

It can be assumed that in this zone, under the conditions of the dynamic impact of the striker, compression along one axis was realized in the target, which was transformed into a volumetric oppositely deformed state, which led to the formation of the indicated oblique cracks.

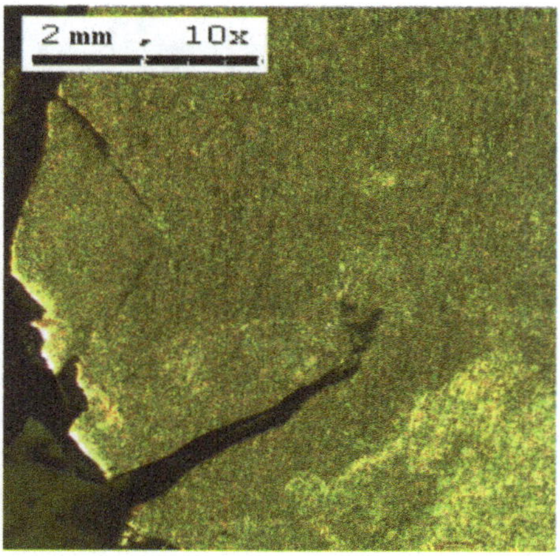

Fig. 2.8 Cracks in the third zone of the aluminum bronze target

2.4 Conclusion

Thus, based on the totality of the studies carried out, the following conclusion can be drawn. After shock loading with the penetration of targets made of different metals with an FCC lattice, signs of the transformation of the deformed state due to the dissipation of mechanical energy were revealed in the following sequence: slip of dislocations, rotational formations and localized adiabatic shears, bands of a fragmented state and dynamic recrystallization, local dynamic polygonization and recrystallization. Structural changes are weakly dependent on the initial structure of the metal and are realized with the participation of interference (wave) effects of shock wave scattering.

Acknowledgements Atroshenko S. A. received financial support for this study under grant No. 22-11-00091 of Russian Science Foundation.

References

1. Orlenko LP (ed) (2002) Physics of the explosion, vol 2. FIZMATLIT, Moscow
2. Kolpakov VI, Savenkov GG, Rudomyotkin KA, Grigoriev AY (2016) Mathematical modeling of the formation of compact projectiles from low-spherical linings. Tech Phys 86(8):21–25
3. Merzhievsky LA, Titov VM (1987) High-speed impact. Phys Combust Explos 23(5):92–108
4. Fomin VM (ed) (1999) High-speed interaction of bodies. Publishing House of SB RAS, Novosibirsk
5. Gladyshev SA, Grigoryan VA (2010) Armor steel. Intermet Engineering, Moscow
6. Barakhtin BK, Prus AA, Savenkov GG (1989) Microstructure anomalies in the deformation of barriers under high-speed penetration by plane jets. J Appl Mech Tech Phys 30(5):820–822

7. Moss JL (1984) Influence of shock waves on the magnitude, rate and temperature of deformation in adiabatic shear bands. Shock waves and phenomena of high-velocity deformation of metals. Metallurgy, Moscow, pp 30–40
8. Rybin VV (1986) Large plastic deformation and destruction of metals. Metallurgy, Moscow

Chapter 3
Decomposition of Solutions of Multicomponent Boundary Value Problems by Solutions of Single-Component Ones for the Purposes of Nanotechnology

Vladimir A. Babeshko, Olga V. Evdokimova, and Olga M. Babeshko

Abstract The integro-differential approach, which is part of the universal method of solving boundary problems proposed by the authors, for the first time builds an exact solution of the boundary problem in a quarter plane for a system of Lame equations under the assumption of setting stress vectors at the boundary. The solution of a vector, two-component, boundary value problem is constructed decomposed by solutions of one-component boundary value problems. Previously, this boundary value problem was considered in a simpler formulation, assuming that displacement vectors were assigned at the boundary. It was suggested that the construction of a solution for the case of stresses is difficult. It is shown that the application of the integro-differential method practically does not complicate the solution of boundary problems with complicated boundary conditions. The research presented in the article is important in studying the behavior of solutions to boundary value problems of complex rheology described by systems of partial differential equations using decompositions by solutions of individual equations. It is proposed to develop this method for studying the behavior of multicomponent nanomaterials in order to build models of their self-organization and self-assembly. They can be used to study their physical and strength properties, as well as the possibility of controlling their parameters.

Keywords Boundary value problems · Block element method · Packed block elements · Lame equations · Helmholtz equations

---

V. A. Babeshko (✉) · O. V. Evdokimova
Scientific Center, Russian Academy of Sciences, Rostov-On-Don, Russia
e-mail: babeshko41@mail.ru

O. V. Evdokimova
e-mail: evdokimova.olga@mail.ru

V. A. Babeshko · O. M. Babeshko
Kuban State University, Krasnodar, Russia
e-mail: babeshko49@mail.ru; olg@kubsu.ru

H. Altenbach et al. (eds.), *Deformation and Destruction of Materials and Structures Under Quasi-static and Impulse Loading*, Advanced Structured Materials 186,
https://doi.org/10.1007/978-3-031-22093-7_3

3.1 Introduction

In the monograph [1] V. Novatsky noted that the application of the Galerkin transform [2, 3] or the representation of solutions using potentials in boundary problems with stresses set at the boundary significantly complicates their solution compared to the given displacements. At the same time, in most boundary problems considered in engineering practice, boundary conditions of the first kind are set, that is, when stress vectors are set at the boundary. The issues of studying and solving boundary value problems for multicomponent materials are complex since it is necessary to consider systems of partial differential equations in arbitrary ones, each of whose components is present in boundary conditions [4].

B.G. Galerkin owns a transformation that allows us to bring systems of linear partial differential equations with constant coefficients to separate differential equations. For a long time, this method was used only for the numerical study of boundary value problems. However, there are tasks, including models, whose solutions are needed in an analytical form. These are, firstly, mixed problems in which solutions may have features [5], and secondly, solutions are needed in an analytical form, for the formation of mutual contacts, adhesion, and decay along the boundaries occurring between the components of multicomponent particles. Models of such interaction are available, for example, in [6]. Of particular interest are the well-known experimental processes of self-organization and self-assembly of nanomaterials [7]. Experiments are known, but there are no strict mathematical models. There is a question of separating and mathematically describing the mechanical aspects of these processes and the physico-chemical ones in order to understand a reasonable way of controlling them and their components.

It is known that the variety of rheologies of multicomponent nanomaterials is huge [8–16]. The more important is the task under consideration.

In this paper, it is shown that the universal method proposed by the authors for solving boundary value problems [17] solves boundary value problems for both of these types of boundary conditions fairly uniformly and without much difficulty. The universality of the method consists in the uniformity of the approach in solving boundary value problems for systems of partial differential equations in non-classical domains, as well as for a number of integral equations, for example, Wiener–Hopf. Boundary value problems can be posed in areas of different dimensions and different rheologies. The available various ways of crushing the carriers of block elements, like the condensation of grids in numerical methods, open up the possibility of using the method for the cases of systems of differential equations with variable coefficients, as it is performed in the finite element method.

The paper gives a solution to a vector boundary value problem decomposed by packed block elements, which are solutions to scalar boundary value problems in a quarter plane. For the first time, this approach has constructed an exact solution in the first quadrant of a plane boundary value problem of the first kind for dynamic Lame equations at the stresses set at the boundary. In the work of the authors [18], an integro-differential method for decomposing solutions of vector boundary value problems by scalar ones was developed, in the work [19], a less general, coordinate method. In both cases, displacements were set on the units below. The integro-differential method is used in the problem under consideration, which is more general than the coordinate method, requiring the use of in-depth properties of solutions to scalar problems, which is not always possible to obtain.

3.2 Basic Equations

Let us consider the plane boundary value problem of the second kind studied in [18], for the system of Lame equations, posed in the first quadrant under harmonic influences at the boundary where displacements are given. Based on it, we formulate a boundary value problem of the first kind by setting stress vectors on the boundary. Previously, it was not possible to obtain an exact solution to it, but the new universal block element method [17] in this paper makes it possible to do this in the form of decomposition of the solution by packed block elements of scalar problems.

In the first quadrant, the dynamic Lame equations, after excluding the time term $exp(-i\omega t)$, have the form

$$\begin{aligned}(\lambda+\mu)\frac{\partial\theta}{\partial x_1}+\mu\Delta u_1+k^2u_1=0,\ \theta=\frac{\partial u_1}{\partial x_1}+\frac{\partial u_2}{\partial x_2},\quad k^2=\rho\omega^2\\(\lambda+\mu)\frac{\partial\theta}{\partial x_2}+\mu\Delta u_2+k^2u_2=0,\ x_1,x_2\in\Omega,\quad \Delta u=\frac{\partial^2u}{\partial x_1^2}+\frac{\partial^2u}{\partial x_2^2}\end{aligned}\tag{3.1}$$

Here $u_n(x_1, x_2)$ are the components of the displacement vectors at the point—$\Omega\Omega$—the area of the first quadrant $x_1 \geq 0$, $x_2 \geq 0$ λ, μ—the Lame parameters—ρ—the density of the material of the deformable body—ω— the frequency of external harmonic influences at the boundary, set by a complex function $exp(-i\omega t)$, where t is time. In a problem of the first kind, the values of normal and tangential stresses at the boundaries of the quadrant are denoted on the abscissa axis by the functions $\sigma_2(x_1, 0)$, $\tau_2(x_1, 0)$ and $\sigma_1(0, x_2)$, $\tau_1(0, x_2)$—on the ordinate axis. The stresses normal to the boundary are indicated by a symbol σ, and tangents—τ. In a problem of the second kind, the components of the vectors of normal and tangential displacements on the boundary of the first quadrant are given the axes $u_1(x_1, 0)$, $u_2(x_1, 0)$ of the abscissa and $u_1(0, x_2)$, $u_2(0, x_2)$ of the ordinate respectively.

3.3 Decomposition of the Boundary Value Problem Solution Using Block Elements

The Lame equations, both in static and dynamic cases, have a long-established property of representing the solution as a sum of potential and vortex components, which are also obtained by using the Galerkin transform [1–3]. Let us use the decomposition of the solution of the Lame equations used in [18] in the following form:

$$\begin{aligned} u_1(x_1, x_2) &= \partial_1 \varphi(x_1, x_2) + \partial_2 \psi(x_1, x_2) \\ u_2(x_1, x_2) &= \partial_2 \varphi(x_1, x_2) - \partial_1 \psi(x_1, x_2) \\ \partial_1 &= \frac{\partial}{\partial x_1}, \quad \partial_2 = \frac{\partial}{\partial x_2} \end{aligned} \tag{3.2}$$

The designations are accepted here

$$\begin{aligned} &(\Delta - p_1^2)\varphi = 0, \ (\Delta - p_2^2)\psi = 0, \ p_1^2 = k_1^2(\lambda + 2\mu)^{-1}, \ p_2^2 = k_1^2 \mu^{-1} \\ &\varphi(x_1, 0) = f_1(x_1, 0), \quad \varphi(0, x_2) = f_2(0, x_2) \\ &\psi(x_1, 0) = g_1(x_1, 0), \quad \psi(0, x_2) = g_2(0, x_2) \end{aligned} \tag{3.3}$$

The functions g_m, $m = 1, 2$ in boundary conditions are arbitrary, satisfying only the conditions of the correctness of the statement of the boundary problem. In particular, they can be taken from the space of slowly growing generalized functions, in which solutions to the boundary value problem in the domain Ω are sought.

The case of the Lame boundary value problem of the first kind is considered. Conditions of the form $\sigma_1(0, x_2)$, $\tau_1(0, x_2)$, $\sigma_2(x_1, 0)$, $\tau_2(x_1, 0)$ are set on the coordinate axes.

Thus, for solutions of the Helmholtz equation, boundary conditions are formed for $x_1 \to 0$ in the form

$$\begin{aligned} \partial_1 \varphi(x_1, x_2) + \partial_2 \psi(x_1, x_2) &= \sigma_1(0, x_2) \\ \partial_2 \varphi(x_1, x_2) - \partial_1 \psi(x_1, x_2) &= \tau_1(0, x_2) \end{aligned} \tag{3.4}$$

Similarly, when $x_2 \to 0$

$$\begin{aligned} \partial_1 \varphi(x_1, x_2) + \partial_2 \psi(x_1, x_2) &= \sigma_2(x_1, 0) \\ \partial_2 \varphi(x_1, x_2) - \partial_1 \psi(x_1, x_2) &= \tau_2(x_1, 0) \end{aligned} \tag{3.5}$$

To solve the boundary problem for the Lame equations with boundary conditions (3.4), (3.5), solutions of boundary problems for the Helmholtz equations are constructed under arbitrary boundary conditions (3.3). The block element method is used, which is described in a number of works by the authors [18, 19]. Packed in the first quadrant in the case of the Dirichlet boundary value problem, the solutions have the form

$$\varphi(x_1, x_2) = \frac{1}{4\pi^2} \iint_{R^2} \frac{\omega_1(\alpha_1, \alpha_2)}{\alpha_1^2 + \alpha_2^2 - p_1^2} e^{-i(\alpha_1 x_1 + \alpha_2 x_2)} d\alpha_1 d\alpha_2$$

$$\psi(x_1, x_2) = \frac{1}{4\pi^2} \iint_{R^2} \frac{\omega_2(\alpha_1, \alpha_2)}{\alpha_1^2 + \alpha_2^2 - p_1^2} e^{-i(\alpha_1 x_1 + \alpha_2 x_2)} d\alpha_1 d\alpha_2$$

$$\varphi(x_1, x_2) = \frac{1}{4\pi^2} \iint_{R^2} \sum_{j=1}^{2} (\alpha_{3-j} - \alpha_{(3-j)1+}) \left[F_j(\alpha_j) - F_j(\alpha_{j1+})\right]$$

$$\times e^{-i(\alpha_1 x_1 + \alpha_2 x_2)} \frac{i d\alpha_1 d\alpha_2}{\alpha_1^2 + \alpha_2^2 - p_1^2} \tag{3.6}$$

$$\psi(x_1, x_2) = \frac{1}{4\pi^2} \iint_{R^2} \sum_{j=1}^{2} (\alpha_{3-j} - \alpha_{(3-j)2+}) \left[G_j(\alpha_j) - G_j(\alpha_{j2+})\right]$$

$$\times e^{-i(\alpha_1 x_1 + \alpha_2 x_2)} \frac{i d\alpha_1 d\alpha_2}{\alpha_1^2 + \alpha_2^2 - p_2^2}$$

$$\alpha_{11+} = i\sqrt{\alpha_2^2 - p_1^2},\ \alpha_{21+} = i\sqrt{\alpha_1^2 - p_1^2},$$

$$\alpha_{12+} = i\sqrt{\alpha_2^2 - p_2^2},\ \alpha_{22+} = i\sqrt{\alpha_1^2 - p_2^2},$$

The sections of multivalued functions are dictated by the requirement to perform automorphisms [18]. In accordance with the construction, the properties (3.3) are valid for the reduced block elements. Using them, we introduce the following notation for solutions of the Helmholtz equations:

$$\phi(x_1, x_2) \equiv \phi[x_1, x_2, f_1(\xi_1, 0), f_2(0, \xi_2)] \to f_1(x_1, 0), \quad 0 < x_2 \ll 1$$
$$\phi(x_1, x_2) \equiv \phi[x_1, x_2, f_1(\xi_1, 0), f_2(0, \xi_2)] \to f_2(0, x_2), \quad 0 < x_1 \ll 1$$
$$\psi(x_1, x_2) \equiv \psi[x_1, x_2, g_1(\xi_1, 0), g_2(0, \xi_2)] \to g_1(x_1, 0), \quad 0 < x_2 \ll 1$$
$$\psi(x_1, x_2) \equiv \psi[x_1, x_2, g_1(\xi_1, 0), g_2(0, \xi_2)] \to g_2(0, x_2), \quad 0 < x_1 \ll 1$$

In earlier works of the authors, the solution of the boundary value problem of the second kind for the Lame equations was constructed by the integro-differential method.

Its exact solution in the first quadrant has the form [18]

$$
\begin{aligned}
&u_1(x_1,x_2)= \\
&\partial_1\Big\langle\varphi_1\Big[x_1,x_2,\frac{1}{2}\partial_1^{(-1)}u_1(\xi_1,0),\frac{1}{2}\partial_1^{(-1)}u_1(0,\xi_2)+\frac{1}{2}\partial_1^{(-1)}F(\xi_2)\Big] \\
&+\varphi_2\Big[x_1,x_2,\frac{1}{2}\partial_2^{(-1)}u_2(\xi_1,0)+\frac{1}{2}\partial_1^{(-1)}D(\xi_1),\frac{1}{2}\partial_2^{-1}u_2(0,\xi_2)\Big]\Big\rangle \\
&+\partial_2\Big\langle\psi_1\Big[x_1,x_2,\frac{1}{2}\partial_2^{(-1)}u_1(\xi_1,0)+\frac{1}{2}\partial_1^{(-1)}(\xi_1),\frac{1}{2}\partial_2^{(-1)}u_1(0,\xi_2)\Big] \\
&-\psi_2\Big[x_1,x_2,\frac{1}{2}\partial_1^{(-1)}u_2(\xi_1,0),\frac{1}{2}\partial_1^{(-1)}u_2(0,\xi_2)+\frac{1}{2}\partial_1^{(-1)}E(\xi_2)\Big]\Big\rangle
\end{aligned}
\tag{3.7}
$$

$$
\begin{aligned}
&u_2(x_1,x_2)= \\
&=\partial_2\Big\langle\varphi_1\Big[x_1,x_2,\frac{1}{2}\partial_1^{(-1)}u_1(\xi_1,0),\frac{1}{2}\partial_1^{-1}u_1(0,\xi_2)+\frac{1}{2}\partial_2^{(-1)}F(\xi_2)\Big] \\
&+\varphi_2\Big[x_1,x_2,\frac{1}{2}\partial_2^{(-1)}u_2(\xi_1,0)+\frac{1}{2}\partial_2^{(-1)}D(\xi_1),\frac{1}{2}\partial_2^{-1}u_2(0,\xi_2)\Big]\Big\rangle \\
&-\partial_1\Big\langle\psi_1\Big[x_1,x_2,\frac{1}{2}\partial_2^{(-1)}u_1(\xi_1,0)+\frac{1}{2}\partial_2^{(-1)}(\xi_1),\frac{1}{2}\partial_2^{(-1)}u_1(0,\xi_2)\Big] \\
&-\psi_2\Big[x_1,x_2,\frac{1}{2}\partial_1^{(-1)}u_2(\xi_1,0),\frac{1}{2}\partial_1^{(-1)}u_2(0,\xi_2)+\frac{1}{2}\partial_2^{(-1)}E(\xi_2)\Big]\Big\rangle
\end{aligned}
\tag{3.8}
$$

$$
\begin{aligned}
C(x_1)&=\partial_2\partial_1^{(-1)}u_1(x_1,0)-\partial_1\partial_2^{(-1)}u_1(x_1,0)\\
D(x_1)&=\partial_2\partial_1^{(-1)}u_2(x_1,0)-\partial_1\partial_2^{(-1)}u_2(x_1,0)\\
E(x_2)&=\partial_1\partial_2^{(-1)}u_2(0,x_2)-\partial_2\partial_1^{(-1)}u_2(0,x_2)\\
F(x_2)&=\partial_1\partial_2^{(-1)}u_1(0,x_2)-\partial_2\partial_1^{(-1)}u_1(0,x_2)
\end{aligned}
$$

Below we will use it to solve the Lame problem of the first kind with boundary conditions (3.4), (3.5).

To do this, we will carry out a number of transformations. Under boundary conditions, we introduce new variables by putting for

$$\sigma_1(x_1,x_2)\to\sigma_1(0,x_2),\quad \tau_1(x_1,x_2)\to\tau_1(0,x_2),\quad x\ll 1$$

$$x_1=(2b_1)^{-1}(z_1+y_1),\quad x_2=(2b_2)^{-1}(z_1-y_1)$$

Similarly, for

$$\sigma_2(x_1,x_2)\to\sigma_2(x_1,0),\quad \tau_2(x_1,x_2)\to\tau_2(x_1,0),\quad x_2\ll 1$$

we will accept

$$x_1=(2b_2)^{-1}(z_2+y_2),\quad x_2=(2b_1)^{-1}(z_2-y_2)$$

As a result, we get representations

$$
\begin{aligned}
&(b_1x_1 - b_2x_2) = y_1, \quad (b_1x_1 + b_1x_2) = z_1, \\
&(b_2x_1 - b_1x_2) = y_2, \quad (b_2x_1 + b_1x_2) = z_2 \\
&b_1 = \sqrt{(\lambda + 2\mu)\mu}, \quad b_2 = \sqrt{\mu\lambda}, \\
&\partial_{y_1} = (b_1\partial_1 - b_2\partial_2), \quad \partial_{z_1} = (b_1\partial_1 + b_2\partial_2), \quad \partial_{y_1}\partial_{z_1} = \left(b_1^2\partial_1\partial_1 - b_2^2\partial_2\partial_2\right), \\
&\partial_{y_2} = (b_2\partial_1 - b_1\partial_2), \quad \partial_{z_2} = (b_2\partial_1 + b_2\partial_2), \quad \partial_{y_2}\partial_{z_2} = \left(b_2^2\partial_1\partial_1 - b_1^2\partial_2\partial_2\right),
\end{aligned}
\tag{3.9}
$$

$$
\begin{aligned}
&\sigma_1\left((2b_1)^{-1}(z_1 + y_1), (2b_2)^{-1}(z_1 - y_1)\right) \equiv \sigma_{10}(y_1, z_1), \\
&\tau_1\left((2b_1)^{-1}(z_1 + y_1), (2b_2)^{-1}(z_1 - y_1)\right) \equiv \tau_{10}(y_1, z_1),
\end{aligned}
\tag{3.10}
$$

$$
\begin{aligned}
&\sigma_2((2b_2)^{(-1)}(z_2 + y_2), \ (2b_1)^{(-1)}(z_2 - y_2)) \equiv \sigma_{20}(y_2, z_2), \\
&\tau_2((2b_2)^{(-1)}(z_2 + y_2), \ (2b_1)^{(-1)}(z_2 - y_2)) \equiv \tau_{20}(y_2, z_2)
\end{aligned}
$$

As in [17], for an arbitrary continuous function $w(\xi, \eta)$ we have the relations

$$
\begin{aligned}
&\partial_{y_n}^{(-1)} w(y_n, z_n) = \int_0^{y_n} w(\xi, \eta)d\xi, \quad \partial_{z_n}^{(-1)} w(y_n, z_n) = \int_0^{z_n} w(\xi, \eta)d\eta, \\
&\partial_{y_n}\partial_{z_n} \int_0^{y_n}\int_0^{z_n} w(\xi, \eta)d\xi d\eta = w(y_2, z_2), \\
&\partial_{y_n}\partial_{z_n}\partial_{y_n}^{(-1)}\partial_{z_n}^{(-1)} w(\xi, \eta)d\xi d\eta = W(y_2, z_2)
\end{aligned}
\tag{3.11}
$$

Note that in order to calculate derivatives or primordial boundary functions with parameters that have turned to zero, it is necessary to use the block element method to construct packed block elements for the Helmholtz equation with their participation, and calculate the required values in the vicinity of the boundary.

Let's add the following relations to (13.7), (13.8) instead of displacements

$$
\begin{aligned}
&u_1(0, x_2) = \mu\partial_1\partial_{y_1}^{(-1)}\partial_{z_1}^{(-1)}\sigma_{10}(y_1, z_1) - \lambda\partial_2\partial_{y_1}^{(-1)}\partial_{z_1}^{(-1)}\tau_{10}(y_1, z_1), \\
&u_2(0, x_2) = -\mu\partial_2\partial_{y_1}^{(-1)}\partial_{z_1}^{(-1)}\sigma_{10}(y_1, z_1) + (\lambda + 2\mu)\partial_1\partial_{y_1}^{(-1)}\partial_{z_1}^{(-1)}\tau_{10}(y_1, z_1), \\
&u_1(x_1, 0) = \mu\partial_1\partial_{y_2}^{(-1)}\partial_{z_2}^{(-1)}\sigma_{20}(y_2, z_2) - (\lambda + 2\mu)\partial_2\partial_{y_2}^{(-1)}\partial_{z_2}^{(-1)}\tau_{20}(y_2, z_2), \\
&u_2(x_1, 0) = -\mu\partial_2\partial_{y_2}^{(-1)}\partial_{z_2}^{(-1)}\sigma_{20}(y_2, z_2) + \lambda\partial_1\partial_{y_2}^{(-1)}\partial_{z_2}^{(-1)}\tau_{20}(y_2, z_2)
\end{aligned}
$$

We prove that the expressions constructed in this way represent the solution of the first boundary value problem for the Lame equation in the first quadrant, decomposed using packed block elements.

To do this, you need to make sure that the boundary conditions (3.4) are met. Let's limit ourselves to the border x_2, the check is performed similarly at the border x_1.

Knowing that in order to verify the fulfillment of the boundary condition (3.4) for the normal voltage on the axis x_2, it is necessary using (3.9), (3.10), (3.11), calculate the expression $\sigma_1(0, x_2) = (\lambda + 2\mu)\partial_1 u_1(0, x_2) + \lambda\partial_2 u_2(0, x_2)$, we get the following sequence of transformations

$$\begin{aligned}
&(\lambda + 2\mu)\partial_1 u_1(0, x_2) + \lambda\partial_2 u_2(0, x_2) \\
&= (\lambda + 2\mu)\partial_1\big[\mu\partial_1\partial_{y_1}^{(-1)}\partial_{z_1}^{(-1)}\sigma_1(0, x_2) - \lambda\partial_2\partial_{y_1}^{(-1)}\partial_{z_1}^{(-1)}\tau_1(0, x_2)\big] \\
&+\lambda\partial_2\big[-\mu\partial_2\partial_{y_1}^{(-1)}\partial_{z_1}^{(-1)}\sigma_1(0, x_2) + (\lambda + 2\mu)\partial_1\partial_{y_1}^{(-1)}\partial_{z_1}^{(-1)}\tau_1(0, x_2)\big] \\
&= (\lambda + 2\mu)\partial_1\mu\partial_1\partial_{y_1}^{(-1)}\partial_{z_1}^{(-1)}\sigma_1(0, x_2) - \lambda\partial_2\mu\partial_2\partial_{y_1}^{(-1)}\partial_{z_1}^{(-1)}\sigma_1(0, x_2) \\
&-(\lambda + 2\mu)\partial_1\lambda\partial_2\partial_{y_1}^{(-1)}\partial_{z_1}^{(-1)}\tau_1(0, x_2) + \lambda\partial_2(\lambda + 2\mu)\partial_1\partial_{y_1}^{(-1)}\partial_{z_1}^{(-1)}\tau_1(0, x_2) \\
&= [(\lambda + 2\mu)\mu\partial_1\partial_1 - \lambda\mu\partial_2\partial_2]\partial_{y_1}^{(-1)}\partial_{z_1}^{(-1)}\sigma_1(0, x_2) \\
&= \partial_{y_1}\partial_{z_1}\partial_{y_1}^{(-1)}\partial_{z_1}^{(-1)}\sigma_1(0, x_2) = \sigma_1(0, x_2)
\end{aligned}$$

The ratio is taken into account here $\partial_{y_1}\partial_{z_1} = ({b_1}^2\partial_1\partial_1 - {b_2}^2\partial_2\partial_2)$.

Consider the case of tangential stresses $\tau_1(0, x_2)$ given at the boundary substitute in the right part

$$\tau_1(0, x_2) = \mu\partial_2 u_1(0, x_2) + \mu\partial_1 u_2(0, x_2)$$

the values $u_1(0, x_2)$, $u_2(0, x_2)$ taken from (3.7), (3.8). As a result, we will have

$$\begin{aligned}
&\mu\partial_2\big[\mu\partial_1\partial_{y_1}^{(-1)}\partial_{z_1}^{(-1)}\sigma_1(0, x_2) - \lambda\partial_2\partial_{y_1}^{(-1)}\partial_{z_1}^{(-1)}\tau_1(0, x_2)\big] + \\
&+\mu\partial_1\big[-\mu\partial_2\partial_{y_1}^{(-1)}\partial_{z_1}^{(-1)}\sigma_1(0, x_2) + (\lambda + 2\mu)\partial_1\partial_{y_1}^{(-1)}\partial_{z_1}^{(-1)}\tau_1(0, x_2)\big] = \\
&= [(\lambda + 2\mu)\mu\partial_1\partial_1 - \lambda\mu\partial_2\partial_2]\partial_{y_1}^{(-1)}\partial_{z_1}^{(-1)}\tau_1(0, x_2) = \\
&= \partial_{y_1}\partial_{z_1}\partial_{y_1}^{(-1)}\partial_{z_1}^{(-1)}\tau_1(0, x_2) = \tau_1(0, x_2)
\end{aligned}$$

The above ratio is used again here.

Thus, it is quite simple to decompose the solution of the boundary value problem of the first kind for the Lame equation in the first quadrant by solutions of boundary value problems for the Helmholtz equations describing vortex and potential processes in the first quadrant.

3.4 Conclusion

Thus, in accordance with the objectives of the task, it is proved that having obtained a solution to a scalar boundary value problem of the second kind, when a displacement vector is set at the boundaries, it is simple enough to construct a solution to a vector boundary value problem with complicated rheology of the first kind, when a stress vector is set at the boundary. The proposed method is easily transferred to the three-dimensional case.

Funding The work was carried out with the financial support of the Russian Science Foundation, project 22-21-00128.

References

1. Novatsky V (1975) Theory of elasticity. Mir, Moscow (In Russian)
2. Galerkin BG (1930) Contribution à la solution générale du problème de la théorie de l'élasticité dans le cas de trois dimensions. C R Acad Sci 190:1047–1048
3. Galerkin BG (1931) Sur l'equilibre élastique d'une plaque rectangulaire épaisse. C R Acad Sci 193:568–571
4. Igumnov, L.A., Grazina, A.V. et al.: Symbolic and numerical modeling of diffusion processes in multi-component solids of finite dimension. Probl Strength Plast 80(3):336–348 (2018) https://doi.org/10.32326/1814-9146-2018-80-3-336-348 (In Russian)
5. Vorovich II, Aleksandrov VM, Babeshko VA (1974) Classical mixed problem of elasticity theory. Nauka, Moscow (In Russian)
6. Babeshko VA, Evdokimova OV, Babeshko OM (2021) Investigation of the three-dimentional Helmholtz equation for a wedge using the block element method. J Appl Mech Tech Phy 62:717–722. https://doi.org/10.1134/S0021894421050023
7. Gordeev SK, Kukushkin SA et al (2000) Self-organization in the formation of a nanoporous carbon material. Phy Solid State 42(12):2314–2317. https://doi.org/10.1134/1.1332156
8. Arghavan S, Singh AV (2011) On the vibrations of single-walled carbon nanotubes. J Sound Vib 330(13):3102–3122. https://doi.org/10.1016/j.jsv.2011.01.032
9. Kang JW, Kwon OK (2012) A molecular dynamics simulation study on resonance frequencies comparison of tunable carbon-nanotube resonators. Appl Surf Sci 258(6):2014–2016. https://doi.org/10.1016/j.apsusc.2011.05.026
10. Yoon JW, Hwang HJ (2011) Molecular dynamics modeling and simulations of a single-walled carbon-nanotube-resonator encapsulating a finite nanoparticle. Comput Mater Sci 50(9):2741–2744. https://doi.org/10.1016/j.commatsci.2011.04.033
11. Yin J, Zhang Z, Li X et al (2014) Waving potential in graphene. Nat Commun 5:3582. https://doi.org/10.1038/ncomms4582
12. Yin J, Li X, Yu J et al (2014) Generating electricity by moving a droplet of ionic liquid along graphene. Nat Nanotech 9:378–383. https://doi.org/10.1038/nnano.2014.56
13. Lei XW, Natsuki T et al (2013) An atomic-resolution nanomechanical mass sensor based on circular monolayer graphene sheet: Theoretical analysis of vibrational properties. J Appl Phys 113:154313. https://doi.org/10.1063/1.4802438
14. Lengiewicz J, Korelc J, Stupkiewicz S (2011) Automation of finite element formulations for large deformation contact problems. Int J Numer Meth Eng 85:1252–1279. https://doi.org/10.1002/nme.3009
15. Roland T, Retraint D et al (2006) Fatigue life improvement through surface nanostructuring of stainless steel by means of surface mechanical attrition treatment. Scr Mater 54:1949–1954. https://doi.org/10.1016/j.scriptamat.2006.01.049
16. Tian J, Villegas J et al (2007) A study of the effect of nanostructured surface layers on the fatigue behaviors of a C-2000 superalloy. Mater Sci Eng: A 468–470:164–170. https://doi.org/10.1016/j.msea.2006.10.150
17. Babeshko VA, Evdokimova OV, Babeshko OM (2021) Fractal properties of block elements and a new universal modeling method. Dokl Phys 66(8):218–222. https://doi.org/10.1134/S1028335821080012

18. Babeshko VA, Evdokimova OV, Babeshko OM (2020) The block element method in expansion of the solutions of complex boundary-value problems in mechanics. Dokl Phys 65(12):431–435. https://doi.org/10.1134/S1028335820120022
19. Babeshko VA, Evdokimova OV, Babeshko OM (2021) On a method for solving boundary value problems of the dynamical theory of elasticity in a quarter plane. Mechanics of Solids 56(7):217–222. https://doi.org/10.3103/S0025654421070037

Chapter 4
Determination of Dynamic Interlayer Strength Properties of Layered Composites Using Measuring Bars

Artem V. Basalin, Anatoly M. Bragov, and Aleksandr Yu. Konstantinov

Abstract The paper presents the results of the development and testing of experimental schemes that allow us to study the characteristics of the interlayer strength of layered composite materials. The schemes are based on the technique of measuring bars. To determine the interlayer strength at separation, a modification of the Kolsky method for direct tension is used. A sample of a special shape is glued to adapters having threaded parts, by means of which the sample is installed in a split measuring bar. To determine the mechanical characteristics of composite materials during interlayer shear, three experimental schemes were proposed and tested: dynamic three-point bending of a short beam, dynamic compression of plate samples with incisions and dynamic extrusion of the middle part of the samples in the form of parallelepipeds. Loading of samples and registration of their deformation processes were carried out using the technique of measuring bars. A numerical simulation was carried out to check the dynamic equilibrium condition of the sample in an experiment on the dynamic bending of a short beam. The schemes were tested on samples of a layered composite material with a polymer matrix reinforced with carbon fabric. The results of a comparative analysis of the schemes for determining shear strength showed that the most preferable is the scheme of extrusion of the middle part of the parallelepiped, since, unlike the bending of the beam, it allows you to vary and control the loading conditions, and unlike the testing of incised samples, it is symmetrical, which eliminates the appearance of bending moments in the sample.

Keywords Layered composites · Experiment · Measuring bar · Strain rate · Dynamics · Beam · Three-point bending · Deformation · Fracture · Numerical simulation · Interlayer strength

---

A. V. Basalin · A. M. Bragov · A. Yu. Konstantinov (✉)
National Research Lobachevsky State University of Nizhny Novgorod, Gagarin Ave. 23, Nizhni Novgorod, Russia 603022
e-mail: konstantinov@mech.unn.ru

A. V. Basalin
e-mail: basalin@mech.unn.ru

A. M. Bragov
e-mail: bragov@mech.unn.ru

H. Altenbach et al. (eds.), *Deformation and Destruction of Materials and Structures Under Quasi-static and Impulse Loading*, Advanced Structured Materials 186,
https://doi.org/10.1007/978-3-031-22093-7_4

4.1 Introduction

The structures made of multilayer composite materials based on glass or carbon fiber and plastic binder (polymer composite materials or PCM) during exploitation can be subjected to dynamic loads of various natures. Interest in the study of the behavior of such materials under intense dynamic loads is determined by the requests of the aerospace and automotive industries, energy technology, construction, etc. Composite materials have a set of properties and features that differ from traditional structural materials (metal alloys) and together open up wide opportunities both for improving existing structures of various purposes and for developing new structures and technological processes. To calculate stress–strain state and assess the strength of structures, modern computing systems are used, for example, ANSYS, LS-DYNA, ABAQUS, LOGOS, etc. The creation of digital models of real structural elements greatly facilitates the optimization of the design under development and significantly reduces the design time, but requires a large amount of reliable experimental information. The development of new materials and the increasing complexity of mathematical models lead to the need to develop experimental research tools. For multilayer composite materials based on glass or carbon fiber and plastic binder (polymer composite materials or PCM), the range of dynamic loads has not been studied much. Available scattered experimental data on the effect of the strain rate on the strength properties of individual classes of composites for specific loading conditions.

Currently, the methods of dynamic testing based on the classical Hopkinson-Kolsky split bar scheme have received the greatest development [14]. The main idea of the method is to use bars measuring indirectly the movement of sample points over time and the history of changes in the force acting on the sample during loading. Due to the small length of the sample compared to the length of the loading pulse, stresses and strains are evenly distributed along its entire length, and deformation occurs under conditions of so-called "dynamic equilibrium". Currently, many modifications of the Split Hopkinson Pressure Bar (SHPB) method are used for compression, tension, shear, torsion, etc. A description of various variants of SHPB can be found in Campbell and Dowling [3–5, 7–9, 12, 15, 18, 19, 21].

In the last decade, methods based on the technique of measuring bars have been widely used for the study of polymers [20] and structural PCM with various reinforcement schemes. For example, studies of the effect of the strain rate on the dynamic behavior of woven PCM under tension [10, 11] and compression [2, 16] have been carried out. The strain rate dependences of the deformation curves of unidirectional carbon fiber plastics under tension [17] and compression [13] in the direction transverse to the direction of the fibers and shear in the plane of the layer [13] are obtained at deformation rates of ~1000 1/s. Cherniaev et al. [6] obtained deformation diagrams of unidirectional carbon fiber in the direction perpendicular to the direction of the fibers in the range of strain rates 10–2000 1/s using the copra test method and the SHPB method. In Akl and Baz [1], experimental studies of the dynamic behavior and damping properties during compression of a thermoplastic polymer filled with carbon nanoparticles using the classical scheme of the SHPB method at strain rates of 2000–7000 1/s were carried out.

The purpose of this work is to develop and test methods for studying the dynamic characteristics of the strength of layered PCM during interlayer fracture: shear and separation.

4.2 Determination of Interlayer Strength at Separation

This section describes an experimental scheme for determining the ultimate strength at the tension of the PCM in the direction perpendicular to the reinforcement plane. A sample from a special-shaped PCM is attached to measuring bars using a snap-in (Fig. 4.1). The system is loaded with a tensile pulse. In order to fix the sample in the measuring bars, special adapters are used. Centering of the sample is performed by drilling holes in the samples into which the adapter guides are inserted. The sample is connected to the adapters by means of an adhesive joint. The "sample-adapter" assembly is screwed into the measuring bars. This mounting scheme allows you to prepare several samples for testing at once, while in the case of gluing the sample directly to the measuring bars, the testing process slows down noticeably.

To determine the ultimate stress at which the fracture of the sample occurs according to the interlayer separation (perpendicular to the reinforcement plane σ_3), the following ratio is used:

$$\sigma_3^+ = max\left(\frac{E_T \cdot S_T \cdot \varepsilon^T(t)}{a^2}\right)$$

where ε^{T}—the transmitted strain pulse registered in the output bar, E_T, S_T—Young's modulus and cross section area of the output bar and a—sample section side.

The left part of Fig. 4.2 shows the characteristic time dependences of the tensile stress in the cross section of the sample (blue line) and the rate of tension (red dotted line). The ultimate value of the stress was determined as the maximum values of σ_{3+}(t). At the same time, the rate of tension at the moment of fracture of the sample was determined. The data obtained during the testing of carbon fabric layered PCM

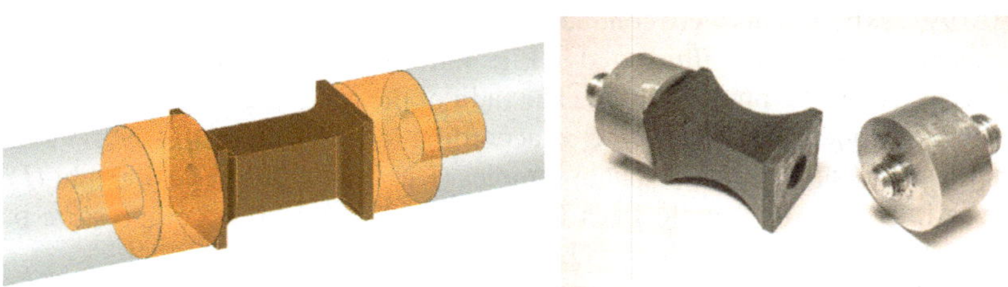

Fig. 4.1 Test scheme for determining the tensile strength of the PCM in the normal direction

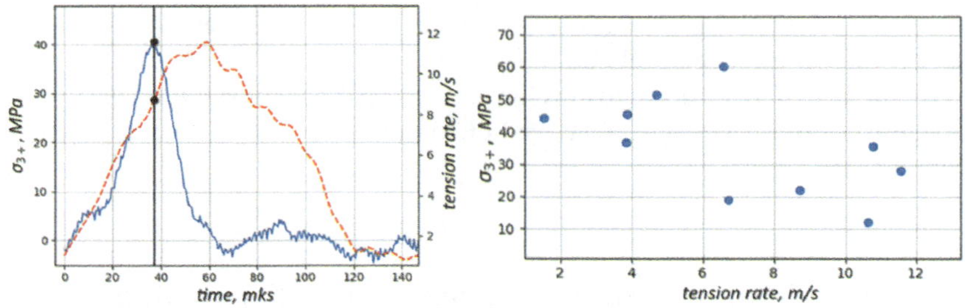

Fig. 4.2 Results of the experiment

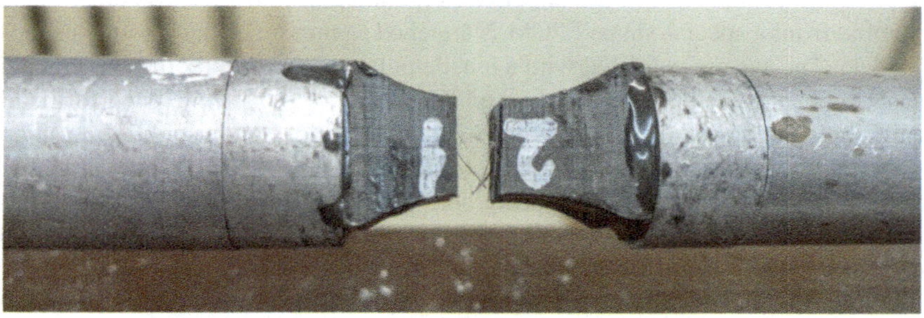

Fig. 4.3 Sample after the test

are shown in the right part of Fig. 4.2. It can be noted that the value of σ_{3+} decreases with increasing loading speed.

The sample after the test is shown in Fig. 4.3.

4.3 Determination of Interlayer Shear Strength by Bending a Short Beam

To determine the strength characteristics of the interlayer shear of the PCM under dynamic loading, by analogy with the standard for static testing of composites ISO 14130:1997, experiments were conducted on the three-point bending of a short beam. The general test scheme is illustrated in Fig. 4.4. In the experiments, measuring bars with a diameter of 20 mm were used. The radii of the rounding bars and the distance between the support bars were selected according to standard ISO 14130:1997. The geometric characteristics of the beam samples (length *L*, width *W* and height *H*) are shown in Fig. 4.4. The sample sizes were 45 × 12 × 6 mm. A compressive pulse through a loading measuring bar loads the sample.

One of the fundamental assumptions and conditions of applicability of the Kolsky or SHPB method for determining the characteristics of materials is the condition of

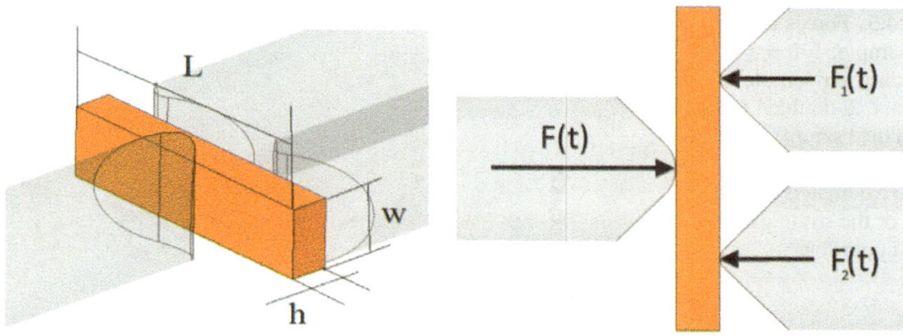

Fig. 4.4 Geometric characteristics of the sample and the forces acting on the beam during the test

dynamic equilibrium of the sample during loading. This means that at each moment of time, the force acting on the sample from the side of the loading measuring bar $F(t)$ must be equal to the sum of the forces acting on the sample from the side of the support bars $F_1(t) + F_2(t)$.

Numerical simulation was carried out to assess the conditions of dynamic equilibrium of the sample-beam in the used configuration of the test facility. The simulation was carried out in LS-DYNA code (Customer number 1069197). The results are shown in Fig. 4.5, which shows the history of changes in the forces acting on the sample from the measuring bars. The numbers indicate 1–the force from the contact "loading bar-sample", 2–the doubled force from the contact "sample-support bar" and 3–the doubled force calculated according to the data of the strain gauges on the support bar. The following conclusions can be drawn: firstly, the force acting on the sample from the side of the loading bar at each moment of time is very close in magnitude to the force acting on the sample from the side of the output-measuring bars. The time difference between the beginning of the action of the first force (F) and the appearance of the force on the support bars (F_1 and F_2) is about 10 microseconds, i.e., in the process of deformation, a dynamic equilibrium condition occurs in the sample. Secondly, the information from the strain gauge located on the support bar allows to determine accurately the force F_1 (F_2).

To process the experimental information obtained during the bending of a composite beam, the following relations are used:

The deflection rate of the beam is equal to

$$V_b(t) = c_I \cdot \left(\varepsilon^I + \varepsilon^R\right) - 0.5 \cdot c_T\left(\varepsilon_1^T + \varepsilon_2^T\right)$$

The deflection is calculated by the following formula:

$$U_b(t) = \int_0^t V_b(\tau)d\tau$$

Force acting on the beam from the input bar is calculated by the following formula:

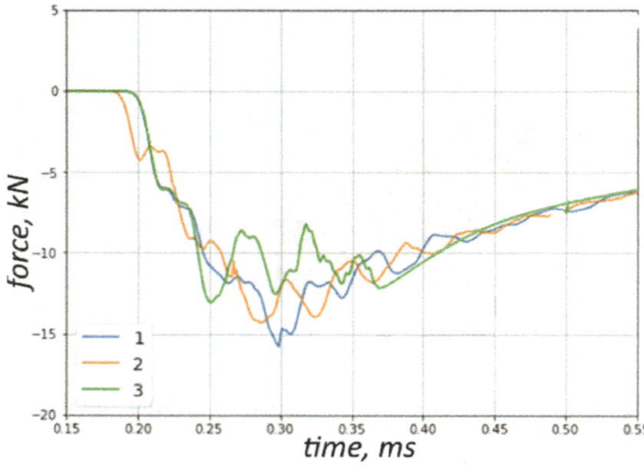

Fig. 4.5 Forces acting on the sample: 1–force from the "loading bar-sample" contact, 2–doubled force from the "sample-support bar" contact and 3–doubled force calculated from the data of the strain gauge on the support bar

$$F(t) = F_1(t) + F_2(t) = E_T \cdot S_T \cdot \left(\varepsilon_1^T + \varepsilon_2^T\right)$$

where ε^I, ε^R—incident and reflected strain pulses measured in the input bar, ε_1^T, ε_2^T—transmitted strain pulses registered in output bars, c_I—the speed of sound of input bar's material, c_T—the speed of sound of material of output bars and E_T, S_T—Young's modulus and cross section area of output bars.

Interlayer shear stress is calculated by the following formula:

$$\tau(t) = \frac{3}{4} \cdot \frac{F(t)}{h \cdot w}$$

The ultimate interlayer shear stress is the maximum value of $\tau(t)$.. The valid tests are only those in which a single or multiple bundle was formed in the sample at the end of the beam (Fig. 4.6).

The scheme was tested on beams made of laminated PCM based on carbon fiber. Photos of the samples after the test are shown in Fig. 4.7. Depending on the amplitude of the loading wave (the velocity of the impactor), different modes of fracture of the sample are observed. On the left side of the photo, a single crack appears on the end of the sample, and on the right, there is an intense multiple delamination in the loading zone.

The characteristic strain pulses recorded in the experiment are shown in Fig. 4.8. It can be noted that the pulses recorded on the support bars are in good agreement with each other, which indicates the correct alignment of the experimental setup and the exact installation of the test sample.

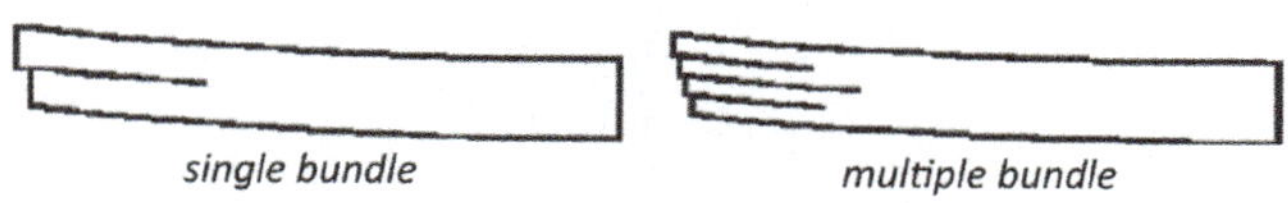

Fig. 4.6 "Valid" types of fracture of the sample

Fig. 4.7 Types of sample fracture

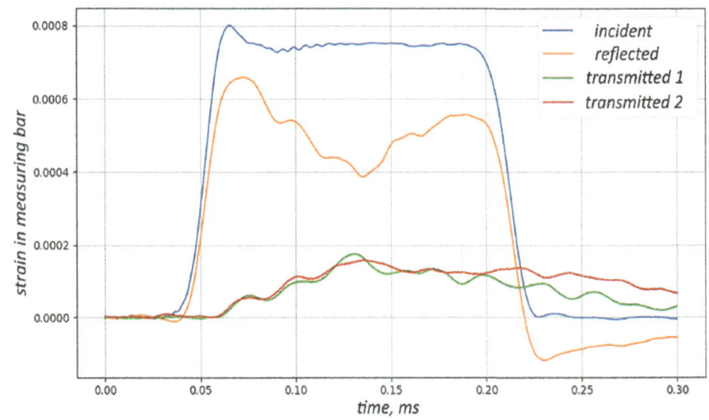

Fig. 4.8 Characteristic strain pulses recorded in the experiment for three-point bending of a short beam

Figure 4.9 illustrates a comparison of the forces acting on the sample from the loading (F) and support ($F_1 + F_2$) bars. It can be noted that these forces correspond quite well, which indicates that the conditions of dynamic equilibrium of the sample-beam during bending are met.

Based on the results of the experiment using the formulas given above, the following parameters were calculated: the deflection rate of the beam, the force F acting on the sample during the test and the magnitude of the shear stress τ. The left part of Fig. 4.10 shows the history of changes in the deflection rate of the beam (blue

Fig. 4.9 Comparison of the forces acting on the sample from the measuring bars in the bending test

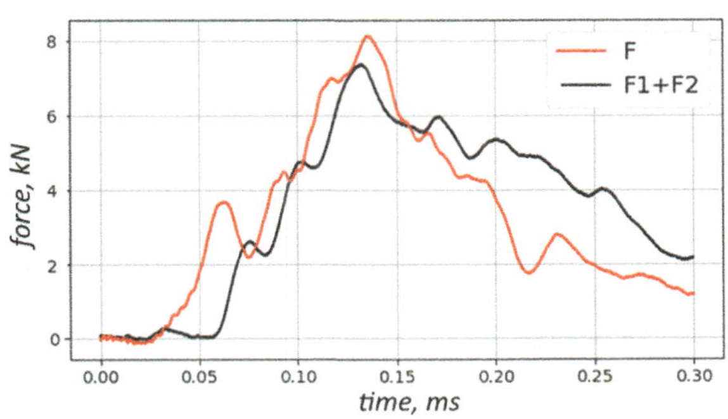

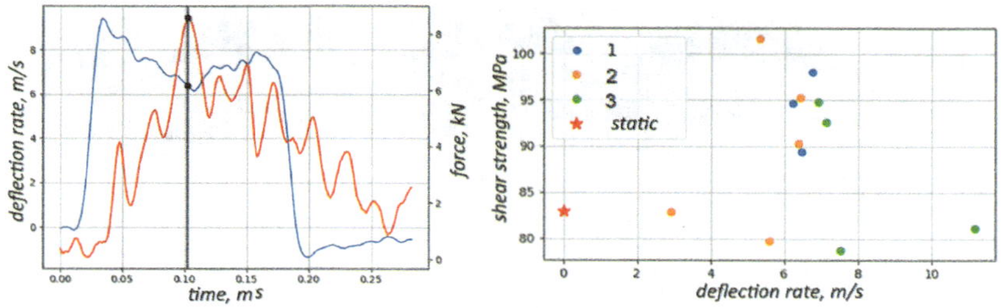

Fig. 4.10 Processing and results of the bending tests

line, left vertical axis) and the force F acting on it (red line, right vertical axis). For each experiment, the maximum value of the force and the corresponding value of the deflection velocity were determined. According to the maximum force F, using the appropriate formula, the value of the ultimate strength of the PCM during interlayer shear was calculated. The velocity dependences of the strength characteristics are presented in the right part of Fig. 4.10. The color of the dots characterizes the fracture modes of the sample: 1—corresponds to the appearance of a single crack at the end of the sample, 2—the appearance of multiple delamination at the end of the sample and 3—intense multiple delamination in the loading zone. It can be concluded that the dynamic strength is 15% higher than the static value.

4.4 Determination of Interlayer Shear Strength by Compression of Samples of Special Shape

In the second scheme, samples with incisions were tested to determine the interlayer shear strength. The configuration of the sample and equipment for fixing the sample in the measuring device is shown in Fig. 4.11.

For processing experimental information, the following relations are used:

The shear rate is equal to

$$V_{sh}(t) = c_I \cdot \left(\varepsilon^I + \varepsilon^R\right) - c_T \cdot \varepsilon^T$$

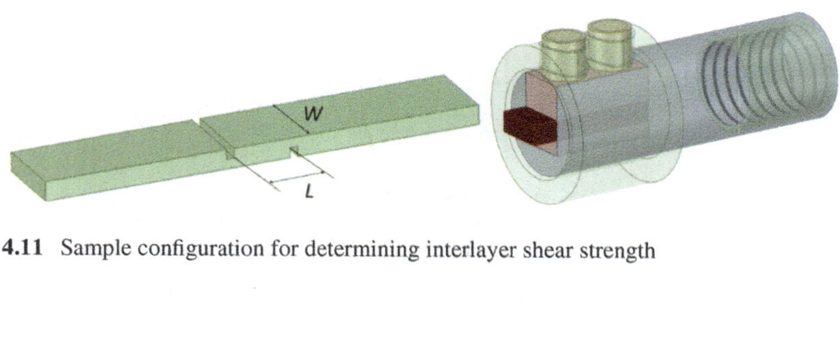

Fig. 4.11 Sample configuration for determining interlayer shear strength

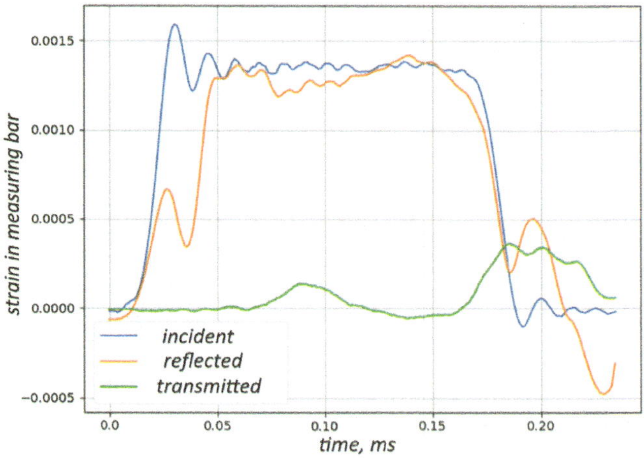

Fig. 4.12 Characteristic strain pulses recorded in the compression experiment of samples with incisions

Shear stress is calculated by the following formula:

$$\tau(t) = \frac{E_T \cdot S_T \cdot \varepsilon^T}{L \cdot w}$$

where ε^I, ε^R—incident and reflected strain pulses measured in the input bar, ε^T—transmitted strain pulse registered in output bar, c_I—the speed of sound of input bar's material, c_T—the speed of sound of material of output bars, E_T, S_T—Young's modulus and cross section area of output bars and L and W—length and width of the working area of the sample.

The scheme was tested on beams made of laminated PCM based on carbon fiber. The characteristic strain pulses recorded in the experiment are shown in Fig. 4.12.

Figure 4.13 illustrates a comparison of the forces acting on the sample from the loading (F_1) and support (F_2) bars calculated according to the Kolsky formulas. The presence of complex equipment for attaching the sample to the measuring bars (forks + tightening clips) distorts the wave pattern. Massive elements lead to additional wave reflections. A drop appears at the beginning of the reflected pulse and, accordingly, an outburst appears on the force, which is determined by the difference between the incident and reflected pulses, which does not reflect the force acting on the sample, but is a superposition of forces acting on the sample and an inert massive rigging. However, during the loading of the sample, the condition of dynamic equilibrium takes place.

Based on the results of the experiment using the formulas given earlier, the following parameters were calculated: the shear rate, the force F acting on the sample during the test and the magnitude of the shear stress τ. The left part of Fig. 4.14 shows the history of changes in the shear rate (blue line, left vertical axis) and the force

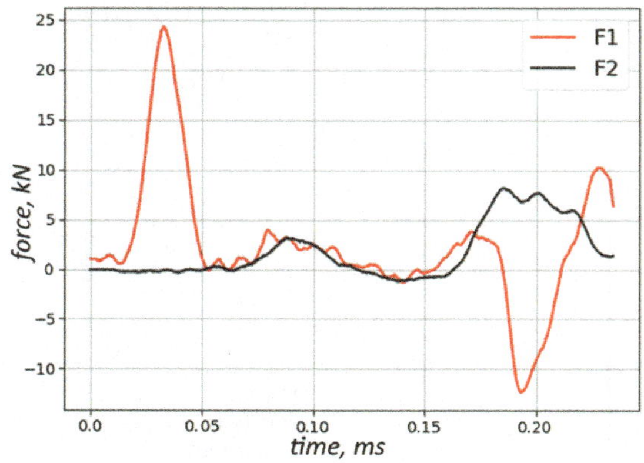

Fig. 4.13 Comparison of the forces acting on the sample from the measuring bars in the shear test

F acting on the sample (red line, right vertical axis). The stress has several peaks. The first corresponds to the destruction of the sample along the cut plane. During the second peak, the parts of the sample are closed when the gap formed by the cut is estimated. For each experiment, the maximum value of the force in the first peak and the corresponding value of the shear rate were determined. According to the maximum force F, the value of the strength of the PCM during interlayer shear was calculated using the appropriate formula. The velocity dependences of the strength characteristics are presented in the right part of Fig. 4.14. The points are grouped by the thickness of the sample. Blue markers correspond to the data obtained on samples with a thickness of 2 mm, and orange—on samples with a thickness of 6.5 mm. The dynamic strength according to the specified test method turned out to be lower than the static one.

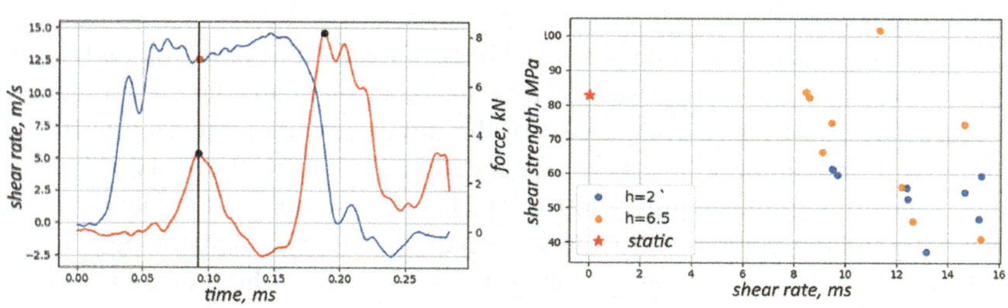

Fig. 4.14 Processing and results of the experiment on compression of samples with incisions

4.5 Determination of the Interlayer Shear Strength of the PCM by Extrusion of the Middle of the Parallelepiped Sample

To determine the interlayer shear strength of the PCM, a scheme was used to extrude the middle of the sample-parallelepiped (two-plane shift). Loading in the RSG system was carried out by a compressive load. The general view of the experimental scheme is shown in Fig. 15a. The equipment for loading sample 5 (Fig. 15b) in the SHPB system (measuring bars 1 and 2) includes parts 3 and 4, as well as a guide 6 for centering parts.

When loading the sample in the described tooling, it shifts along the planes highlighted in red in Fig. 4.16. Samples of 25 × 25x10 mm were tested.

The following relations are used to process experimental information:

Shear rate:

$$V_{sh}(t) = c_I \cdot \left(\varepsilon^I + \varepsilon^R\right) - c_T \cdot \varepsilon^T$$

Shear stress:

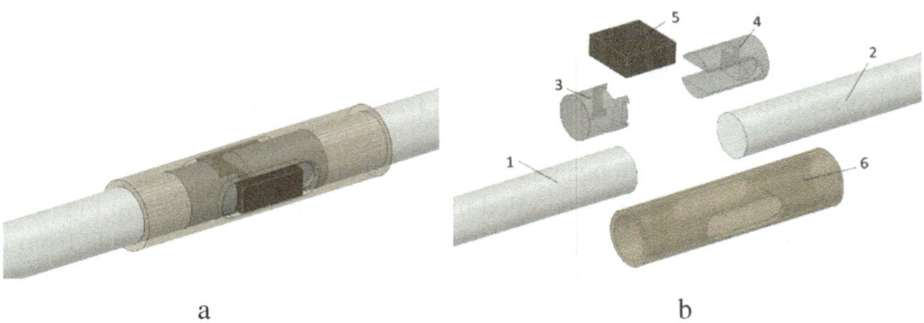

Fig. 4.15 General view of the experimental scheme

Fig. 4.16 Geometric characteristics of the sample

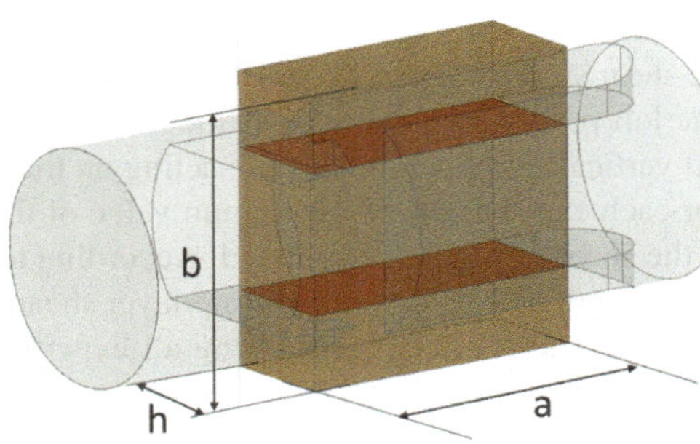

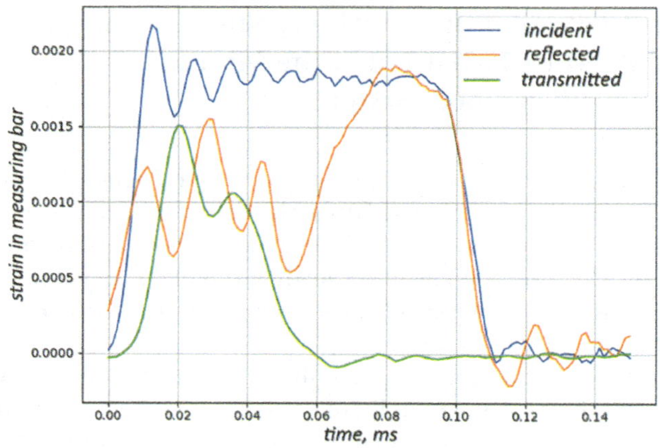

Fig. 4.17 Characteristic strain pulses recorded in the extrusion experiment

$$\tau(t) = \frac{E_T \cdot S_T \cdot \varepsilon^T}{2 \cdot a \cdot h}$$

where ε^I, ε^R—incident and reflected strain pulses measured in input bar, ε^T—transmitted strain pulse registered in output bar, c_I—the speed of sound of input bar's material, c_T—the speed of sound of material of output bars, E_T, S_T—Young's modulus and cross section area of output bars and a and h are the length and width of the shear zone.

The scheme was tested on beams made of laminated PCM based on carbon fiber. The characteristic strain pulses recorded in the experiment are shown in Fig. 4.17.

Figure 4.18 illustrates a comparison of the forces acting on the sample from the loading (F_1) and support (F_2) bars calculated according to the Kolsky formulas. It is seen that the presence of adapters used to implement extrusion introduces distortions into the wave pattern, as a result of which the conditions of dynamic equilibrium are not met.

Based on the results of the experiment, the following parameters were calculated: the shear rate, the force F acting on the sample during the test and the magnitude of the shear stress τ. The force is determined by the strain in the output-measuring bar. The left part of Fig. 4.19 shows the history of changes in the shear rate (blue line, left vertical axis) and the force F acting on the sample (red line, right vertical axis). For each experiment, the maximum value of the force and the corresponding value of the shear rate were determined. According to the maximum force F, the value of the strength of the PCM during interlayer shear was calculated using the appropriate formula. The results of processing all experiments are illustrated in the right part of Fig. 4.19. The asterisk corresponds to the static value of shear strength during interlayer shear. The values of interlayer shear strength obtained in the dynamic range were on average 25% lower than the static characteristic.

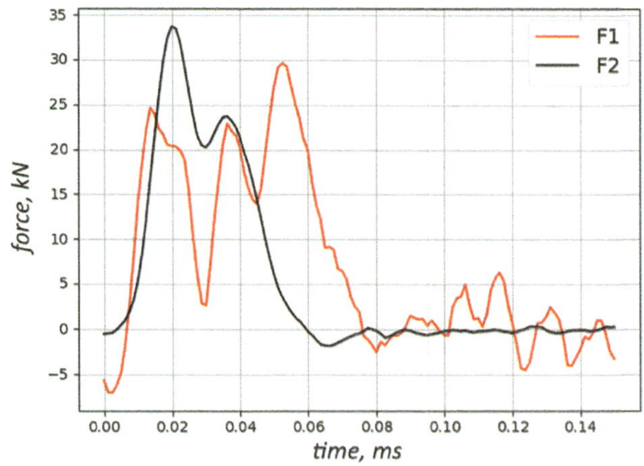

Fig. 4.18 Comparison of forces acting on the sample from the measuring rods in the extrusion experiment

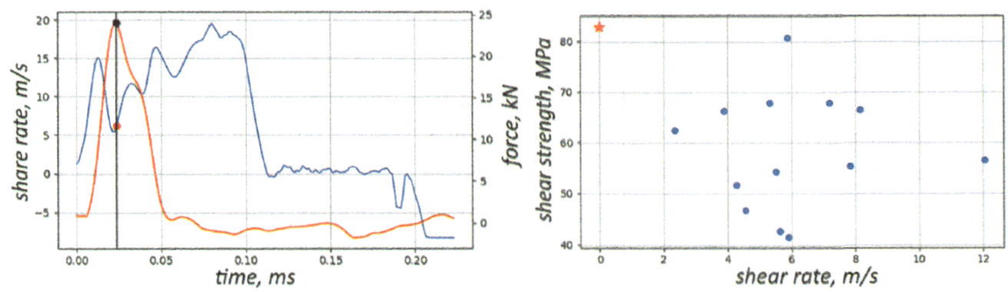

Fig. 4.19 Processing and results of the extrusion experiments

4.6 Comparative Analysis of Schemes for Interlayer Shear

Figure 4.20 shows a comparison of the values of the strength of the PCM during interlayer shear obtained by different methods. It can be noted that the data obtained by extrusion (red triangles) are in good agreement with the data determined by the dynamic compression of samples with incisions (orange and blue squares). The strength characteristic determined by the method of bending a short beam turned out to be noticeably higher.

4.7 Conclusions

As a result of the work performed, a number of experimental schemes and corresponding experimental installations were created based on the technique of

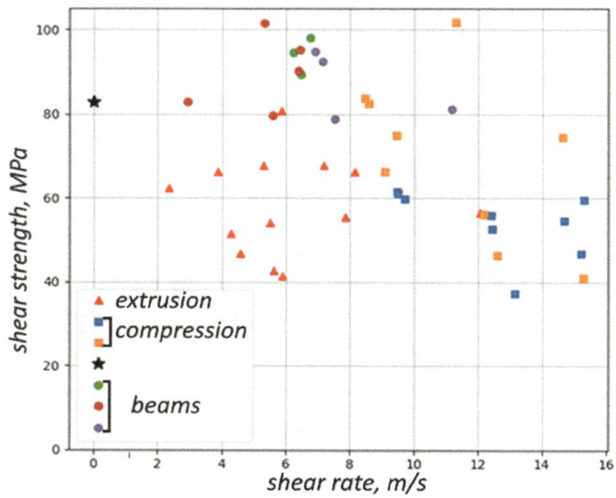

Fig. 4.20 Comparison of data obtained by different methods

measuring bars, which allow testing samples from PCM to determine the strength characteristics of interlayer shear and separation depending on the loading speed. The method of dynamic bending of a short beam does not allow varying the loading conditions in a sufficiently wide range, since the shape of the fracture of the sample changes depending on the intensity of the load. In addition, in this type of test, it is impossible to assess the conditions of destruction by the shape under study (for example, the shear rate). The scheme for testing a sample with incisions is asymmetric. Despite the small thickness of the sample, due to geometric features, a bending moment occurs, which can disrupt the isolation of the fracture mode by interlayer shear. The most informative is the scheme with the extrusion of the middle part of the sample in the form of a parallelepiped. The approbation of the schemes was carried out on the example of a layered composite of woven reinforcement. A comparative analysis of the data obtained according to the described schemes is performed.

Acknowledgements The theoretical study was done with financial support from the Ministry of Science and Higher Education of the Russian Federation (Project 0729-2020-0054). Experimental studies were carried out with the financial support of the Russian Science Foundation (project 21-19-00283).

References

1. Akl W, Baz AM (2018) Dynamic behavior and damping characteristics of carbon black polymer composites at high strain rates. Adv Polym Technol 37(2):3364–3375. https://doi.org/10.1002/ADV.22120

2. Bandaru AK, Chouhan H, Bhatnagarc N (2020) High strain rate compression testing of intraply and inter-ply hybrid thermoplastic composites reinforced with Kevlar/basalt fibers. Polymer Testing 84, Article No 106407. https://doi.org/10.1016/j.polymertesting.2020.106407.
3. Bol'shakov AP, Novikov SA, Sinitsyn VA (1979) Dynamic uniaxial tension and compression curves of copper and alloy AMg6. Strength Mater 11(10):1159–1161. https://doi.org/10.1007/BF00768612
4. Bragov AM, Gandurin VP, Grushevsky GM, Lomunov AK (1995) New potentials of Kol'skii's method for studying the dynamic properties of soft soils. J Appl Mech Tech Phys 36(3):476–481. https://doi.org/10.1007/BF02369791
5. Campbell JD, Dowling AR (1979) The behaviour of materials subjected to dynamic incremental shear loading. J Mech Phys Solids 18(1):43–63. https://doi.org/10.1016/0022-5096(70)90013-X
6. Cherniaev A, Zenga Y, Cronina D, Montesanoa J (2019) Quasi-static and dynamic characterization of unidirectional non-crimp carbon fiber fabric composites processed by HP-RTM. Polym Testing 76:65–375. https://doi.org/10.1016/J.POLYMERTESTING.2019.03.036
7. Dharan CKH, Hauser FE (1970) Determination of stress-strain characteristics at very high strain 18 rates. Exp Mech 10(9):370–376. https://doi.org/10.1007/BF02320419
8. Doerner MF, Nix WD (1986) A method for interpreting the data from depth-sensing indentation instruments. J Mater Res 1(4):601–609. https://doi.org/10.1557/JMR.1986.0601
9. Duffy JA, Campbell JD, Hawley RH (1971) On the use of a torsional split Hopkinson bar to study rate effects in 1100-0 aluminum. J Appl Mech 38(1):83–91. https://doi.org/10.1115/1.3408771
10. Elmahdy A, Verleysen P (2020) Mechanical behavior of basalt and glass textile composites at high strain rates: a comparison. Polym Test 81, Article No. 106224. DOI:https://doi.org/10.1016/j.polymertesting.2019.106224.
11. Elmahdy A, Verleysen P (2019) Tensile behavior of woven basalt fiber reinforced composites at high strain rates. Polym Test 76:207–221. https://doi.org/10.1016/j.polymertesting.2019.03.016
12. Klepaczko JR (1982) Discussion of a new experimental method in measuring fracture toughness initiation at high loading rates by stress waves. Transactions of the ASME. Journal of Engineering for Industry. 104(1):29–35
13. Koerber H, Xavier J, Camanho PP (2010) High strain rate characterisation of unidirectional carbon-epoxy IM7-8552 in transverse compression and in-plane shear using digital image correlation. Mech Mater 42(11):1004–1019. https://doi.org/10.1016/j.mechmat.2010.09.003
14. Kolsky H (1949) An investigation of the mechanical properties of material at very high rates of loading. Proc Phys Soc B 62(11):676–700
15. Lewis JL, Goldsmith W (1973) A biaxial split Hopkinson bar for simultaneous torsion and compression. Rev Sci Instrum 44(7):811–813. https://doi.org/10.1063/1.1686253
16. Liu T, Sun B, Gu B (2018) Size effects on compressive behaviors of three-dimensional braided composites under high strain rates. J Compos Mater 53(28):3895–3908. https://doi.org/10.1177/0021998318771459
17. Melin L, Asp L (1999) Effects of strain rate on transverse tension properties of a carbon/epoxy composite: Studied by moiré photography. Compos Part A Appl Sci Manuf 30(3):305–316. https://doi.org/10.1016/S1359-835X(98)00123-7
18. Muzychenko VP, Kashchenko SI, Guskov VA (1986) Application of the Hopkinson composite rod method in the study of the dynamic properties of materials (review). Head. Laboratory 1:58–66 (in Russ.)
19. Nicholas T (1981) Tensile testing of materials at high rates of strain. Exp Mech 21(5):177–195. https://doi.org/10.1007/BF02326644
20. Siviour CR, Jordan JL (2016) High strain rate mechanics of polymers: a review. J Dyn Behav Mater 2(1):15–32. https://doi.org/10.1007/s40870-016-0052-8
21. Zukas JA, Nicholas T, Swift H, Greszczuk LB, Curran DR (1982) Impact dynamics. New York. Chichester. Brisbane. Toronto. Singapore

Chapter 5
Durability of High-Load Structures

Valentin S. Bondar and Dmitry R. Abashev

Abstract The article describes the constitutive model based on the applied inelasticity theory—one of the combined hardening flow theories. The authors identify the material functions closing the applied inelasticity theory and formulate the fundamental experiment. The life of structural materials in the case of non-isothermal cyclic loading is predicted by analyzing the durability of the air-cell diesel's edge and the uncooled conical nozzle tip in the case of thermal cycling. Life estimates based on the applied inelasticity theory are compared to experimental data and conservative life estimation methods. The authors also consider examples of estimating the life of a durable power generation system's structure. Loading modes resulting in considerable life reduction are described.

Keywords The combined hardening flow theory · Life of structural materials · Experiment · Thermoviscoplasticity

5.1 Introduction

Reliable information on the damage accumulation processes in structural materials is required to ensure long-term (up to several decades) safe use of critical facilities (nuclear and thermal power generation systems, chemical, gas, oil, aerospace aviation facilities, etc.). As structural elements reach the end of their nominal service life, there appears a problem of extending such life and ensure safe use throughout it.

These goals may be reached only by fulfilling operational monitoring methods, the primary objectives whereof include estimation of the worked-out service life and prediction of the extendable life of the materials making up critical areas of the most loaded structural elements on the basis of mathematical modeling of damage accumulation processes using contemporary thermoviscoplasticity or inelasticity theories.

V. S. Bondar · D. R. Abashev (✉)
Department of Technical Mechanics, Moscow Polytechnic University, Moscow 107023, Russian Federation
e-mail: tm@mospolytech.ru

H. Altenbach et al. (eds.), *Deformation and Destruction of Materials and Structures Under Quasi-static and Impulse Loading*, Advanced Structured Materials 186,
https://doi.org/10.1007/978-3-031-22093-7_5

A large number of publications are dedicated to the issues of generating alternative variants of thermoviscoplasticity or inelasticity theories. See monographs, reviews, and separate publications for the main theory-building trends and extensive bibliography [1–34].

The applied variants of the combined hardening-based flow theory have become the most widely used for practical calculations. The best evidence-based and widely used for estimating the life of structural elements exposed to thermoviscoplastic deformation characterized by repeated long-term thermomechanical stresses are Bondar [2–6], Korotkih [16, 17] and Chaboche [7–12] theory variants. It ought to be mentioned that Korotkih and Chaboche distinguish between elastic, plastic, and creep deformations, whereas Bondar—between elastic and inelastic ones. That is why the former variants belong to the thermoviscoplasticity theory, whereas the latter one—to the inelasticity theory. Accordingly, evolutionary equations for the yield surface radius, displacement (backstresses), and damage accumulation are in the former variants formulated separately for plasticity and creep, whereas in the latter one—only for inelasticity. It ought to be mentioned that the division of irreversible deformations into plastic and creep ones is conventional in nature. Rapid processes are characterized by smaller creep deformations; these are neglected in favor of plastic deformations. Slower processes, though, are characterized by both plastic and creep deformations, and their division is conventional in nature, as in that case the irreversible deformation is one. Rapid deformation processes are usually characterized by viscous fractures following damage accumulation in the grain body, whereas very slow processes—by brittle fractures following damage accumulation along grain edges. That is why in the former variants damage accumulation processes are divided and described with different kinetic equations. The main drawback of the first approach consists in the absence of interconnection between plasticity and creep processes characterizing both deformations and fractures. This comes in contrast to experimental data [1, 13, 16, 17, 23, 32, 34]. The second approach takes the deformation process history and interconnection of plasticity and creep into account a priori—a rapid process affects the subsequent slow one and vice versa. As for fractures, the presence of the kinetic equation to describe embrittlement among the theory's equations makes both viscous (in case of significant deformations and short-term processes, or if the number of cycles is low) and brittle (in case of smaller deformations and longer-term processes, or high-cycle fatigue) fractures possible. Therefore, this theory variant allows simulating mixed types of deformations and fractures.

This article describes the main provisions and equations of the applied inelasticity theory—one of the combined hardening flow theories. The authors identify the material functions closing the applied inelasticity theory and formulate the fundamental experiment. The life of structural materials in the case of non-isothermal cyclic loading is predicted by analyzing the durability of the air-cell diesel's edge and the uncooled conical nozzle tip in the case of thermal cycling. Life estimates based on the applied inelasticity theory are compared to experimental data and conservative life estimation methods. The authors also consider examples of estimating the life of a durable power generation system's structure. Loading modes resulting in considerable life reduction are described.

5.2 Main Provisions and Equations of the Applied Inelasticity Theory

The material is homogenous and initially isotropic. The article describes small deformations of polycrystalline structural steel grades and alloys taking place at temperatures not characterized by phase transitions and at such deformation rates that dynamic effects may be neglected. Inelastic deformation may result only in the material's deformation-induced anisotropy. Below is a summary of the basic equations for the applied inelasticity theory.

$$\dot{\varepsilon}_{ij} = \dot{\varepsilon}_{ij}^{e} + \dot{\varepsilon}_{ij}^{p}$$

$$\dot{\varepsilon}_{ij}^{e} = \frac{1}{E}\left[\dot{\sigma}_{ij} - \nu\left(3\dot{\sigma}_0\delta_{ij} - \dot{\sigma}_{ij}\right)\right] + \alpha_{ij}^{eT}\dot{T},$$

$$\alpha_{ij}^{eT} = \alpha_T\delta_{ij} - \frac{1}{E^2}\left[\sigma_{ij} - \nu\left(3\sigma_0\delta_{ij} - \sigma_{ij}\right)\right]\frac{dE}{dT} - \frac{1}{E}\left(3\sigma_0\delta_{ij} - \sigma_{ij}\right)\frac{d\nu}{dT},$$

$$f\left(\sigma_{ij}\right) = \frac{3}{2}\left(s_{ij} - a_{ij}\right)\left(s_{ij} - a_{ij}\right) - C^2 = 0,$$

$$\dot{C} = q_\varepsilon\dot{\varepsilon}_{u*}^{p} + q_T\dot{T} - q_R, \quad \dot{\varepsilon}_{u*}^{p} = \left(\frac{2}{3}\dot{\varepsilon}_{ij}^{p}\varepsilon_{ij}^{p}\right)^{1/2},$$

$$\dot{a}_{ij} = \frac{2}{3}g\dot{\varepsilon}_{ij}^{p} + \left(\frac{2}{3}g_\varepsilon\varepsilon_{ij}^{p} + g_a a_{ij}\right)\dot{\varepsilon}_{u*}^{p} + \left(\frac{2}{3}g_\varepsilon^{T}\varepsilon_{ij}^{p} + g_a^{T}a_{ij}\right)\dot{T} - \left(\frac{2}{3}g_\varepsilon^{R}\varepsilon_{ij}^{p} + g_a^{R}a_{ij}\right),$$

$$\dot{\varepsilon}_{ij}^{p} = \frac{\partial f}{\partial\sigma_{ij}}\dot{\lambda} = \frac{3}{2}\frac{s_{ij}^{*}}{\sigma_u^{*}}\dot{\varepsilon}_{u*}^{p}, \quad s_{ij}^{*} = s_{ij} - a_{ij}, \quad \sigma_u^{*} = \left(\frac{3}{2}s_{ij}^{*}s_{ij}^{*}\right)^{1/2},$$

$$\dot{\varepsilon}_{u*}^{p} = \frac{1}{E_*}\left[\frac{3}{2}\frac{s_{ij}^{*}\dot{\sigma}_{ij}}{\sigma_u^{*}} - B^T\dot{T} + B^R\right],$$

$$\dot{\varepsilon}_{u*}^{p} = \frac{1}{E_* + 3G}\left[3G\frac{s_{ij}^{*}\dot{\varepsilon}_{ij}}{\sigma_u^{*}} - B^T\dot{T} + B^R\right],$$

$$E_* = q_\varepsilon + g + g_\varepsilon\varepsilon_u^{p*} + g_a a_u^{*},$$

$$B^T = q_T + g_\varepsilon^{T}\varepsilon_u^{p*} + g_a^{T}a_u^{*},$$

$$B^R = q_R + g_\varepsilon^{R}\varepsilon_u^{p*} + g_a^{R}a_u^{*},$$

$$\varepsilon_u^{p*} = \frac{s_{ij}^{*}\varepsilon_{ij}^{p}}{\sigma_u^{*}}, \quad a_u^{*} = \frac{3}{2}\frac{s_{ij}^{*}a_{ij}}{\sigma_u^{*}},$$

$$\sigma_u^* < C \cup \dot{\varepsilon}_{u*}^p \le 0 \quad - \quad \text{elasticity } \left(\dot{\varepsilon}_{u*}^p = 0\right),$$
$$\sigma_u^* = C \cap \dot{\varepsilon}_{u*}^p > 0 \quad - \quad \text{inelasticity } \left(\dot{\varepsilon}_{u*}^p \ne 0\right),$$

$$\dot{\omega} = \alpha\,\omega^{\frac{\alpha-1}{\alpha}}\,\frac{1}{W}\left(a_{ij}\dot{\varepsilon}_{ij}^p\right) - g_\omega\omega,$$

$$\dot{W} = g_W^T\dot{T} - g_W W.$$

Here ε_{ij}, ε_{ij}^e, ε_{ij}^p—total, elastic and inelastic tensors; T—temperature; $\sigma_0 = \sigma_{ii}/3$—average stress; δ_{ij}—Kronecker delta; σ_{ij}, s_{ij}—stress tensor and deviator; a_{ij}—backstress deviator; C—yield surface radius (size); ε_{u*}^p—accumulated inelastic deformation; s_{ij}^*—active stress deviator; G—shear modulus; ω—damage ($\omega \in [0;1]$); W—fracture power.

Defining functions q_ε, q_T, q_R, g, g_ε, g_a, g_ε^T, g_a^T, g_ε^R, g_a^R, g_ω, g_W^T, g_W, α are expressed with material functions as follows [2–6]:

$$q_\varepsilon = \frac{\partial C_p}{\partial \varepsilon_{u*}^p},\quad q_T = \frac{C}{C_p}\frac{\partial C_p}{\partial T},\quad q_R = q_\varepsilon P_c,$$

$$g = E_a + \beta\sigma_a,\quad g_\varepsilon = \beta E_a,\quad g_a = -\beta,$$

$$g_\varepsilon^T = \frac{dE_a}{dT} - \frac{E_a}{\sigma_a}\frac{d\sigma_a}{dT},\quad g_a^T = \frac{1}{\sigma_a}\frac{d\sigma_a}{dT},$$

$$g_\varepsilon^R = g_\varepsilon P_a,\quad g_a^R = (g + g_a a_u)\frac{P_a}{a_u},\quad a_u = \left(\frac{3}{2}a_{ij}a_{ij}\right)^{1/2},$$

$$g_\omega = \lambda,\quad g_W = \rho,\quad g_W^T = \frac{W}{W_0}\frac{dW_0}{dT},$$

$$\alpha = \left(\frac{\sigma}{\sigma_{au}}\right)^{n_\alpha},\quad \sigma_{au} = \left[\frac{3}{2}\left(a_{ij} - \frac{2}{3}E_a\varepsilon_{ij}^p\right)\left(a_{ij} - \frac{2}{3}E_a\varepsilon_{ij}^p\right)\right]^{1/2},$$

$$P_c = \exp(b_c)\,|C - C_{P0}|^{n_C}(1-\omega)^{-m_\omega},\quad C_{P0} = C_P(T,\ 0),$$

$$P_a = \exp(b_a)\,(a_u)^{n_a}(1-\omega)^{-m_\omega},$$

$$\lambda = \begin{cases} 0, & \text{if } \sigma_{ii} \ge 0 \\ \exp(b_\lambda)\,|\sigma_{ii}|^{n_\lambda}, & \text{if } \sigma_{ii} < 0 \end{cases},$$

$$\rho = \exp\left(b_\rho\right)(\sigma_u)^{n_\rho},\quad \sigma_u = \left(\frac{3}{2}s_{ij}s_{ij}\right)^{1/2}.$$

The equations above belong to the primary fundamental version of the inelasticity theory—applied inelasticity theory [2–6]—one of the single-surface flow theories for combined (translation/isotropic) hardening characterized by isotropic expansion or narrowing of the loaded surface and displacement of the loaded surface's center. For the yield surface radius, we formulated the evolutionary equation to express isotropic hardening, non-isotermal transition, and annealing softening. The evolutionary equation for the backstress deviator features three terms, i.e. the yield surface displacement rate deviator may be divided into inelastic deformation rate, displacement, and inelastic deformation deviators. The first three addenda of this evolutionary equation express an anisotropic hardening, and the subsequent one—non-isothermal transition and recrystallization (softening). It ought to be mentioned that this theory does not distinguish between plastic and creep deformations; all deformations are inelastic. We also formulated kinetic damage accumulation equations to describe damage accumulation processes, where we adopt the yield of backstresses characterizing inelastic deformations as the power expended for damaging the material. Apart from the addendum to express damage accumulation caused by backstresses, kinetic equations contain the addenda to ensure non-isothermal transition, softening, and healing.

5.3 Material Functions and Fundamental Experiment

The applied inelasticity theory is closed by the following material functions:

- $E\,(T),\ \ \nu\,(T),\ \ \alpha_T\,(T\,)$—elastic parameters;
- $E_a\,(T),\ \ \sigma_a\,(T),\ \ \beta\,(T)$—moduli of anisotropic hardening;
- $C_p\left(T\,,\ \varepsilon^P_{u*}\right)$—isotropic hardening function;
- $W_0\,(T\,)$—initial fracture energy;
- $n_\alpha(T)$—damage accumulation process's non-linearity parameter equal to 1.5 for almost all structural steel grades and alloys;
- $b_c\,(T),\ \ b_a\,(T)\,,\ \ n_c(T)\,,\ \ n_a\,(T),\ \ m_\omega(T)$—isotropic and anisotropic creep parameters;
- $b_\lambda\,(T)\,,\ \ b_\rho\,(T),\ \ n_\lambda(T)\,,\ \ n_\rho\,(T)$—healing and softetning parameters.

The following general (fundamental) experimental data set for different temperature levels is sufficient to define material functions:

- elastic parameters $E,\ \nu,\ \alpha_T$ as defined by conservative methods;
- uniaxial extension $\sigma_1(\varepsilon)$ plastic deformation (up to 0.05–0.1) diagram;
- uniaxial extension $\sigma_2(\varepsilon)$ plastic deformation (up to 0.05–0.1 after compression to deformation of 0.01–0.02) diagram;
- cyclic uniaxial extension/compression data at unchanging plastic deformation range: number of cycles to failure (macrocrack) N_f and dependence of maximum stress values σ_N^+ during a cycle (in the end of a cycle) on a number of loading cycles N (plastic deformation range $\Delta\varepsilon^P$ of ca. 0.01–0.02);

- creep data for uniform extension stress: dependence of the minimal creep rate on strain throughout the whole range of strain change from short-term to rather long creep;
- long-term durability data: a long-term extension durability curve that includes all three segments and a long-term compression durability curve that corresponds to the second segment only.

The computational-experimental method of defining material functions on the basis of the fundamental experiment is detailed in Bondar publications [2–6], which list material functions for a range of structural steel grades and alloys.

5.4 Air-Cell Diesel's Durability

Operational experience demonstrates that increasing diesel power leads to the premature breakdown of forces due to cracking. Extension of cracks perpendicular to the air cell diesel's edge is typical of deep chambers. Variable thermal stresses caused by transient diesel operation modes have a prevailing effect over fractures of forcer combustion chambers. It appears reasonable to perform accelerated testing of test samples on engineless rigs at the new forcer design development stage. An engineless thermal rig heats forcers with radiant energy (halogen lamps), while compressed air blow-off ensures cooling of forcer combustion chamber edges.

Analysis of computed stressed-deformed states of diesel forcers demonstrated a virtually uniaxial stressed state on the combustion chamber edges where cracks appear. It ought to be mentioned that hoop stresses are one hundred times larger than other stressed state components.

All the aforesaid helped to develop a computational-experimental method to predict the durability of combustion chamber edges. It consists in the experimental determination of the edge's temperature and radial displacement (hoop deformation) as functions of time in the thermal loading cycle, as well as in the calculation of the stressed-deformed state kinetics and the number of cycles to fracture (macrocrack) on the basis of the applied inelasticity theory.

We analyzed two thermal loading programs for an *AL25* aluminum alloy-made forcer differing in heating and cooling durations. The heating and cooling duration in the first program was $47 + 47 = 94$ s. In the second program, it was $72 + 72 = 144$ s. Figure 5.1 representes experimental changes of the edge's temperature T and hoop deformation ε_θ, as well as calculated changes of the force hoop deformation $\varepsilon_{\theta S}$ (the whole deformation excluding the temperature one) and of the inelastic hoop deformation ε_θ^P for the second thermal loading program. The stabilized cyclic loop for the second thermal loading program is given in Fig. 5.2. A back loop in the extension stress area is caused by intensive heat transfer in the forcer cooling mode. Inelastic deformation ranges for the first and the second programs are 0.00047 and 0.00121, respectively. The estimated number of cycles for the first program was 1,870, whereas the experimental one varied from 1,900 to 2,200; the estimated number of cycles for the second program was 260, whereas the experimental one was 300. The comparison of estimated and experimental numbers of cycles to the cracks caused by non-isothermal loading of the combustion chamber edge shows that they are satisfactory. It ought to be mentioned that it is rather problematic to use Coffin's test to predict durability, as the available plasticity of the AL25 aluminum alloy strongly depends on the temperature.

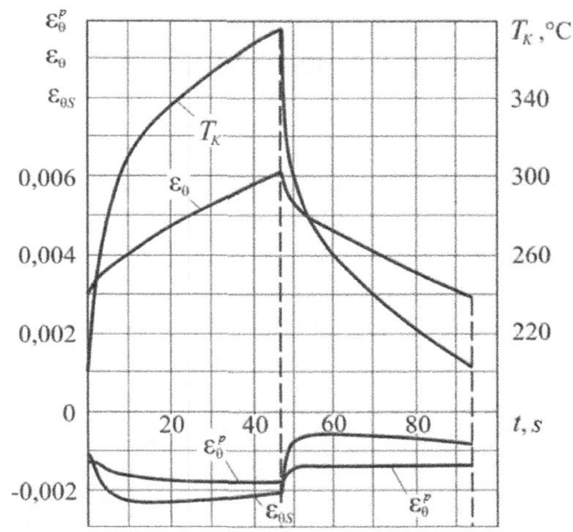

Fig. 5.1 Experimental changes of the combustion chamber edge's temperature and hoop deformation

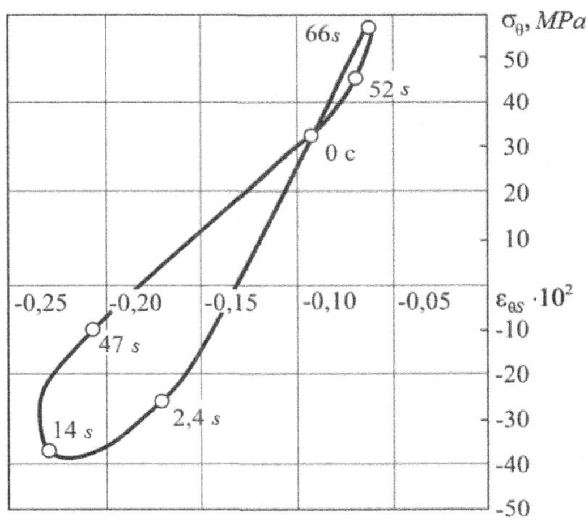

Fig. 5.2 AL25 aluminum alloy. Stabilized cyclic loop

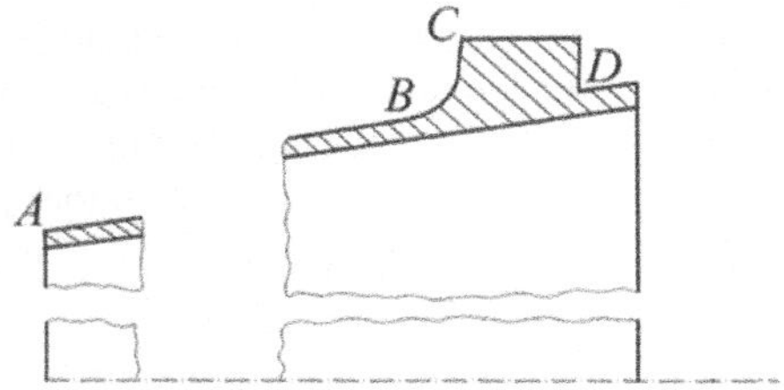

Fig. 5.3 Geometric properties of the liquid rocket engine's nozzle tip

5.5 Low-Cycle Durability of the Conical Nozzle Tip

The study object is represented by a single-layer truncated conical shell (Fig. 5.3). The shell's back edge is supported by a stiffening ring to ensure the required stiffness. Shell wall's thickness is 1.0 mm, structural material—BZh100 heat-resistant steel. The structure was exposed to multiple intensive heating instances (immersion into melted salt) and subsequent cooling (by means of air blow-off and immersion into liquid nitrogen) in a laboratory. Figure 5.4 contains experimental temperature change curves for the shell (*AB* segment) and the stiffening ring (*CD* segment) throughout the full thermal cycle. There is a significant alternating temperature gradient of the generating structure in the *BC* segment of the shell/stiffening ring transition. The applied inelasticity theory was used to calculate the kinetics of the stressed-deformed state. We also evaluated the low-cycle durability of the nozzle tip. Figure 5.5 contains a calculated trajectory of the inelastic deformations throughout three thermal cycles for point *B* of the structure where the fracture took place (separation of the stiffening ring from the shell). The estimated number of thermal cycles to structural fracture was 50, whereas experimental fractures took place after 46–53 thermal cycles. The number of cycles to fracture estimated on the basis of the Novozhilov-Rybakina test is 33, whereas on the basis of the deformation kinetic test it is 250. The aforegiven studies demonstrate that the loads in the analyzed point of the structure are complex and non-isothermal, and that durability estimates (of the number of cycles to fracture) based on the Novozhilov-Rybakina test and the deformation kinetic test are either understated (almost two times) or overestimated (almost five times). At the same time, applied inelasticity theory-based calculations satisfactorily correspond with the experimental data.

5.6 Prediction of Life of a Power Generation System

We calculated the life of the power generation system's material in the worst loaded point of the shelled structure made of structural stainless steel for two complex non-isothermal loading modes. Laws of changes for two stress components (soft loading) and the temperature corresponding to the first mode are given in Fig. 5.6. In this loading mode, the structure is taken to the heavy-rate mode that lasts for one year

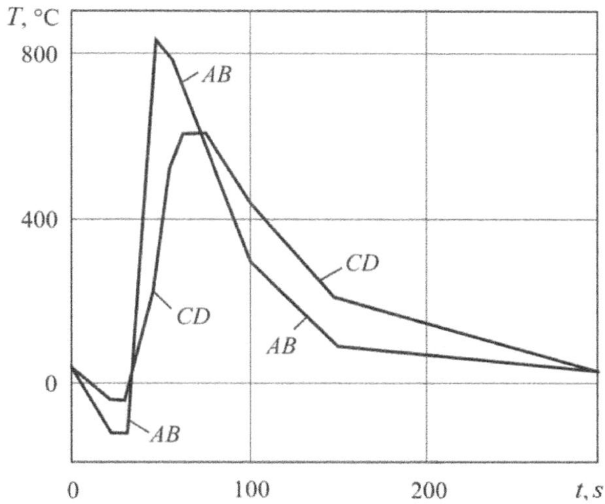

Fig. 5.4 Experimental temperature changes of tip's segments

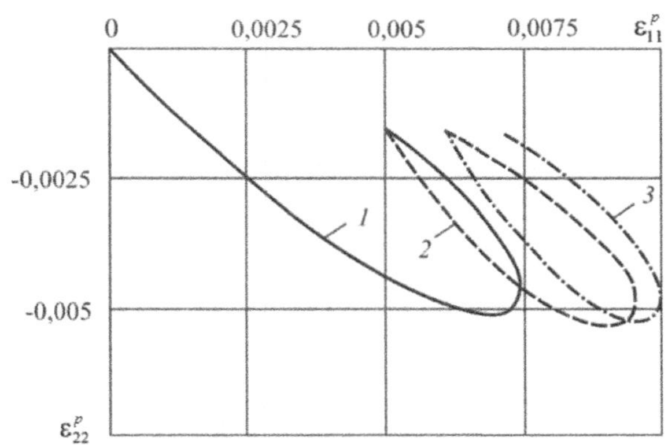

Fig. 5.5 BZh100 heat-resistant alloy. Calculated trajectory of inelastic deformations

within one hour; after that, the stress and the temperature decrease within two hours, and the calculation continues with the nominal mode until the structural material's fracture. The total time to fracture was 99,000 h (11.3 years).

The second mode is determined by the laws of stress and temperature given in Fig. 5.7. The structural material remains at the nominal loading mode for five years; after that it is taken to the heavy-rate mode that lasts for one year within one hour; after that, the stress is taken back to the nominal mode. The structural material fractured after 119,000 h (13.6 years), i.e. the structure's life increased significantly.

The calculation demonstrates that the material accumulates damage non-linearly and that this process depends on the sequence of heavy-rate and nominal modes. It is obvious that linear addition-based life estimates in the case of the loading according to the laws given in Figs. 5.6 and 5.7 will be the same, as the only difference between them is the sequence of heavy-rate and nominal modes. The significant difference

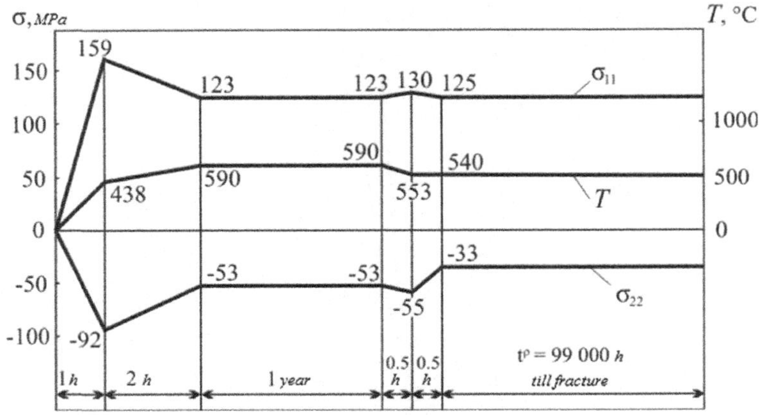

Fig. 5.6 First loading mode

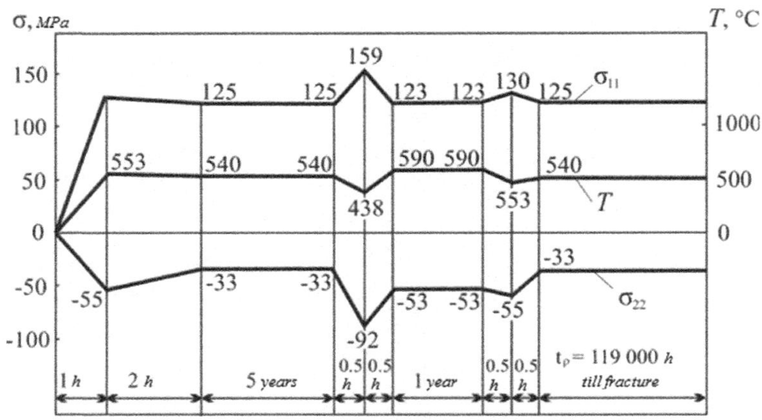

Fig. 5.7 Second loading mode

in the time to fracture obtained using different loading programs is caused by the non-linear nature of the structural material's damage accumulation.

5.7 Conclusion

Having analyzed estimates and actual life of materials and structures exposed to repeated and long-term thermomechanic loading causing inelastic deformation, we may distinguish between the following several peculiarities of life-reducing loads:

- the non-isothermal nature of loading results in significant uncertainty of life estimates, especially when the available material plasticity largely depends on the temperature;
- complex non-isothermal loading makes life the least predictable. A sufficiently significant life prediction requires adequate simulation of deformation and damage accumulation processes.
- the life of durable power generation systems in the nominal mode following the heavy-rate mode in the very beginning of operation is considerably lower than when a power generation system is taken to the heavy-rate mode after having operated in the nominal mode.

The applied inelasticity theory variant analyzed herein allows for adequately predicting the life of materials and structures exposed to repeated and long-term thermomechanical loading.

References

1. Birger IA, Shorr BE et al (1975) Termoprochnost' detalej mashin: Spravochnik. Mashinostroenie, Moscow
2. Bondar VS (1990) Neuprugoe povedenie i razrushenie materialov i konstrukcii pri slozhnom neizotermicheskom nagruzhenii: dis d-ra fiz.-mat.nauk. Izd-vo MAMI, Moscow
3. Bondar VS (2004) Neuprugost'. Varianty teorii. FIZMATLIT, Moscow
4. Bondar VS, Danshin VV (2008) Plastichnost'. Proporcional'nye i neproporcional'nye nagruzhenija. FIZMATLIT, Moscow
5. Bondar VS (2013) Inelasticity. Variants of the theory. Begell House, New York
6. Bondar VS, Abashev DR (2018) Applied theory of inelasticity. PNRPU Mech Bull 4:147–162
7. Chaboche JL, Rousselier G (1983) On the plastic an viscoplastic constitutive equations. ASME J Pres Vessel Techn 105:153–164
8. Chaboche JL (1989) Constitutive equation for cyclic plasticity and cyclic viscoplasticity. Int J Plast 5(3):247–302
9. Chaboche JL (1991) Thermodynamically based viscoplastic constitutive equations: theory versus experiment. In: ASME winter annual meeting, Atlanta, GA (USA), pp. 1–20
10. Chaboche JL (1993) Cyclic viscoplastic constitutive equations, parts I and II. ASME J Appl Mech 60:813–828
11. Chaboche JL (2008) A review of some plasticity and viscoplasticity constitutive theories. Int J Plast 24:1642–1692
12. Besson J, Cailletaud G, Chaboche J-L, Forest S, Blétry M (2010) Non-linear mechanics of materials. Springer, Heidelberg
13. Kachanov LM (1960) Teorija polzuchesti. FIZMATLIT, Moscow
14. Il'jushin AA (1963) Plastichnost'. Osnovy obshchei matematicheskoi teorii. Izd. AN SSSR, M.
15. Il'jushin AA (1990) Mehanika sploshnoj sredy. Izd-vo MGU, Moscow
16. Korotkih JuG, Volkov IA (2008) Uravnenija sostojanija vjazkouprugoplasticheskih sred s povrezhdenijami. FIZMATLIT, Moscow
17. Korotkih JuG, Volkov IA, Igumnov LA (2015) Prikladnaja teorija vjazkoplastichnosti. Monografija. Izd-vo Nizhegorodskogo gos.universiteta, Nizhnij Novgorod
18. Krempl E (1974) The influence of state of stress on low-cycle fatigue of structural materials: a literature survey and interpretive report. Am Soc Test Mater Spec, Techn Publ, no 549, pp 1–46

19. Krempl E, Lu H (1984) The hardening and dependent behavior of fully annealed AISI Type 304 stainless steel under biaxial in phase and out-of-phase strain cycling at room temperature. ASME J Eng Mater Technol 106:376–382
20. Krieg RD (1975) A. Practical Two Surface plasticity Theory. J Appl Mech 42:641–646
21. Krieg RD, Swearengen JC, Rhode RW (1978). A physicallybased internal variable model for rate-dependent plasticity. In: Proc. ASME/CSME PVP Conference, pp 15–27
22. Lindholm US, Chan KS, Bodner SR, Weber RM, Walker KP, Cassenti BN (1985) Constitutive modeling for isotropic materials (HOST). Second annual contract report, NASA CR, 174980
23. Malinin NN (1975) Prikladnaja teorija plastichnosti i polzuchesti. Mashinostroenie, M., 400 p
24. Miller AK (1978) A unified approach to predicting interactions among creep, cyclic plasticity, and recovery. Nucl Eng Des 51:35–43
25. Miller KJ, Brown MW (1984) Multiaxial fatigue: a brief review. In: Adv. Fract. Res. Proc. 6th Int. Conf. New Delhi 4–10, vol I, pp 31–56
26. Miller AK, Tanaka TG (1988) NONSS: a new method for integrating unified constitutive equations ander complex histories. Trans ASME: J Eng Mater and Technol 110(3):205–211
27. Novozhilov VV, Kadashevich JuI (1990) Mikronaprjazhenija v konstrukcionnyh materialah. Mashinostroenie, Leningrad
28. Ohno N (1982) A constitutive model of cyclic plasticity with a nonhardening strain region. J Appl Mech 49:721–727
29. Ohno N (1990) Recent topics in constitutive modeling of cyclic and viscoplasticity. Appl Mech rev 43:283–295
30. Ohno N, Wang JD (1991) Transformation of a nonlinear kinematics hardening rulle to a multisurface form under isothermal and nonisothermal conditions. Int J Plast 7:879–891
31. Ohno N, Wang JD (1993) Kinematics hardening rule with critical state of dynamic recovery. Parts I and II. Int J Plast 9:375–403
32. Rabotnov JN (1966) Polzuchest' jelementov konstrukcij. Fizmatgiz, M.
33. Temis JM (2005) Modelirovanie plastichnosti i polzuchesti konstrukcionnyh materialov GTD. Materialy 49-oj Mezhdunarodnoj nauchno-tehnicheskoj konferencii AAI «Prioritety razvitija otechestvennogo avtotraktorostroenija i podgotovki inzhenernyh nauchnyh kadrov». Shkola-seminar «Sovremennye modeli termovjazkoplastichnosti». Chast' 2. MAMI, Moscow, pp 25–76
34. Vasin RA (1987) Jeksperimental'no-teoreticheskoe issledovanie opredeljajushhih sootnoshenij v teorii uprugoplasticheskih processov. In: Avtoref. diss. d.f-m.n. MGU, Moscow

Chapter 6
Monotonic and Cyclic Loading Processes

Valentin S. Bondar and Dmitry R. Abashev

Abstract Experimental analysis of 12X18H10T stainless steel specimens subjected to strain-controlled cyclic loading that comprises sequential monotonic and cyclic loading under uniaxial tension-compression and standard temperature is used to identify some features and dissimilarities of isotropic and anisotropic hardening processes that occur during monotonic and cyclic loading. In order to describe these features in terms of the plasticity theory (the Bondar model), which can be classified as a combined-hardening flow theory, plastic-strain redirection criterion and the memory surface concept are introduced in the plastic-strain tensor space so as to separate monotonic and cyclic strain. Evolution equations for isotropic and anisotropic hardening processes are derived to describe the monotonic-to-cyclic and cyclic-to-monotonic evolutions in transients. The basic experiment used to determine the material functions consists of three stages: cyclic loading, monotonic loading, and subsequent cyclic loading until fracture. The results of the basic experiment are fundamental to the proposed method for identifying the material functions. Basic-experiment results and the identification method are used to identify the room-temperature material functions of 12X18H10T stainless steel. The paper compares the computational analysis and the experimental analysis of stainless steel subjected to a strain-controlled fatigue test (loading) in five stages: cyclic, monotonic, cyclic, monotonic, and cyclic loading until fracture. It further compares the computational and experimental kinetics of the stress-strain state throughout the deformation process. Changes in the amplitude and mean cycle stress during the cyclic stress stages are subsequently analyzed. These stages are characterized by hysteresis loop stabilization. Computational and experimental results fit reliably. The theory adequately describes the processes of how the kinetics, the amplitudes, and the mean cycle stress alter when subjecting a specimen to strain-controlled loading, which enables a more adequate description of stress-controlled loading, especially when loading is non-stationary and non-symmetric.

V. S. Bondar · D. R. Abashev (✉)
Department of Technical Mechanics, Moscow Polytechnic University, Moscow 107023, Russian Federation
e-mail: tm@mospolytech.ru

H. Altenbach et al. (eds.), *Deformation and Destruction of Materials and Structures Under Quasi-static and Impulse Loading*, Advanced Structured Materials 186,
https://doi.org/10.1007/978-3-031-22093-7_6

Keywords Experiment · Strain-controlled cyclic loading · Plasticity · Combined-hardening flow theory

6.1 Introduction

Non-stationary asymmetric cyclic strain is a deformation process that is a sequence of monotonic and cyclic loadings. It is a very complex problem to model such processes mathematically when subjecting a specimen to strain-controlled cyclic loading, even more so in the case of stress-controlled loading. Besides, such loadings are associated with the hard-to-model hysteresis loop ratcheting and stabilization. As for the assessment and prediction of the resource under non-stationary and asymmetric cyclic loading conditions, fatigue damage accumulation must be determined throughout the deformation process given the significant non-linearity of such damage.

Mathematical modeling of strain and damage accumulation when subjecting a specimen to cyclic loading is mainly based on variants of plasticity theories belonging to the class of combined (isotropic and anisotropic) hardening plastic-flow theories as reviewed and analyzed in [1, 2, 4–12, 14–34, 37–40]. In this paper, such modeling is based on the Bondar model, a version of plasticity theory [7–9, 11, 12] (Bondar et al., 2013) which, as shown in [13], is the most adequate version for describing cyclic loading-induced strain and fracture, as compared to the Korotkikh [21, 25, 33–35] or Chaboche [6, 14, 18] models. This paper presents the basic equations of the Bondar Model.

In order to identify the features of strain induced by non-stationary and asymmetric cyclic loading, strain-controlled loading is analyzed by subjecting 12X18H10T stainless steel specimens to tension–compression tests in a sequence of five stages: cyclic, monotonic, cyclic, monotonic, and cyclic loading until fracture. Analysis of the cyclic-to-monotonic and monotonic-to-cyclic transients shows the need to separate the monotonic and the cyclic deformation processes. To that end, a plastic-strain redirection criterion and the memory surface concept for separating the monotonic and cyclic deformation processes are introduced in the plastic-strain space. Evolution equations of isotropic and anisotropic hardening parameters for monotonic and cyclic loading are further introduced in the Bondar plasticity theory equations.

Separation of the monotonic and cyclic strain is also a feature of the Korotkikh model [34], where it is only used to describe the evolution of isotropic hardening. The memory surface in this model is constructed in the backstress deviator space while determining the maximum backstress intensity value in the deformation process. In [25, 34], the evolution of anisotropic hardening in a plastic-strain deviator space is described by introducing a memory surface while determining the maximum plastic-strain intensity amplitude in the deformation process. The paper [36] uses the same memory surface to describe the anisotropic hardening evolution as in the case of isotropic hardening. All these approaches [25, 34, 36] have one significant drawback: the resulting memory surface size can potentially decrease and increase at the end of the cycle, resulting in a chance of it either both monotonic or cyclic loading at the end of each cycle. Besides, the evolution equation for the maximum cyclic loading backstress intensity means that this value is always diminishing, although it should remain constant in a stabilized cycle. In conclusion, it should also be noted that there is no documented adequate rationale for the considered approaches [25, 34, 36].

Taking into account the identified features of monotonic and cyclic loading for the refined equations of the modified Bondar plasticity theory, this research has defined the basic experiment as well as the method for identifying the material functions. The material functions of 12H18N10T stainless steel at room temperature are obtained. This paper compares the computational analysis and experimental analysis of 12H18N10T stainless steel subjected to strain-controlled loading that is a sequence of monotonic and cyclic loadings. The kinetics of the stress–strain state is analyzed, and changes in the amplitude and mean stress of the cycle during the cyclic loading stages are taken into account.

6.2 Basic Equations of the Plasticity Theory

A simplified version of the plasticity theory [10, 11, 13], which is a partial version of the theory of inelasticity [7, 9], is considered. This version is a single-surface combined-hardening flow theory. Its applicability is limited to small strains of initially isotropic metals at temperatures that entail no phase transformations, at such strain rates where dynamic and rheological effects are negligible.

Below is a summary of the basic equations for this plasticity theory version.

1. $\dot{\varepsilon}_{ij} = c + \dot{\varepsilon}^p_{ij}$.
2. $\dot{\varepsilon}^e_{ij} = \frac{1}{E}\left[\dot{\sigma}_{ij} - \nu\left(3\dot{\sigma}_0\delta_{ij} - \dot{\sigma}_{ij}\right)\right]$.
3. $f\left(\sigma_{ij}\right) = \frac{3}{2}\left(s_{ij} - a_{ij}\right)\left(s_{ij} - a_{ij}\right) - C^2 = 0$.
4. $\dot{C} = q_\varepsilon \dot{\varepsilon}^p_{u*}, \quad \dot{\varepsilon}^p_{u*} = \left(\frac{2}{3}\dot{\varepsilon}^p_{ij}\ \dot{\varepsilon}^p_{ij}\right)^{\frac{1}{2}}$.
5. $\dot{a}_{ij} = \sum\limits_{m=1}^{M} \dot{a}^{(m)}_{ij}$.
6. $\dot{a}^{(1)}_{ij} = \frac{2}{3}g^{(1)}\dot{\varepsilon}^p_{ij} + g^{(1)}_a a^{(1)}_{ij}\dot{\varepsilon}^p_{u*}$.
7. $\dot{a}^{(2)}_{ij} = \frac{2}{3}g^{(2)}\dot{\varepsilon}^p_{ij} + g^{(2)}_a a^{(2)}_{ij}\dot{\varepsilon}^p_{u*}$.
8. $\dot{a}^{(m)}_{ij} = \frac{2}{3}g^{(m)}\dot{\varepsilon}^p_{ij} \quad (m = 3, \ldots, M)$.
9. $\dot{\varepsilon}^p_{ij} = \frac{\partial f}{\partial \sigma_{ij}}\lambda = \frac{3}{2}\frac{s^*_{ij}}{\sigma^*_u}\dot{\varepsilon}^p_{u*}, \quad s^*_{ij} = s_{ij} - a_{ij}, \qquad \sigma^*_u = \left(\frac{3}{2}s^*_{ij}\ s^*_{ij}\right)^{\frac{1}{2}}$.
10. $\dot{\varepsilon}^p_{u*} = \frac{1}{E_*}\frac{3}{2}\frac{s^*_{ij}\ \dot{\sigma}_{ij}}{\sigma^*}, \quad E_* = q_\varepsilon \sum\limits_{m=1}^{M} g^{(m)} + \sum\limits_{m=1}^{2} g^{(m)}_a a^{(m)*}_u, \quad a^{(m)*}_u = \frac{3}{2}\frac{s^*_{ij}a^{(m)}_{ij}}{\sigma^*_u}$
11. $\dot{\varepsilon}^p_{u*} = \frac{1}{E_* + 3G}\, 3\, G\frac{s^*_{ij}\ \dot{\varepsilon}_{ij}}{\sigma^*}, \quad G = \frac{E}{2(1+\nu)}$.
12. $\sigma^*_u < C \cup \dot{\varepsilon}^p_{u*} \leq 0$—elasticity,
    $\sigma^*_u = C \cap \dot{\varepsilon}^p_{u*} > 0$—elastoplasticity.
13. $\dot{\omega} = \alpha\omega^{\frac{\alpha-1}{\alpha}}\frac{a^{(2)}_{ij}\dot{\varepsilon}^p_{ij}}{W_a}, \qquad \alpha = \left(\sigma^{(2)}_a / a^{(2)}_u\right)^{n_\alpha}, \quad a^{(2)}_u = \left(\frac{3}{2}a^{(2)}_{ij}\ a^{(2)}_{ij}\right)^{\frac{1}{2}}$.

Here, $\dot{\varepsilon}_{ij}$, $\dot{\varepsilon}^e_{ij}$, $\dot{\varepsilon}^p_{ij}$ are the total, elastic, and plastic-strain rate tensors; σ_{ij}, s_{ij}, s^*_{ij}, a_{ij} is the stress tensor, stress, active-stress, and backstress deviators; ε^p_{u*} is the accumulated plastic strain; ω is the damage; E, ν are Young's modulus and Poisson's ratio; C is the radius (size) of the yield surface; $a^{(1)}_{ij}$, $a^{(2)}_{ij}$, $a^{(m)}_{ij}$ are Type I, II, and III backstresses (yield surface center displacement deviator); and q_ε, $g^{(m)}$, $g^{(m)}_a$ are the defining functions, the relationship whereof to the material functions is described below.

6.3 Monotonic and Cyclic Loading of 12X18H10T Stainless Steel

The paper presents the results of experimenting with 12X18H10T stainless steel subjected to uniaxial strain-controlled loading, which is a sequence of monotonic and cyclic loading stages. The experiment consists of five loading stages:

- Stage 1 involves cyclic loading at $\varepsilon^{(1)}_m = 0$, $\Delta\varepsilon^{(1)} = 0.016$, and $N^{(1)} = 20$ cycles;
- Stage 2 involves monotonic tension test up to $\varepsilon^{(2)} = 0.05$;
- Stage 3 involves cyclic loading at $\varepsilon^{(3)}_m = 0.05$, $\Delta\varepsilon^{(3)} = 0.012$, and $N^{(3)} = 200$ cycles;
- Stage 4 involves monotonic tension up to $\varepsilon^{(4)} = 0.1$;
- Stage 5 involves cyclic loading at $\varepsilon^{(5)}_m = 0.1$, $\Delta\varepsilon^{(5)} = 0.012$, and $N^{(5)} = N_f$ cycles until fracture.

Here, $\varepsilon^{(i)}_m$ is the mean cycle strain; $\Delta\varepsilon^{(i)}$ is the cycle strain amplitude; $\varepsilon^{(i)}$ is the final monotonic strain; $N^{(i)}$ is the number of cycles.

Figure 6.1 shows the experimental diagram of the 12X18H10T steel strain that covers all five loading stages. The cyclic diagrams of Stages I, II, and III show the loops for the first cycle and the last cycle. Experimental results are analyzed below.

Cyclic deformation at Stage I entails a cyclical hardening of 12X18H10T steel at the initial stage, which slows down to insignificant levels $\left(dC_p/d\varepsilon^p_{u*} \approx 1\,\mathrm{M}\prod \mathrm{a}\right)$; then the steel becomes virtually cyclically stable.

Stages III and V feature stabilization of the hysteresis loop. These stages are identical stabilization-wise, as if there was no pre-history of strain. Thus, the modulus E_a, which is part of the Type I backstress evolution equation and is necessary for loop stabilization, must have the same initial value $E_a = E_{a0}$. That said, during monotonic post-cyclic loading, when E_a is reduced to nearly zero, the modulus E_a must quickly return to its initial value E_{a0}.

Hardening is constant at Stages II and IV of monotonic loading. Here, hardening is determined by the modulus E_{a0} and to a lesser extent by the modulus of monotonic isotropic hardening.

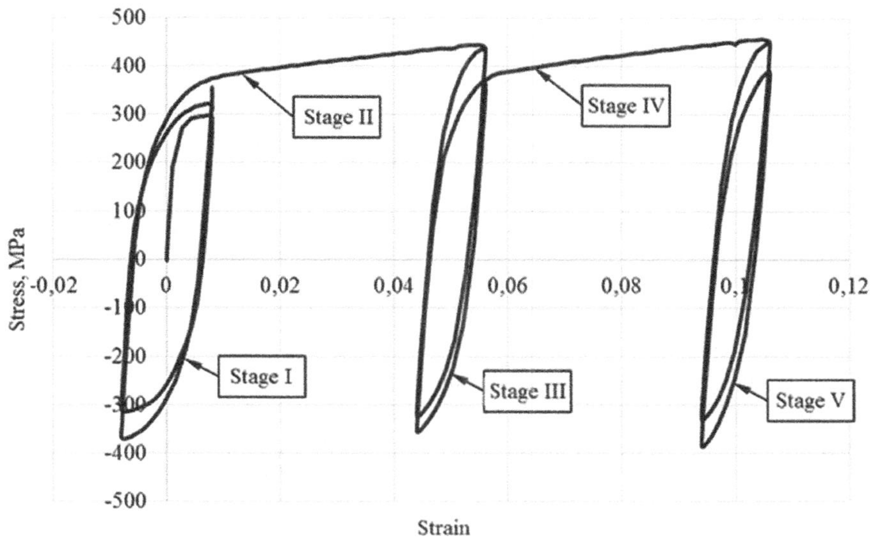

Fig. 6.1 Stress–strain diagram of 12X18H10T steel

Thus, the behavior of the modulus E_a that describes the anisotropic hardening, and therefore the behavior of the isotropic hardening parameters, will depend significantly on whether the strain is cyclic or monotonic.

Memory surface that limits the cyclic deformation area is introduced in the plastic-strain tensor space ε_{ij}^{p} in order to separate monotonic and cyclic strain. The surface is determined by the position of its center ξ_{ij} and its radius (size) C_{ε}. To compute the center and size of the surface, two plastic-strain tensors $\varepsilon_{ij}^{p(1)}$ and $\varepsilon_{ij}^{p(2)}$ are introduced to define the surface boundaries. These variables are zero as strain begins. The displacement and size of the memory surface are determined at the time plastic strain is redirected. The following condition is assumed as the redirection criterion:

$$\dot{\varepsilon}_{ij(t-0)}^{p}\dot{\varepsilon}_{ij(t)}^{p} < 0, \tag{6.1}$$

where $\dot{\varepsilon}_{ij(t)}^{p}$ is the current plastic-strain rate tensor; $\dot{\varepsilon}_{ij(t-0)}^{p}$ is the plastic-strain rate tensor at the preceding time point.

At this moment, the change in the boundaries, center, and size of the yield surface is described based on the following relationships:

$$\varepsilon_{ij}^{p(2)} = \varepsilon_{ij}^{p(1)}, \tag{6.2}$$

$$\varepsilon_{ij}^{p(1)} = \varepsilon_{ij}^{p}, \tag{6.3}$$

$$\xi_{ij} = \frac{\varepsilon_{ij}^{p(1)} + \varepsilon_{ij}^{p(2)}}{2}, \tag{6.4}$$

$$C_\varepsilon = \left[\frac{2}{3}\left(\frac{\varepsilon_{ij}^{p(1)} - \varepsilon_{ij}^{p(2)}}{2}\right)\left(\frac{\varepsilon_{ij}^{p(1)} - \varepsilon_{ij}^{p(2)}}{2}\right)\right]^{\frac{1}{2}}. \tag{6.5}$$

Then the condition of cyclic strain is the strain within the memory surface

$$\left[\frac{2}{3}\left(\varepsilon_{ij}^{p} - \xi_{ij}\right)\left(\varepsilon_{ij}^{p} - \xi_{ij}\right)\right]^{\frac{1}{2}} \le C_\varepsilon \tag{6.6}$$

Outside the memory surface, the strain is monotonous.

Based on the above peculiarities of monotonic and cyclic loading, the following equations are derived for the modulus E_a and backstress defining functions:

$$g^{(1)} = E_a,\ g^{(2)} = \beta^{(2)}\sigma_a^{(2)},\ g_a^{(2)} = -\beta^{(2)}, \tag{6.7}$$

$$g^{(m)} = \begin{cases} \beta^{(m)},\ \sigma_a^{(m)}, \\ 0, \text{ если } a_u^{(m)} \ge \sigma_a^{(m)} \cap a_{ij}^{(m)} s_{ij}^{*} > 0, \end{cases} \tag{6.8}$$

$$a_u^{(m)} = \left(\frac{3}{2} a_{ij}^{(m)} a_{ij}^{(m)}\right)^{\frac{1}{2}} \quad m = 3, \dots, M$$

$$\dot{E}_a = \begin{cases} -K_E\left(\frac{E_a}{E_{a0}}\right)^{n_E} \dot{\varepsilon}_{u*}^{p} & \text{cyclic loading,} \\ M_E\left(\frac{E_{a0} - E_a}{E_{a0}}\right) \dot{\varepsilon}_{u*}^{p} & \text{monotonic loading,} \end{cases} \tag{6.9}$$

$$g_a^{(1)} = \begin{cases} \frac{1}{E_a}\frac{d E_a}{d \varepsilon_{u*}^{p}} & \text{cyclic loading,} \\ 0 & \text{monotonic loading.} \end{cases} \tag{6.10}$$

Therefore, the following material functions need to be derived in order to describe backstresses:

$E_{a0},\ \sigma_a^{(m)},\ \beta^{(m)}$ are the moduli of anisotropic hardening;

$K_E,\ n_E,\ M_E$ are the parameters of anisotropic hardening when subjecting the material to cyclic and monotonic strain.

The results of the experiment (Fig. 6.1) are used to define these material functions.

Anisotropic hardening modulus E_{a0} is found by the formula

$$E_{a0} = \frac{\sigma_m^{(3)}}{\varepsilon_m^{p(3)}}, \tag{6.11}$$

where $\sigma_m^{(3)}$ is the mean stress in the first Stage III cycle; $\varepsilon_m^{p(3)}$ is the mean plastic strain at the first cycle Stage III cycle.

The moduli of anisotropic hardening $\sigma_a^{(m)}$ and $\beta^{(m)}$ are found by processing the cyclic diagram of the last Stage I semi-cycle as per the procedure described in [10, 11].

The anisotropic hardening parameters K_E and n_E are found based on the results of the hysteresis loop stabilization at Stages III and V. To that end, the dependence in the coordinates

$$Y_E = \ln\left[\frac{\sigma_m(N-1) - \sigma_m(N)}{2\Delta\varepsilon^p\, \varepsilon_m^p}\right], \tag{6.12}$$

$$X_E = \ln\left[\frac{\sigma_m(N)}{\varepsilon_m^p\, E_{a0}}\right] \tag{6.13}$$

is constructed, where N is the cycle number; $\sigma_m(N)$ is the mean stress of the Nth cycle; $\Delta\varepsilon^p$ is the plastic-strain amplitude; ε_m^p is the mean plastic strain. The dependence obtained is approximated by the linear function

$$Y_E = a_E\, X_E + b_E. \tag{6.14}$$

Thus,

$$K_E = \exp(b_E), \quad n_E = a_E. \tag{6.15}$$

The parameter of anisotropic hardening M_E of a specimen subjected to monotonic loading is determined from the considerations of restoring the parameter E_a from 0 to the value E_{a0}, whereby plastic strain changes under monotonic loading over ε_{st}^p. Thus, the parameter M_E shall be determined by the formula

$$M_E = \frac{E_{a0}}{\varepsilon_{st}^p}. \tag{6.16}$$

Having found the backstresses over the entire process from Stage I to Stage V, one can determine the behavior of the yield surface size (radius), i.e. the change in isotropic hardening in cyclic-to-monotonic and monotonic-to-cyclic strain transients.

Figure 6.2 shows the change in the yield surface size (functional C) throughout the deformation process from Stage I to Stage V.

The dotted line in Fig. 6.2 shows the function of isotropic hardening $C = C_p\left(\varepsilon_{u*}^p\right)$ induced by cyclic loading. Analysis of the results, presented in Fig. 6.2, shows that the transition from cyclic to monotonic strain (Stages II and IV) is associated with an increase in the intensity of isotropic hardening. The transition from monotonic to cyclic strain (Stages III and V) is associated with a slowdown in such isotropic hardening, as it tends to be isotropic $C = C_p\left(\varepsilon_{u*}^p\right)$ when subjecting the specimen to cyclic strain.

Based on the above peculiarities of how isotropic hardening is altered by cyclic or monotonic loading, the following dependence is assumed for the defining function of isotropic hardening:

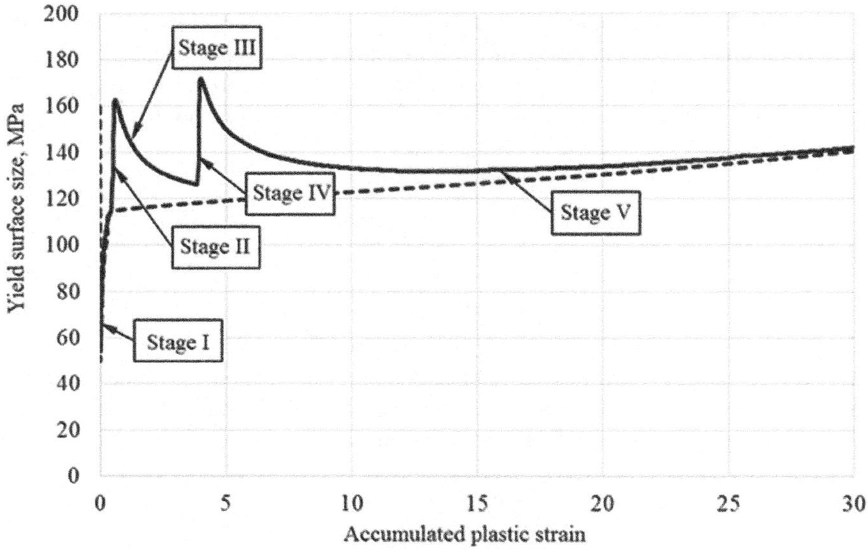

Fig. 6.2 Yield surface-size change

$$q_\varepsilon = \begin{cases} \left[\frac{dC_p}{d\varepsilon_{u*}^p} - K_C\left(\frac{C-C_p}{C_p}\right)^{n_C}\right] & \text{cyclic loading,} \\ \left[\frac{dC_p}{d\varepsilon_{u*}^p} + M_C\right] & \text{monotonic loading.} \end{cases} \quad (6.17)$$

Thus, to describe such isotropic hardening, the following material functions must be defined:

$C_p(\varepsilon_{u*}^p)$ is the function of isotropic hardening induced by cyclic loading;

K_C, n_C, M_C are the moduli of isotropic hardening induced by cyclic and monotonic loading.

These material functions are defined using the experiment results, see Fig. 6.2. The function of isotropic hardening induced by cyclic loading $C_p(\varepsilon_{u*}^p)$ is determined based on the surface-size changes at Stages III and V; see the dotted curve in Fig. 6.2 and I.

Cyclic loading isotropic hardening parameters K_C and n_C are found from the results of decreasing the yield surface size at Stages III and V. To that end, a dependence is constructed in the coordinates

$$Y_C = \ln\left[\frac{d\left(C_p - C\right)}{d\varepsilon_{u*}^p}\right], \quad (6.18)$$

$$X_C = \ln\left[\frac{(C - C_p)}{C_p}\right]. \quad (6.19)$$

Table 6.1 Material functions of 12X18H10T steel

E, MPa	ν	E_{a0}, MPa	$\sigma_a^{(2)}$,MPa	$\beta^{(2)}$	K_E, MPa	n_E	M_E, MPa	K_C, MPa	n_C	M_C, MPa
$2 \cdot 10^5$	0.3	800	140	260	$1,4 \cdot 10^4$	3.5	$4 \cdot 10^4$	148	1.4	960

$\sigma_a^{(3)}$, MPa	$\beta^{(3)}$	$\sigma_a^{(4)}$, MPa	$\beta^{(4)}$	$\sigma_a^{(5)}$, MPa	$\beta^{(5)}$	W_a, MJ/M^3	n_α
45	5000	41	2000	36	1100	1830	1.5

Table 6.2 Isotropic hardening function of 12X18H10T steel

ε_{u*}^p	0	0.0003	0.0006	0.0014		0.0045	0.006	0.01	0.025
C_p, MPa	160	125	110	100		65	50	51	57
ε_{u*}^p	0.1	0.15	0.3	0.45	0.6	1	8	25	45
C_p, MPa	85	90	105	110	115	115	121	135	159

The dependence obtained is approximated by the linear function

$$Y = a_C X_C + b_C \ . \tag{6.20}$$

Thus

$$K_C = \exp(b_C), \qquad n_C = a_C. \tag{6.21}$$

The parameter of isotropic hardening M_C induced by monotonic strain is found from the slope of the strain curve at Stages II and IV using the formula

$$M_C = \frac{d\sigma}{d\varepsilon^p} - E_{a0} - \frac{dC_p}{d\varepsilon^p} \tag{6.22}$$

6.4 Material Functions of 12X18H10T Stainless Steel

Material functions have been derived based on the results of room-temperature experiments with 12X18H10T stainless steel; see Tables 6.1 and 6.2.

6.5 Verification of the Modified Plasticity Theory

To verify the modified plasticity theory, the researchers have computed the kinetics of the stress–strain state of 12X18H10T stainless steel subjected to strain-controlled cyclic and monotonic loading according to the five-stage program described in

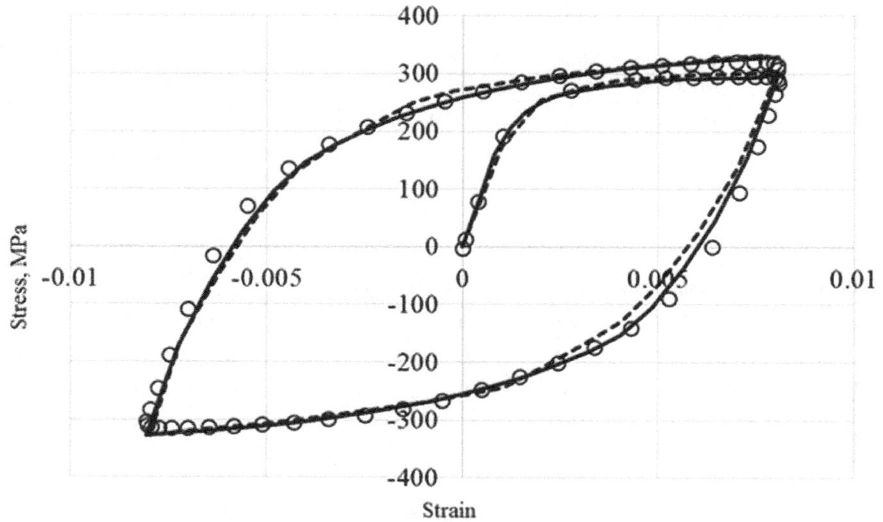

Fig. 6.3 First Stage I cycle

Section 6.2. Computation uses the material functions per Section 6.3. Figures 6.3, 6.4, 6.5, 6.6, and 6.7 present a comparison of the computed (solid curves) and experimental (open circles) results. The dotted curves show the results based on the variant [13] of the modified Bondar model. Figure 6.3 shows the cyclic diagram of the first Stage I cycle; Fig. 6.4 shows the 20th (last) Stage I cycle, the monotonous loading at Stage II, and the first Stage III cycle; Fig. 6.5 shows the 200th (last) Stage III cycle, the monotonous loading at Stage IV, and the first Stage V cycle. Variations in the stress amplitude and mean cycle stress at Stages I, III, and V are shown in Figs. 6.6 and 6.7.

There is a significant improvement in the description of the stress–strain state kinetics based on the variant proposed herein, as compared to the previously [13] modified model. As for the changes in the amplitude and mean stress of the cycles, the proposed version adequately describes these rather complex processes.

6.6 Conclusions

Analysis of the stainless steel experiments leads to a conclusion that the processes of isotropic and anisotropic hardening vary significantly depending on whether the strain is monotonic or cyclic. Monotonic-to-cyclic and cyclic-to-monotonic strain transitions are associated with hardening transients.

In the light of the identified features of monotonic and cyclic loading, the equations of the modified Bondar plasticity theory have been refined. The researchers

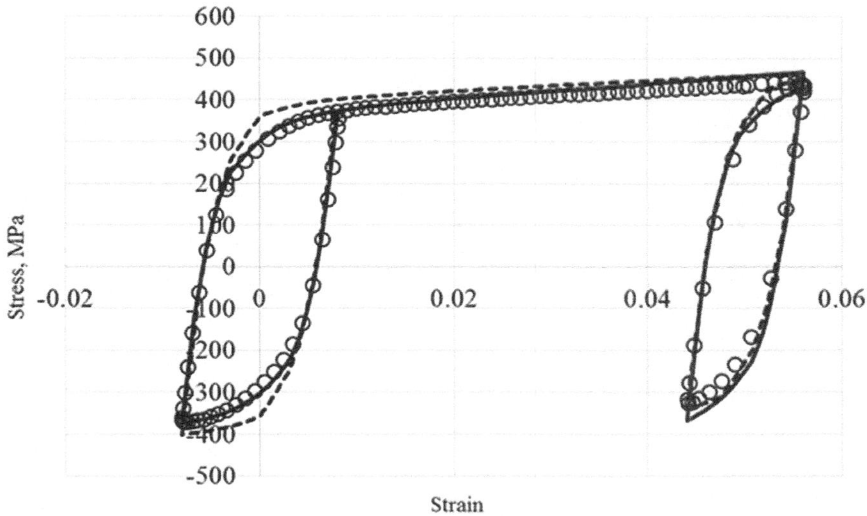

Fig. 6.4 Last Stage I cycle, Stage II, and first Stage III cycle

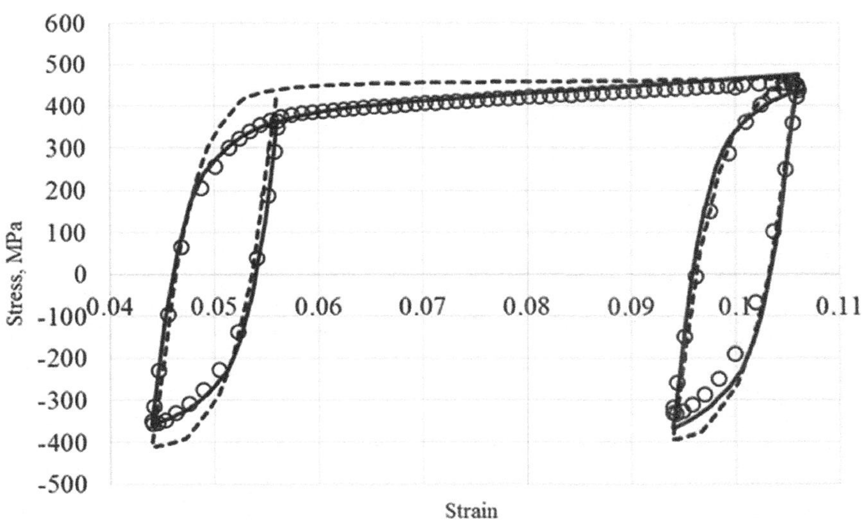

Fig. 6.5 Last Stage III cycle, Stage IV, and first Stage V cycle

have defined the basic experiment, derived the material-function identification method, and obtained such material functions for 12X18H10T stainless steel at room temperature.

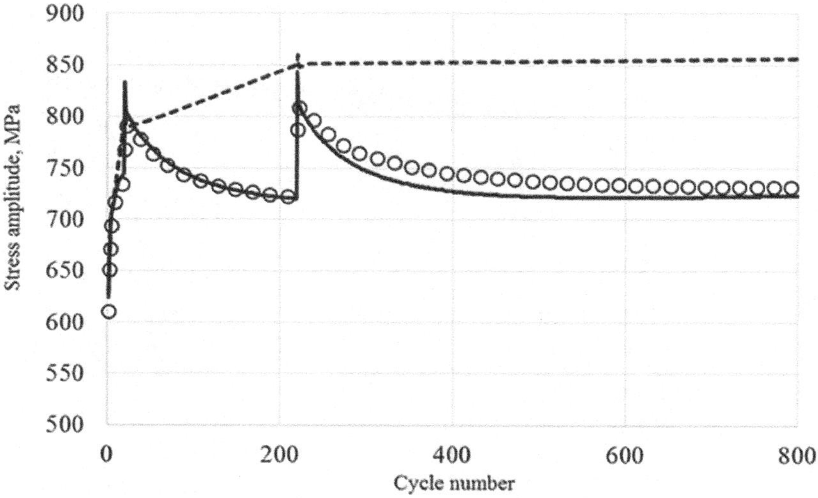

Fig. 6.6 Stress peak-to-peak amplitude

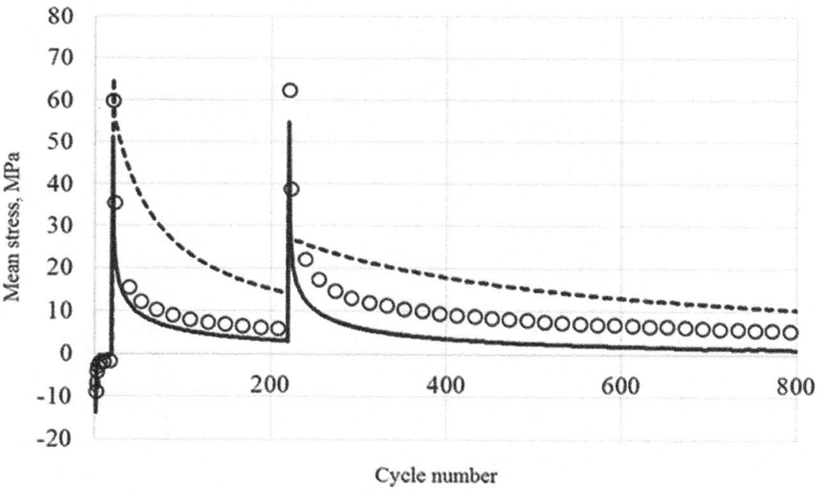

Fig. 6.7 Mean stress

The paper compares the results of computational and experimental studies of 12X18H10T stainless steel subjected to strain-controlled loading, a process consisting of a sequence of monotonic and cyclic loadings. Stress–strain state kinetics has been analyzed. Changes in the amplitude and mean stress of the cycle during cyclic loading have been dwelled upon. Computational and experimental results fit reliably.

The theory adequately describes the processes of how the kinetics, the amplitudes, and the mean cycle stress alter when subjecting a specimen to strain-controlled loading, which enables a more adequate description of stress-controlled loading, especially when loading is non-stationary and non-symmetric.

References

1. Abdel-Karim M (2009) Modified kinematic hardening rules for simulations of ratchetting. Int J Plast 25:1560–1587. https://doi.org/10.1016/j.ijplas.2008.10.004
2. Abdel-Karim M (2010) An evaluation for several kinematic hardening rules on prediction of multiaxial stress-controlled ratchetting. Int J Plast 26:711–730. https://doi.org/10.1016/j.ijplas.2009.10.002
3. Abdel-Karim M (2010) An extension for the Ohno-Wang kinematic hardening rules to incorporate isotropic hardening. Int J Plast 87:170–176. https://doi.org/10.1016/j.ijpvp.2010.02.003
4. Abdel-Karim M (2011) Effect of elastic modulus variation during plastic deformation on uniaxial and multiaxial ratchetting simulations. Int J Mech 30:11–21. https://doi.org/10.1016/j.euromechsol.2010.08.002
5. Bari S, Hassan T (2002) An advancement in cyclic plasticity modeling for multiaxial ratcheting simulation. Int J Plast 18:873–894. https://doi.org/10.1016/S0749-6419(01)00012-2
6. Besson J, Cailletaud G, Chaboche J-L, Forest S, Blétry M (2010) Non-linear mechanics of materials. Springer, Heidelberg
7. Bondar VSM (2004) Neuprugost'. Varianty teorii [Inelasticity. Variants of the theory]. Moscow, FIZMATLIT
8. Bondar VS, Danshin VV (2008) Plastichnost'. Proporcional'nye i neproporcional'nye nagruzhenija [Plasticity. Proportional and disproportionate loading]. Moscow, FIZMATLIT.
9. Bondar VS (2013) Inelasticity. Variants of the theory. New York, Begell House
10. Bondar VS, Danshin VV, Makarov DA, (2014) Mathematical modelling of deformation and damage accumulation under cyclic loading. PNRPU Mech Bull (2):125–152
11. Bondar VS, Danshin VV, Kondratenko AA (2015) Version of the theory of thermoplasticity. PNRPU Mech Bull (2):21–35. https://doi.org/10.15593/perm.mech/2015.2.02
12. Bondar VS, Danshin VV, Kondratenko AA (2016) Variant of thermoviscoplasticity theory. PNRPU Mech Bull (1):39–56.https://doi.org/10.15593/perm.mech/2016.1.03
13. Bondar VS, Abashev DR, Petrov VK (2017) Comparative analysis of variants of plasticity theories under cyclic loading. PNRPU Mech Bull 2:23–44. https://doi.org/10.15593/perm.mech/2017.2.02
14. Chaboche J-L (2008) A review of some plasticity and viscoplasticity constitutive theories. Int J Plast 24:1642–1692. https://doi.org/10.1016/j.ijplas.2008.03.009
15. Chaboche J-L, Kanouté P, Azzouz F (2012) Cyclic inelastic constitutive equations and their impact on the fatigue life predictions. Int J Plast 35:44–66. https://doi.org/10.1016/j.ijplas.2012.01.010
16. Hassan T, Taleb L, Krishna S (2008) Influence of non-proportional loading on ratcheting responses and simulations by two recent cyclic plasticity models. Int J Plast 24:1863–1889. https://doi.org/10.1016/j.ijplas.2008.04.008
17. Chang K-H, Jeon J-T, Lee C-H (2016) Effects of residual stresses on the uniaxial ratcheting behavior of a girth-welded stainless steel pipe. Int J Steel Struct 16:1381–1396. https://doi.org/10.1007/s13296-016-0072-1
18. Huang ZY, Chaboche JL, Wang QY, Wagner D (2014) Bathias C. Effect of dynamic strain aging on isotropic hardening in low cycle fatigue for carbon manganese steel. Mater Sci Eng A589:34–40. https://doi.org/10.1016/j.msea.2013.09.058

19. Kan Q, Kang G (2009) Constitutive model for uniaxial transformation ratcheting of super-elastic NiTi shape memory alloy at room temperature. Int J Plast 26(3):441–465. https://doi.org/10.1016/j.ijplas.2009.08.005
20. Kang G, Kan Q (2007) Contitutive modeling for uniaxial time-dependent ratcheting of SS304 stainless steel. Mech Mater 39:488–499. https://doi.org/10.1016/j.mechmat.2006.08.004
21. Kapustin SA, Churilov YA, Gorohov VA (2015) Simulation of nonlinear deformation and fracture of structures under conditions of multifactorial effects based on FEM. N. Novgorod, Izd-vo NNGU
22. Kim JH, Kim D, Lee YS, Lee M-G, Chung K, Kim H-Y, Wagonere RH (2013) A temperature-dependent elasto-plastic constitutive model for magnesium alloy AZ31 sheets. Int J Plast 50:66–93. https://doi.org/10.1016/j.ijplas.2013.04.001
23. Lee C-H, Van Do VN, Chang K-H (2014) Analysis of uniaxial ratcheting behavior and cyclic mean stress relaxation of a duplex stainless steel. Int J Plast 62:17–33. https://doi.org/10.1016/j.ijplas.2014.06.008
24. Lee J-Y, Lee M-G, Barlat F, Bae G (2017) Piecewise linear approximation of nonlinear unloading-reloading behaviors using a multi-surface approach. Int J Plast 93:112–136. https://doi.org/10.1016/j.ijplas.2017.02.004
25. Mitenkov FM., Volkov IA, Igumnov LA, Kaplienko AV, Korotkikh IG, Panov VA (2015) Prikladnaia teoriia plastichnosti. Moscow, FIZMATLIT
26. Muhhamad W, Mohammadi M, Kang J, Mishra RK, Inal K (2015) An elasto-plastic constitutive model for evolving asymmetric/anisotropic hardening behavior of AZ31B and ZEK100 magnesium alloy sheets considering monotonic and reverse loading paths. Int J Plast 70:30–59. https://doi.org/10.1016/j.ijplas.2015.03.004
27. Qiao H, Agnew SR, Wu PD (2015) Modeling twinning and detwinning behavior of Mg alloy ZK60A during monotonic and cyclic loading. Int J Plast 65:61–84. https://doi.org/10.1016/j.ijplas.2014.08.010
28. Rahman SM, Hassan T, Corona E (2008) Evaluation of cyclic plasticity models in ratcheting simulation of straight pipes under cyclic bending and steady internal pressure. Int J Plast 24:1756–1791. https://doi.org/10.1016/j.ijplas.2008.02.010
29. Smith BD, Shih DS, McDowell DL (2018) Cyclic plasticity experiments and polycrystal plasticity modeling of three distinct Ti alloy microstructures. Int J Plast 101:1–23. https://doi.org/10.1016/j.ijplas.2013.10.004
30. Taleb L, Cailletaud G (2011) Cyclic accumulation of the inelastic strain in the 304L SS under stress control at room temperature: ratcheting or creep. Int J Plast 27(12):1936–1958. https://doi.org/10.1016/j.ijplas.2011.02.001
31. Taleb L (2013) About the cyclic accumulation of the inelastic strain observed in metals subjected to cyclic stress control. Int J Plast 43:1–19. https://doi.org/10.1016/j.ijplas.2012.10.009
32. Taleb L, Cailletaud G, Saï K (2014) Experimental and numerical analysis about the cyclic behavior of the 304L and 316L stainless steels at 350 °C. Int J Plast 61:32–48. https://doi.org/10.1016/j.ijplas.2014.05.006
33. Volkov IA, Igumnov LA (2007) Vvedenie v kontinualnuyu mehaniku povrezhdennoj sredy. [Introduction to conrinual mechanics of damaged media]. Moscow, FIZMATLIT
34. Volkov IA, Korotkikh JuG (2008) Uravnenija sostojanija vjazkouprugoplasticheskih sred s povrezhdenijami [The equation of state viscous elastoplastic media with injuries]. FIZMATLIT, Moscow
35. Volkov I.A., Igumnov L.A., Korotkikh Iu.G., 2015. Prikladnaia teoriia viazkoplastichnosti. N. Novgorod, Izd-vo NNGU.
36. Volkov IA, Igumnov LA, Tarasov IS, Shishulin DN, Markova MT (2018) Simulation of the fatigue life of polycrystalline structural alloys with block non-symmetrical low-cycle loading. Problemy Prochnosti i Plastichnosti 80(1):15–30
37. Yu C, Guozheng K, Song D, Kan Q (2015) Effect of martensite reorientation and reorientation-induced plasticity on multiaxial transformation ratchetting of super-elastic NiTi shape memory alloy: New consideration in constitutive model. Int J Plast 67:69–101. https://doi.org/10.1016/j.ijplas.2014.10.001

38. Zecevic M, Knezevic M (2015) A dislocation density based elasto-plastic self-consistent model for the prediction of cyclic deformation: application to AA6022-T4. Int J Plast 72:200–217. https://doi.org/10.1016/j.ijplas.2015.05.018
39. Zhu Y, Guozheng K, Yu C (2017) A finite cyclic elasto-plastic constitutive model to improve the description of cyclic stress-strain hysteresis loops. Int J Plast 95:191–215. https://doi.org/10.1016/j.ijplas.2017.04.009
40. Zhu Y, Kang G, Kan Q, Bruhns OT (2014) Logarithmic stress rate based constitutive model for cyclic loading in finite plasticity. Int J Plast 54:34–55. https://doi.org/10.1016/j.ijplas.2013.08.004

Chapter 7
Experimental and Computational Study of Deformation and Fracture of Pine Under Dynamic Three-Point Bending of Beams

Anatoly M. Bragov, Mikhail E. Gonov, Leonid A. Igumnov, Aleksandr Yu. Konstantinov, Andrey K. Lomunov, and Tatiana N. Yuzhina

Abstract The paper presents the results of an experimental study, as well as numerical modeling of deformation and fracture of pine beams under dynamic loading. Experiments are carried out on an installation implementing a dynamic three-point bending scheme. To create a load and register the forces acting on the beams during loading, the technique of measuring bars is used. Deflections are calculated according to the Kolsky formulas based on data from measuring bars, as well as by direct measurement using the digital image correlation method based on high-speed video recording. A procedure for determining the ultimate strain of pine in perpendicular to the fiber direction is proposed according to experiments on three-point bending of beams. Modeling of dynamic three-point bending of beams in LS-DYNA is carried out. To describe the behavior of pine, the MAT_WOOD material model is used. The use in the model of the value of ultimate strain determined during the experimental study allowed us to obtain a good coincidence of the crack formation time in full-scale and numerical experiments.

A. M. Bragov · M. E. Gonov · L. A. Igumnov · A. Yu. Konstantinov (✉) · A. K. Lomunov · T. N. Yuzhina
National Research Lobachevsky State University of Nizhny Novgorod, Gagarin ave. 23, 603950 Nizhny Novgorod, Russia
e-mail: konstantinov@mech.unn.ru

A. M. Bragov
e-mail: bragov@mech.unn.ru

M. E. Gonov
e-mail: gonov@mech.unn.ru

L. A. Igumnov
e-mail: igumnov@mech.unn.ru

A. K. Lomunov
e-mail: lomunov@mech.unn.ru

T. N. Yuzhina
e-mail: yuzhina@mech.unn.ru

H. Altenbach et al. (eds.), *Deformation and Destruction of Materials and Structures Under Quasi-static and Impulse Loading*, Advanced Structured Materials 186,
https://doi.org/10.1007/978-3-031-22093-7_7

Keywords Wood · Pine · Experiment · High-speed video recording · Measuring bar · Strain rate · Dynamics · Beam · Three-point bending · Deformation · Fracture · Numerical simulation

7.1 Introduction

In recent years, the number of shipments of nuclear energy waste, components of nuclear weapons, a wide range of toxic substances, etc. has increased. The requirements for their safety during transportation have increased. Calculations of the stress–strain state and strength of containers in which the above materials are transported are of great importance. The problems of analyzing possible emergency situations have particular relevance. Situations, accompanied by intense dynamic impacts, are possible during transportation, falling containers, terrorist acts, man-made and natural disasters. Due to the increasing requirements for environmental safety, the problem of creating reliable aviation containers for the air transportation of radioactive materials becomes very relevant and attracts the attention of researchers [1–5]. Its complexity is due to the high levels of impacts characteristic of an aviation accident.

Wood of different tree species, having a relatively low density with sufficient strength, is used as one of the damping materials. It is able to mitigate the results of high-speed impacts on containers and their contents.

To date, wood is considered to be a material whose properties have orthogonal anisotropy. When calculating wooden structures, an approximation of a transversally isotropic material is usually used. The properties differ along and across the fibers for the wood. For reliable calculation of the behavior of containers under impact, data on the dynamic properties of wood are needed as well as reliable verified mathematical models describing the behavior of wood under impact loads.

Wood in its structure is a complex natural composite cellular material similar to cellular structures, metal ring systems, polymer foams, etc. Such materials, due to their structure, have a good ability to absorb the energy of an impact or explosion [6]. In [7] has given a detailed historical overview of the use of wood in the 80 s of the last century. Since the mechanical properties of wood strongly depend on the place of growth, its age, and the place of sample cutting, the results obtained by different authors may differ quite a lot from each other. In this connection, many scientists around the world continue to study the dynamic properties of various types of wood.

The physical reaction of wood to dynamic loading is no different from other materials, such as metals or rigid foams. However, for many years, the application of the principles of impact mechanics to wood testing has been very limited, and empirical approaches have often been preferred. Systematic studies of the dynamic properties of five types of wood (balsa, pine, mahogany, American oak, and yucca) at impact speeds up to 300 m/s are given in the works of Reid and co-authors [8–10]. In these studies, the direct impact method was used, in which a Hopkinson measuring bar was used to measure the forces. The tests were carried out both under uniaxial strain and under uniaxial stress conditions. The samples were loaded both along and across the fibers. As a result, the values of ultimate stresses were obtained and the fracture energy was determined. It turned out that the ultimate dynamic stresses are several times higher than the static values, and their magnitude increases with increasing impact velocity. The authors explained the growth of ultimate stress under dynamic loading, based on the results of Ashby [11], by the influence of inertial effects on mechanical properties at the micro level. The authors noted a significant difference in the mechanical properties of wood samples tested along and across the fibers: the strength along the fibers was an order of magnitude higher than across. A large amount of dynamic tests of sequoia, birch, pine, and aspen was performed by S. A. Novikov and his collaborators [12]. Cylindrical samples of sequoia and birch with a diameter and height of 25 mm were cut at angles of 0°, 5°, 10°, 15°, 30°, 45°, and 90° relative to the direction of the fibers and tested for uniaxial compression by the Kolsky method at temperatures of −30 °C, +20 °C, and +65 °C at a fixed humidity of 6–7%. Y. Byukhar and his collaborators [6, 13, 14] conducted studies of the behavior of coniferous and deciduous wood species (spruce, pine, oak, beech, birch) under intense dynamic impacts during explosive loading of wooden beams and slabs. In their research, the authors used both experimental methods and numerical modeling methods using the LS-DYNA software package. Static strength was determined by tension, compression, and bending. At average strain rates, a three-point bending of the beam under shock loading was used (the Sharpie test). For a strain rate of ~103 s^{-1}, the Kolsky method was used. In addition, experiments were carried out on loading plates with a thickness of 50 mm using a cumulative explosive device. The results of the conducted experiments on explosive loading and their numerical modeling using experimentally obtained mechanical properties of wood in mathematical models showed a good qualitative and satisfactory quantitative correspondence.

In [15], dynamic tests of the European beech were carried out using the Kolsky method in all the main directions of loading. In [16], the behavior under compression of the seaside pine in the transverse (radial-tangential) plane under quasi-static and high-speed deformation modes is considered. For tests at high strain rates, the Kolsky method was used in combination with the digital image correlation method for reconstruction of deformation fields. Quasi-static compression tests were also carried out in order to compare the results. In [17], the behavior of fir under compression in two orthotropic directions (longitudinal and transverse) was studied in a wide range of strain rates from $2.2 \cdot 10^{-3}$ s^{-1} to $1 \cdot 10^{-3}$ s^{-1}. In [18], the SHPB (Split Hopkinson Pressure Bar) system was used to study the dynamic fracture of dry maple wood. To study the influence of the geometric dimensions of the samples on the behavior during fracture with a high strain rate, samples of two different thicknesses were made. In [19], a comprehensive experimental program was carried out studying the mechanical behavior of maple and ash wood for a range of densities used for the manufacture of basic baseball bats. The experimental program included a four-point bending test to determine the elastic modulus and breaking force and a Sharpie test to determine the deformation to failure depending on the strain rate and density of wood. Then the material parameters were calibrated by modeling using the finite element method of the Sharpie experiment in the LS-DYNA software package using the MAT_WOOD material model.

The paper [20] presents an overview of the mechanical properties and material models for wood.

The literature describes many applications in which constitutive models of wood were required, considering the strain rate. These may include wooden aircraft [21], impacts during an atomic explosion [22], rams, warships, hammered wooden piles for coastal structures, ballistic strikes [7], road fence posts and wooden road fences [23–29], protective structures of bus stops from the impact of various objects in a strong hurricane wind [30], enclosing structures of stables [31], wooden buildings [32, 33], wooden pallets for transporting goods, baseball bats and handles for working tools are all examples in which medium and high–speed deformations can occur.

This paper presents the results of experimental and computational studies of the behavior of wood beam samples under dynamic three-point bending.

7.2 Test Method and Proceeding Procedure

For dynamic loading generation and measuring of the dynamic deformation processes of beams, the measuring bar technique was used. Initially, Hopkinson developed his shock rod technique for measuring the pressure created by explosives [34]. This technique was further developed by Davis [35] and Kolsky [36], therefore the Split Hopkinson Pressure Bar (SHPB) is also known as the Kolsky setup. More detailed information about the SHPB technique can be found in the review by [37] and in the article by [38]. To determine the strength of the material during bending in static, a scheme for three-point or four-point bending of the beam is used. A similar technique has been tested for the case of dynamic loading of wooden beams. A general view of the sample-loading scheme is shown in Fig. 7.1.

In the experiments, measuring bars with a diameter of 20 mm were used. The ends of bars were cut into a wedge. Loading of the samples was carried out by a striker with a diameter of 20 mm and a length of 400 mm through a steel measuring bar. To increase the level of the strain gauge signal, duralumin measuring bars were used as supports.

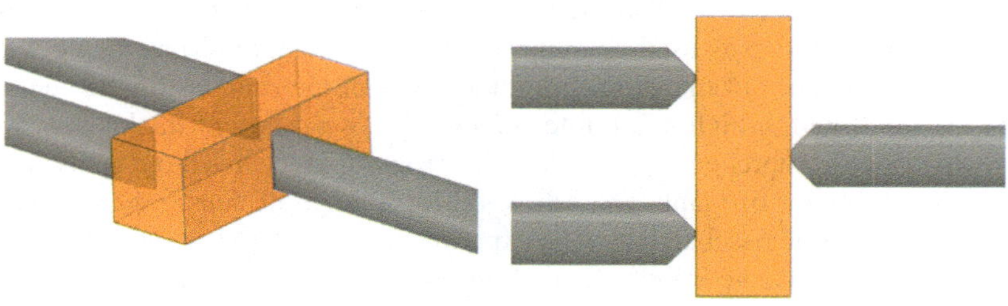

Fig. 7.1 The sample-loading scheme

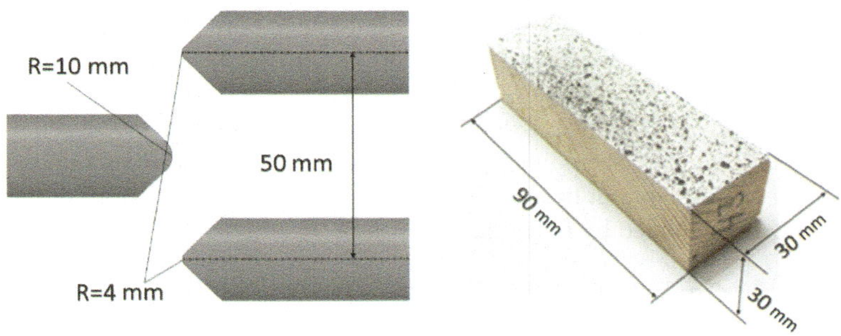

Fig. 7.2 Geometric parameters of the setup and the beam

The geometric parameters of the setup (the radii of the rounded bars and the distance between the output bars) are shown in the left part of Fig. 7.2. The geometric parameters of the sample beam are shown on the right side of Fig. 7.2.

For interpretation of the experimental data from the dynamic three-point bending experiments the following formulas were used:

Bending velocity was calculated by formula:

$$V_b(t) = c_I \cdot \left(\varepsilon^I + \varepsilon^R\right) - 0.5 \cdot c_T\left(\varepsilon_1^T + \varepsilon_2^T\right)$$

Beam deflection was calculated by formula:

$$U_b(t) = \int_0^t V_b(\tau)d\tau$$

Force acting on beam from input bar was calculated by formula:

$$F(t) = E_I \cdot S_I \cdot \left(\varepsilon^I - \varepsilon^R\right)$$

Forces acting on beam from output bars were calculated by formula:

$$F_1(t) = E_T \cdot S_T \cdot \varepsilon_1^T$$

$$F_2(t) = E_T \cdot S_T \cdot \varepsilon_2^T$$

here $\varepsilon^I, \varepsilon^R$—the incident and reflected strain pulses in the input measuring bar, $\varepsilon_1^T, \varepsilon_2^T$—the transmitted strain pulses, measured in the output bars, c_I—the sound speed in the input bar material, c_T—the sound speed in the output bar material, E_I, S_I—the Young modulus and the cross section area of the input bar, E_T, S_T—the Young modulus and the cross section area of the output bar.

In addition to the standard to the measuring bar techniques strain gauge measurements, the high-speed video registration was used to qualify the process of dynamic deformation of the beams (Fig. 7.3). The digital image correlation (DIC) technique was used to determine the displacement and strain fields in the samples during experiment.

The deflection of the beam was determined using the displacements of some points in the sample (Fig. 7.4). The point P0 was chosen near the incident bar and the points P1 and P2—near the output bars as shown in Fig. 7.4. The time history of the beam deflection was calculated by formula:

$$U_b^{DIC}(t) = U^{P0}(t) - 0.5 \cdot \left(U^{P0}(t) + U^{P1}(t)\right)$$

where $U^{P0}(t)$, $U^{P1}(t)$ и $U^{P2}(t)$—vertical displacements of point P0, P1, and P2.

The time histories of the vertical displacements of the points P0, P1, and P2 (solid lines) and the deflection (dotted line) of the beam during the dynamic loading determined using the DIC method are shown in Fig. 7.5.

Fig. 7.3 Video registration tools

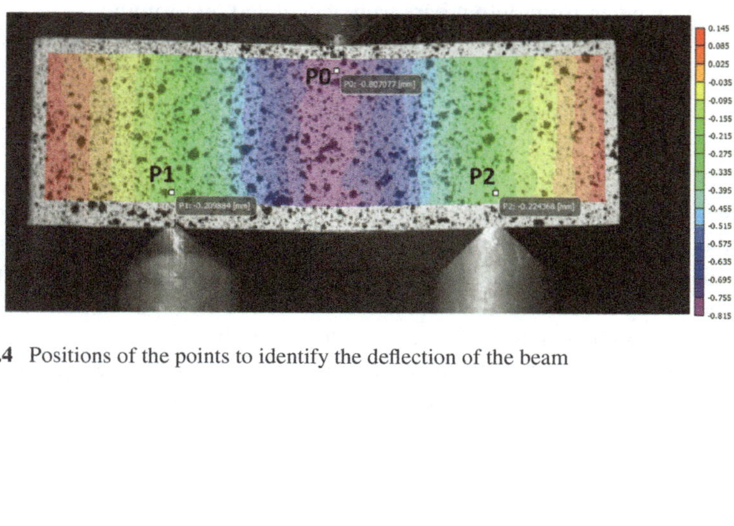

Fig. 7.4 Positions of the points to identify the deflection of the beam

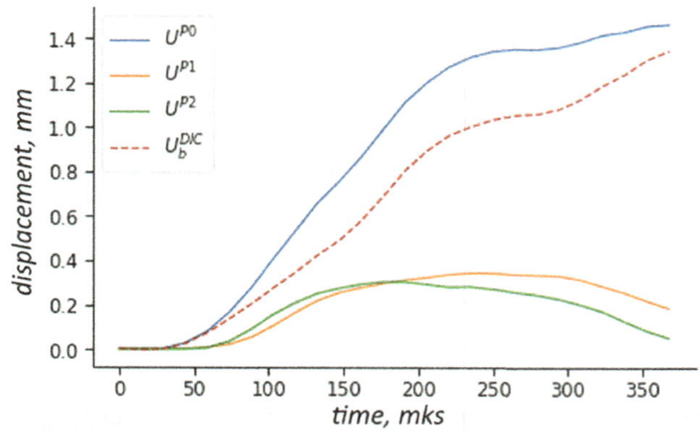

Fig. 7.5 The displacements and deflection determined using the DIC

Figure 7.6 shows a comparison of the deflections of the beam calculated according to the above Kolsky formulas (blue solid line) and determined using the DIC method based on high-speed video registration (orange dotted line). It can be seen that the deflection obtained by the DIC method is noticeably smaller. This is due to the fact that indirect displacement measurements using measuring bars do not actually register displacements of beam points, but only displacements of the bars themselves. Since wood is a soft material, there are local deformations of beams in places where metal bars affect them. Due to imperfections, gaps are also sampled. This leads to a difference in the values of deflections measured in different ways. It should be noted that the maximum deflection difference is about 0.2 mm.

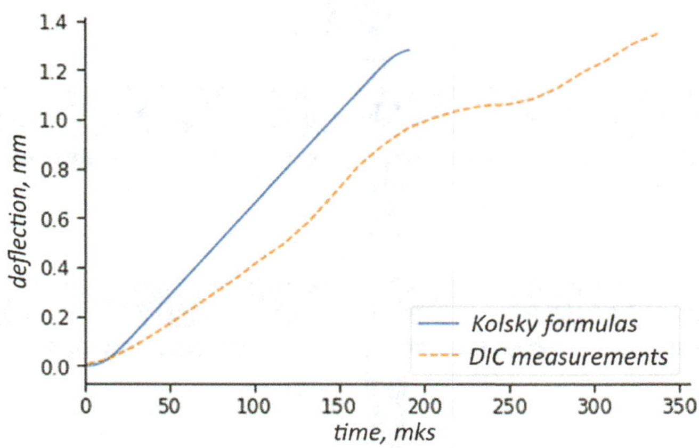

Fig. 7.6 Comparison of beam deflections calculated according to the Kolsky formulas and measured using DIC

To assess the ultimate strength characteristics of the wood, according to high-speed video recording data, deformation fields (normal components of the strain tensor in the direction of the beam axis) were determined at each moment of time. An example of deformation field, in the frame preceding the initiation of the crack, is shown in Fig. 7.7. It can be seen that compression strains take place on the loaded surface, and tension strains take place on the opposite surface. For further analysis and evaluation of the fracture strain, the distribution of strains along the line shown in Fig. 7.7 was determined.

Figure 7.8 shows the process of crack generation and growth during dynamic beam bending.

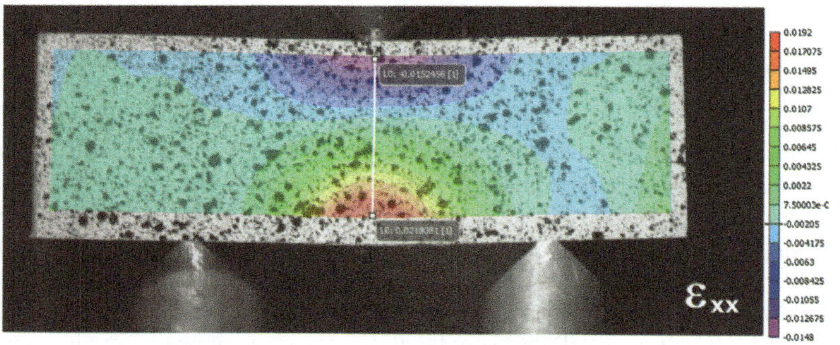

Fig. 7.7 Strain field in the beam

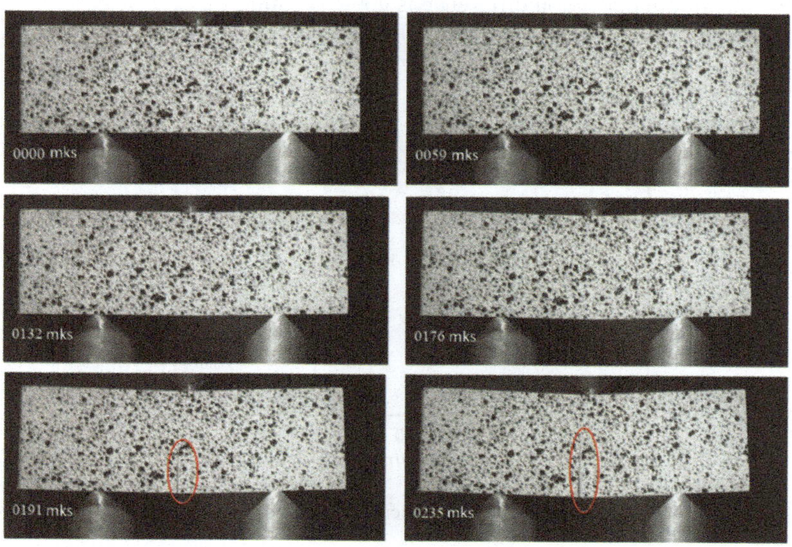

Fig. 7.8 Kinogram of the beam deformation process

Experimental studies have shown that the distributions of strains along the thickness of the sample are linear up to the moment preceding the fracture initiation, and the neutral axis of the beam coincides with its geometric center.

The specificity of the digital image correlation method is that the displacement and deformation fields are not determined for the entire surface of the sample. These values are not calculated on the part of the surface near the boundary (as can be seen in Fig. 7.7).

This is because deformations in the DIC method are calculated at the central points of elementary patterns (yellow rectangles in Fig. 7.9), which are used to track movements. The patterns cannot be too small, since the pattern must be unique and recognizable for the algorithm. As a result, a zone is formed at the border of the sample, the width of which is equal to half the size of the elementary pattern. The deformation parameters in this zone are unknown. Since the fracture originates on the surface of the sample, extrapolation is necessary for a correct assessment of the fracture strain value.

Figure 7.10 shows the result of extrapolation of strains from the area processed by the DIC method to the sample boundary.

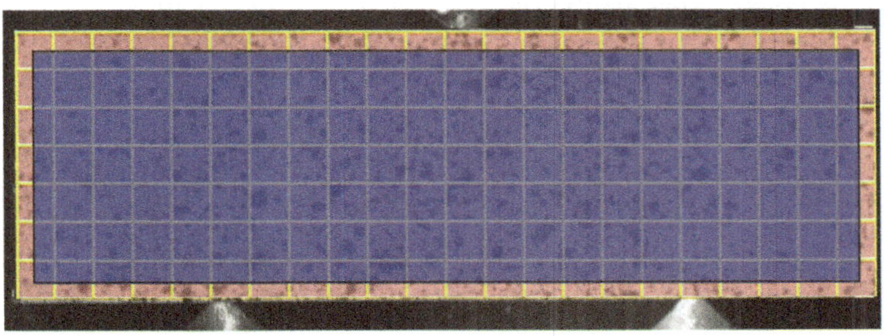

Fig. 7.9 Partitioning of the sample surface into zones when using the DIC method

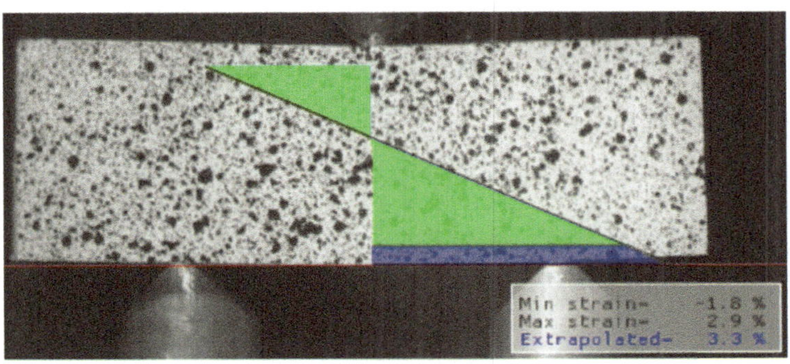

Fig. 7.10 Extrapolations of strains to the sample boundary

It can be noted that the maximum tensile strain in the processed region is 2.9%, and the extrapolated value is 3.3%. By repeating a similar procedure for each frame, it is possible to calculate the time dependence of the extrapolated strain value for each moment of time. To assess the loading conditions, the value of the strain rate was determined by differentiating the time history of strain.

7.3 The Results of Experiments

Experiments according to the scheme described above were carried out on pine beam samples. The samples were cut in such a way that the fiber was oriented along the width (b) of the sample (Fig. 7.11). The loading scheme, the designations of the sample sizes and the orientation of the fibers are shown in Fig. 7.11. The velocity of the striker was about 8 m/s.

Figure 7.12 shows the time dependences of the deflection of beams obtained under close loading conditions. Solid lines show the data calculated according to the Kolsky formulas, and dotted lines correspond to the dependencies determined by the DIC method. As noted earlier, deflections calculated from signals from measuring bars turn out to be larger than deflections determined from digital images.

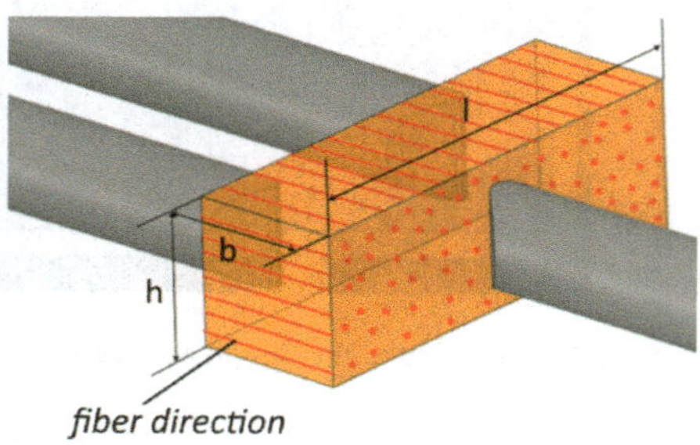

Fig. 7.11 Beam testing scheme

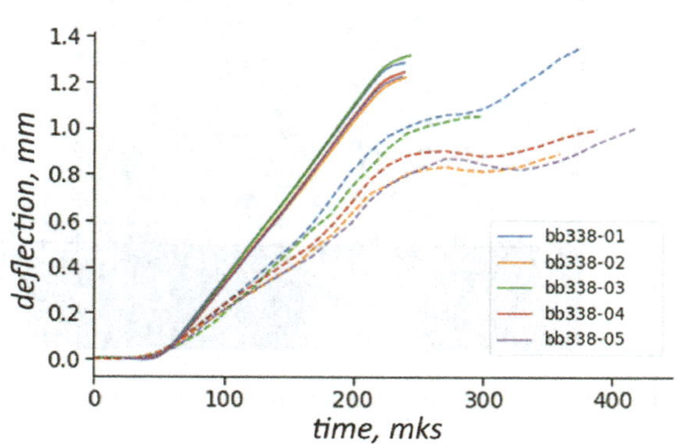

Fig. 7.12 Time dependences of deflection of beams

Figure 7.13 illustrates the time dependences of the forces acting on the beam during the test. The forces were determined by strain gauge signals from output measuring bars. It is possible to note the good repeatability of the test results carried out under similar conditions. The maximum value of the force was 5 kN.

Figure 7.14 shows the results of measurement of the history of tensile strain across the fibers (left) and the corresponding strain rate (right) obtained using the DIC method. When testing the beams, a tensile strain rate of about 700 s^{-1} was obtained. The values of the fracture strain are ~3.5%.

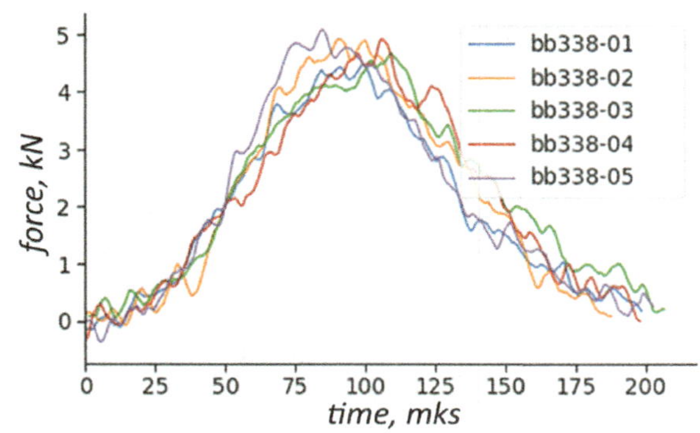

Fig. 7.13 Time dependencies of forces

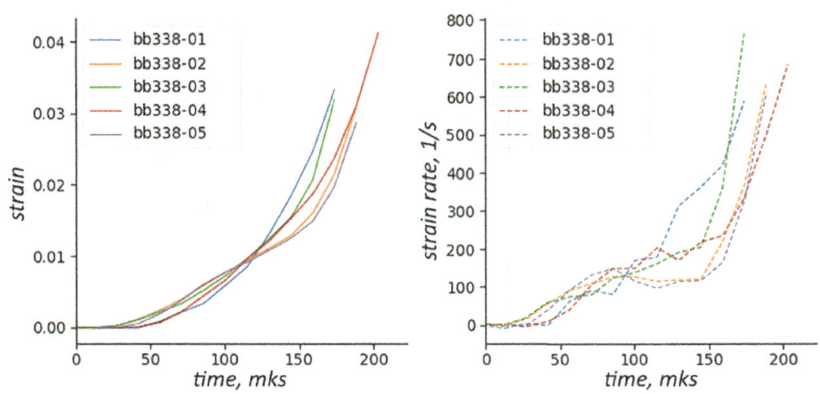

Fig. 7.14 Results of direct measurement of strain (left) and strain rate (right) by DIC method

7.4 Material Model

To model the behavior of wood in code LS-DYNA, there is a MAT_WOOD model, which is described in [39–41].

Within the framework of this model the following features of the behavior of wood are laid down:

- The transversally isotropic behavior of the material is assumed.
- Irreversible deformation of the material is taken into account.
- The deformation hardening of the material is taken into account.
- The change in the properties and fracture of the material is laid within the framework of the theory of the damaged medium.
- It is possible to take into account the influence of the strain rate on the strength characteristics.

A distinctive feature of the MAT_WOOD model is the predefined set of parameters embedded in it for two types of wood: yellow pine and Douglas fir.

The theoretical foundations of the MAT_WOOD model are described in [41]. It is noted that wood is a fairly diverse material, however, a number of features can be distinguished that differ it from other materials. The stiffness and strength characteristics of wood depend on the direction and differ for the longitudinal, radial, and tangential directions. The direction of the fiber is taken as the longitudinal direction. It is also noted that for modeling purposes, differences in properties in radial and tangential directions are insignificant, therefore, the behavior of wood is usually described by a transversally isotropic model, and the terms "parallel" and "perpendicular" are used to classify directions.

The strength characteristics of wood also differ for different types of loading: compression, tension, and shear. The behavior of the material when tensioned in the "parallel" and "perpendicular" directions, as well as when sheared close to linear up to fracture (brittle behavior). When compressed, the wood behaves non-linearly, there is a visco-plastic flow.

The wood model implemented in LS-DYNA includes the following components:

- Linear equation of state for elastic region;
- The criterion of fracture;
- The law of plastic flow;
- The law of hardening;
- The law of softening (degradation of properties) as damage accumulates;
- An equation describing the effect of the strain rate on the fracture stresses.

7.5 Modeling of Impact Bending of a Wooden Beam

Using the MAT_WOOD_PINE model, the simulation of the process of high-speed bending of beams was carried out on the basis of the experimental scheme described earlier. The simulation was carried out using LS-DYNA with the use of

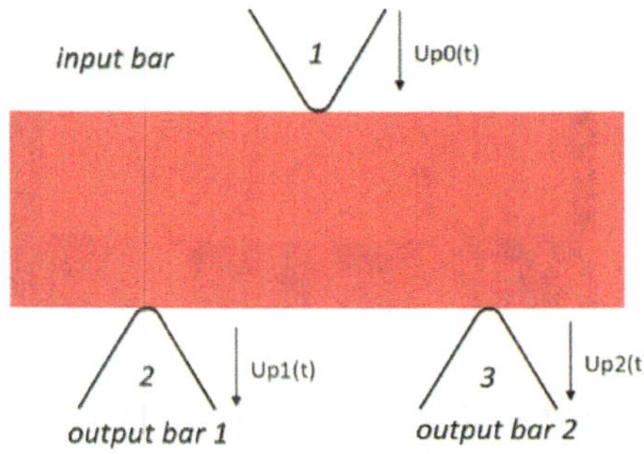

Fig. 7.15 Setting up a dynamic beam bending simulation

an explicit scheme for integrating equations in time and the finite element method in a Lagrangian formulation for space discretization. The problem was solved in a three-dimensional formulation. The geometric formulation of the problem and the boundary conditions are shown in Fig. 7.15. The loading (1) and supporting (2 and 3) bars were modeled by non-deformable bodies, for which time dependences of vertical displacements were applied. Those time histories were determined by the digital image correlation method according to the data of a full-scale test.

In the full-scale experiment, crack initiation began at the time between frames corresponding to 309 and 324 microseconds (in the time frame in which the boundary conditions were calculated). When using the MAT_WOOD model with default strength parameters, no beam fracture was observed in the calculation.

As an experimental study has shown, a crack initiates with a tensile strain of about 3.5%. This corresponds to a value of ultimate stress of about 8.6 MPa. The results of modeling the impact three-point bending of the beam after making appropriate adjustments to the model are shown in Fig. 7.16. The red color corresponds to the fractured material.

When the size of the final element decreases, multiple cracks appear in the fracture zone (Fig. 7.17), but the moment of the initiation of the first fracture practically does not change.

7.6 Conclusions

Modern tools for recording fast-flowing processes, such as the digital image correlation method based on high-speed video recording, allow us to obtain comprehensive information about the process of deformation of samples under shock loading. Based on this method, a scheme for determining the ultimate deformation of fracture during dynamic three-point bending of beams is proposed and implemented. The determination of the ultimate deformation is carried out by extrapolation of the strain fields determined by the DIC method. The loading conditions (strain rate) are estimated by differentiating the histories of deformation changes at the crack initiation point. The value of the ultimate tensile strain in perpendicular direction of pine is determined

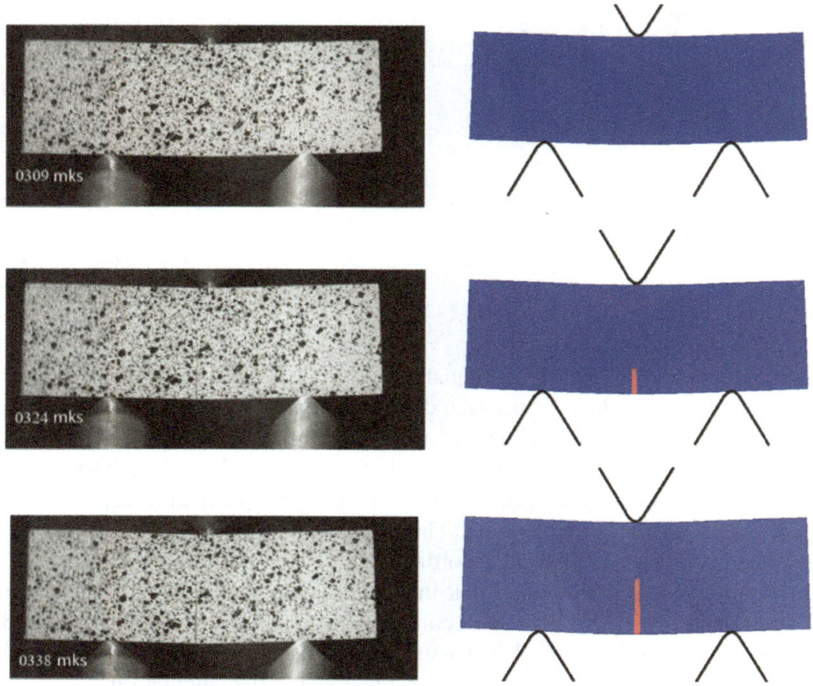

Fig. 7.16 Fracture of the beam during dynamic bending

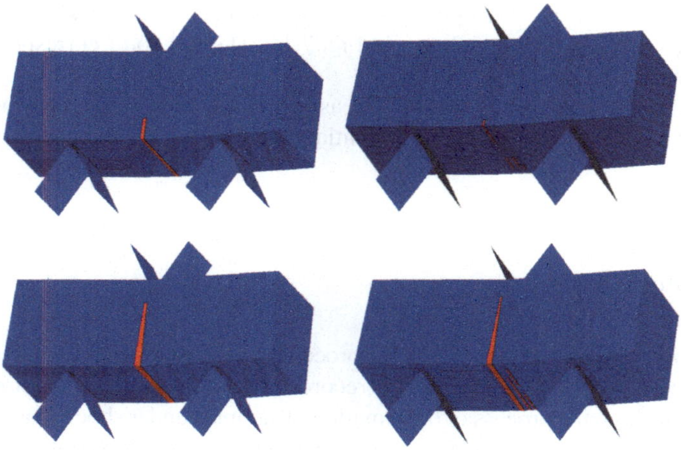

Fig. 7.17 The influence of the grid size on the simulation results

from experiments on three-point bending. Obtained value was about 3.5%. The strain rate at the moment of fracture was about 700 1/s. The use of the specified value of the ultimate strain in the model made it possible to reliably predict the destruction of the beam in a numerical experiment for dynamic three-point bending.

Acknowledgements The theoretical study was done with financial support from the Ministry of Science and Higher Education of the Russian Federation (Project 0729-2020-0054). The experimental investigations were conducted with financial support from the Russian Science Foundation (project 21-19-00283).

References

1. Adalian C, Morlier P (1998) Modeling the behaviour of wood during the crash of a cask impact limiter. PATRAM'98. Conference Proceedings, vol 1. Paris, France
2. Neumann M (2009) Investigation of the behavior of shock-absorbing structural parts of transport casks holding radioactive substances in terms of design testing and risk analysis. PhD thesis, Bergische Universität Wuppertal, Germany
3. Ryabov AA, Romanov VI, Kukanov SS, Spiridonov VF, Tsiberev KV (2016) Numerical analysis of impact and thermal resistances of air transport package PAT-2. Problems of Strength and Plasticity J 78(1):101–111 (in Russian)
4. Ryabov AA, Romanov VI, Kukanov SS, Skurikhin SG (2006) Numerical simulations of dynamic deformation of air transport package PAT-2 in accidental impacts. Proceedings of 9th International LS-DYNA Users Conference. Dearborn, USA, pp 43–51
5. Eisenacher G et al (2013) Crushing characteristics of spruce wood used in impact limiters of type B packages. Packaging and transportation of radioactive materials PATRAM. San Francisco, USA
6. Buchar J, Severa L, Havlicek M, Rolc S (2000) Response of wood to the explosive loading. J. Phys. IV France 10:529–534
7. Johnson W (1986) Historical and present-day references concerning impact on wood. Int J Imp Eng 4(3):161–174
8. Harrigan JJ, Reid SR, Tan PJ, Reddy TY (2005) High rate crushing of wood along the grain. Int J Mech Sci 47:521–544
9. Reid SR, Peng C (1997) Dynamic Uniaxial Crushing of Wood. Int. J. Imp. Eng. 19:531–570
10. Reid SR, Reddy TY, Peng C (1993) Dynamic compression of cellular structures and materials. In: Wierzbicki T (ed) Jones N. Structural crashworthiness and failure. Taylor & Francis Publ, London-New York, pp 257–294
11. Gibson LJ, Ashby MF (1997) Cellular solids-structure and properties, 2nd edn. Cambridge University Press
12. Bolshakov AP, Gerdyukov NN, Novikov SA et al (2001) Damping properties of Sequoia, Birch, Pine, and Aspen under shock loading. J Appl Mech Tech Phys 42(2):202–210
13. Buchar J, Adamik V (2001) Wood strength evaluation under impact loading. 39th international conference experimental stress analysis. Tabor, Czech Republic
14. Buchar J, Krivanek I, Severa L (2001) High rate behaviour of wood. In: Nowacki WK, Klepaczko JR (eds) New experimental methods in material dynamics and impact, trends in mechanics of materials. Warsaw, pp 357–362
15. Sebeka F, Kubika P, Brabecb M, Tippner J (2020) Modelling of impact behaviour of European beech subjected to split Hopkinson pressure bar test. Compos Struct 245:112330
16. Gomesa F, Xavierb J, Koerber H (2019) High strain rate compressive behaviour of wood on the transverse plane. Procedia Struct Integr 17:900–905

17. Zhou SC, Demartino C, Xiao Y (2020) High-strain rate compressive behavior of Douglas fir and glubam. Constr Build Mater 258:119466
18. Allazadeh MR, Wosu SN (2012) High strain rate compressive tests on wood. Strain 48(2):101–107
19. Fortin-Smith J, Sherwood J, Drane P, Kretschmann D (2016) Characterization of maple and ash material properties as a function of wood density for bat/ball impact modeling in LS-DYNA. Procedia Engineering 147:413–418
20. Zhao S, Zhao JX, Han GZ (2016) Advances in the study of mechanical properties and constitutive law in the field of wood research. IOP Conf. Series: Materials Science and Engineering 137:012036
21. Liska JA (1950) Effect of rapid loading on the compressive and flexural strength of wood. USDA for serv report no. 1767. USDA For Serv Forest Products Laboratory, Madison, WI, United States
22. Keeton JR (1968) Dynamic properties of small, clear specimens of structural-grade timber. Technical report R-573, Y-F011-05-04-003. U.S. Navy Civ Eng Lab, Port Hueneme, CA, United States
23. Gatchell C, Michie J (1974) Pendulum impact tests of wooden and steel highway guardrail posts. USDA for serv research paper NE-311. Upper Darby, PA, United States
24. Leijten AJM (2000) Literature review of impact strength of timber and joints. World conference on timber engineering, Whistler, Canada
25. Bocchio N, Paola R, van de Kuilen JWG (2001) Impact loading tests on timber beams. In: IABSE, vol 85. Lahti, Finland, pp 19–24
26. Botting JK (2003) Development of an FRP reinforced hardwood glulam guardrail. Master thesis, The University of Maine, Orono, ME, United States
27. Kubojima Y, Ohsaki H, Kato H, Tonosaki M (2006) Fixed-fixed flexural vibration testing method of beams for timber guardrails. J Wood Sci 52(3):202–207
28. Gutkowski RM, Shigidi A, Abdallah MT, Peterson ML (2007) Dynamic impact load tests of a bridge guardrail system. MPC report no. 07–188. Mountain-Plains Consortium, Fargo, ND, p 37
29. Polocoşer T, Stöckel F, Kasal B (2016) Low-velocity transverse impact of small, clear spruce and pine specimens with additional energy absorbing treatments. J Mater Civ Eng. 28(8): 04016048
30. Turnbull-Grimes C, Charlie WA, Gutkowski RM, Balogh J (2010) Bus-stop shelters–improved safety. Report for North Dakota State University, Fargo, ND, United States
31. Benthien JT, Georg H, Maikowski S, Ohlmeyer M (2012) Infill planks for horse stable constructions: Thoughts about kick resistance determination and alternative material development. Landbauforsch Appl Agric For Res 62(4):255
32. Jacques E, Lloyd A, Braimah A, Saatcioglu M, Doudak G, Abdelalim O (2014) Influence of high strainrates on the dynamic flexural material properties of spruce–pine–fir wood studs. Can J Civ Eng 41(1):56–64
33. Kasal B, Guindos P, Polocoser T, Heiduschke A, Urushadze S, Pospisil S (2014) Heavy laminated timber frames with rigid three-dimensional beam-to-column connections. J Perform Constr Facil. 28(6): A4014014
34. Hopkinson B (1914) A method of measuring the pressure in the detonation of high explosives or by the impact of bullets. Philos. Trans. R. Soc. London Ser A 213:437–456
35. Davies RM (1948) A critical study of the Hopkinson pressure bar. Philos Trans R Soc London, Sect. B 240(821):375–457
36. Kolsky H (1949) An investigation of the mechanical properties of materials at very high rates of loading. Proc Phys Soc London, Sect B 62(II-B):676–700.
37. Gama BA (2004) Hopkinson bar experimental technique: a critical review. Appl Mech Rev 57(4):223–250
38. Gray III GT (2000) Classic split-Hopkinson pressure bar testing. In: Vol 8 Mechanical testing and evaluation, ASM Handbook. ASM International, pp 462–476

39. LS-DYNA Keyword User's Manual (2001) Volume2, Version 960. Livermore Software Technology Corporation, Livermore, CA
40. Murray YD (2004) Manual for LS-DYNA wood material model 143, report no. FHWAHRT-04–097. Federal Highway Administration
41. Murray YD, Reid JD, Faller RK, Bielenberg BW, Paulsen TJ (2005) Evaluation of LS-DYNA wood material model 143, report no. FHWA-HRT-04-096. Federal Highway Administration

Chapter 8
Study of Dynamic Properties of Fiber Concrete with Polymeric, Steel, and Combined Fiber under High-Speed Compression

Anatoly M. Bragov, Mikhail E. Gonov, Aleksandr Yu. Konstantinov, and Andrey K. Lomunov

Abstract An experimental study of the dynamic properties of three types of fiber-reinforced concrete under dynamic uniaxial compression relative to the original fine-grained concrete was carried out. Three types of fiber-reinforced concrete were produced: with polymer fiber, steel fiber, and with a combination of two types of fiber. Static and dynamic tests were carried out. Dynamic compression tests were carried out using the Kolsky method at strain rates from 10^2 to 10^3 s^{-1}. The tests were carried out using a FASTCAM Mini UX100 high-speed camera. The paper presents the compositions of the studied materials, test parameters, as well as a comparative analysis of the data obtained. The introduction of a reinforcing fiber into the original fine-grained concrete increased the dynamic strength of the material. The highest strength under dynamic uniaxial compression was shown by fiber-reinforced concrete with steel fiber. The dependences obtained demonstrate that the maximum breaking stresses achieved in the experiments grow linearly with an increase in the strain rate, as do the corresponding limiting strains. The time before the onset of fracture decreases with increasing strain rate according to a power law.

Keywords Dynamic strength · Stress · Deformation · Strain rate · Compression · Concrete · Fiber reinforced concrete · Dynamic tests · Kolsky method · Mechanics

---

A. M. Bragov · M. E. Gonov · A. Yu. Konstantinov (✉) · A. K. Lomunov
National Research Lobachevsky State University of Nizhny Novgorod, Gagarin Ave. 23, 603022 Nizhny Novgorod, Russia
e-mail: constantinov.al@yandex.ru

A. M. Bragov
e-mail: bragov@mech.unn.ru

M. E. Gonov
e-mail: gonov_mikhail@mech.unn.ru

A. K. Lomunov
e-mail: lomunov@mech.unn.ru

H. Altenbach et al. (eds.), *Deformation and Destruction of Materials and Structures Under Quasi-static and Impulse Loading*, Advanced Structured Materials 186,
https://doi.org/10.1007/978-3-031-22093-7_8

8.1 Introduction

In parallel with the growth of the technogenicity of civilization, the number of emergency situations is growing, exposing the structures of buildings and structures for civil, industrial, or military purposes to high-speed dynamic effects in the form of shocks and explosions. At the same time, structural materials such as concrete and its varieties (fiber-reinforced concrete, reinforced concrete, special types of concrete) are subjected to high-intensity dynamic loads. In addition to man-made disasters, the increasing frequency of natural disasters, which are associated with climate change on our planet, is coming to the fore. Various natural disasters are expressed by hurricanes, tornadoes, floods, and strong earthquakes that have become more frequent in recent decades. Another factor in the occurrence of dynamic loads is the terrorist threat to modern civilization in the form of detonation of various explosive and combustible substances. The main danger of such phenomena is the loss of human lives. The threat to life arises both from the loss of the bearing capacity of buildings and structures, and from the occurrence of many shrapnel wounds caused by brittle materials.

Ordinary concrete has relatively low tensile strength and low ductility, so it is prone to cracking. The formation of cracks leads to the early onset of the destruction of concrete or reinforced concrete. It has been proven that the addition of steel or non-metallic fibers increases the tensile strength and plasticity of fiber-reinforced concrete [1–5]. At present, steel reinforcement is widely used, however, along with this, the technology of reinforcing concrete with fiber is becoming more widespread. Various types of fibers are used as reinforcing elements in concrete. Fiber-reinforced concrete is a composite material that contains fibers to increase its resistance to tensile loads. Recently, new types of fiber-reinforced concrete have been developed with various improved properties [6–9]. Today, fibers of various cross sections (flat, round, rectangular, etc.) are used. Traditionally, fibers are classified by material and, as a rule, metal, glass, synthetic, and natural fibers are used in fiber-reinforced concrete. According to the length of the fiber, microfibers shorter than 20 mm with an equivalent diameter of 0.005–0.2 mm and macrofibers 20–80 mm long with a length-to-diameter ratio of 40–120 are distinguished.

An important issue affecting the physical and mechanical properties of fiber-reinforced concrete is the shape, volume fraction, and geometry of the fiber. The influence of these factors was studied by various authors in [10–16]. The article [17] considered the influence of the loading rate on the delamination of high-strength steel fibers in heavy-duty fiber-reinforced concrete. The research focus is on experimental strength characteristics at various strain rates to understand the dynamic properties of heavy-duty fiber-reinforced concrete. Various mechanisms of detachment of straight and geometrically bound fibers have been studied in detail. The experimental study considers four types of high-strength fibers such as straight brass with a diameter of 0.2 and 0.38 mm, with half-hooks at the ends with a diameter of 0.38 mm, and twisted helical fibers with an equivalent diameter of 0.3 mm. The materials were tested at four different strain rates. The test results show that the semi-hooked fibers show the highest strain rate sensitivity compared to the rest of the fibers used in this study, which may be related to cracking of the stressed cement backing. In addition, the influence of the angles of the location of the fibers to the axis of the sample was studied at various strain rates, which affect the delamination of the fibers. Three fiber angles are considered, 0°, 20°, and 45°. The results show that there is a relationship between fiber angle and strain rate sensitivity, which influences fiber delamination.

In [18], the effect of various fiber shapes on the strength characteristics of fiber-reinforced concrete in compression during impact loading is presented. This study examined five commercially available types of fibers with different shapes and material properties, namely synthetic, wavy, cold rolled, flattened, and hooked fibers. In addition, two new types of helical fibers, called Helix I and Helix II, have been proposed and tested to minimize fiber-reinforced concrete fiber spalling under dynamic loading. The results showed that the properties of the fiber and its geometry play an important role in determining the dynamic characteristics of fiber-reinforced concrete in compression. For example, the impact resistance of the helical fiber samples was better than other commercially available fibers.

In [19], a comparative analysis of the response of plain concrete and fiber-reinforced concrete to impact loading was carried out. In this study, hook-end steel fibers with volume fractions of 0.5% and 1% and polypropylene fibers with volume fractions of 0.2%, 0.3%, and 0.5% were used. Experimental and numerical analysis has shown that steel fibers can increase impact resistance to a greater extent than polypropylene fibers due to their longer length, higher tensile strength, and better cohesiveness. In addition, tensile and compressive strength, energy absorption degree can be increased by increasing the fiber volume fraction.

Today, there are many types of fiber-reinforced concrete with steel fibers, among which there are four main types, depending on the volume fraction of the fiber and the initial strength properties of the concrete. The first type is fiber-reinforced concrete with discontinuous discrete steel fibers. As indicated in [20], the volume fraction of the fiber used in this type of fiber-reinforced concrete should be in the range of 0.5–1.5%, since exceeding this value reduces the workability of the concrete mixture and causes clumping, which will be extremely difficult to overcome during vibrocompaction. However, high fiber percentages can be achieved with special fiber addition techniques and even alignment procedures. The second type is high-strength fiber-reinforced concrete, which is generally defined as high-strength concrete with a compressive strength above 40 MPa, with the addition of short randomly distributed fibers with a volume fraction of 2.0% or more. The addition of steel fibers to high-strength concrete forms high-strength fiber-reinforced concrete and improves its impact and fracture resistance. The third type—special fiber-reinforced concrete—is a special type of high-strength concrete reinforced with high-strength fibers (150–400 MPa) and a large volume fraction of steel fibers (5–10%). The cement base of special fiber-reinforced concrete is extremely plastic due to the high content of steel fibers. This feature makes it possible to effectively use fiber reinforcement without the formation of large cracks under operating conditions. Quadruple type—ultra-high strength fiber-reinforced concrete—is a new class of materials that combine a very strong and dense binder cement base with a high fiber content. This material has a very high compressive strength (about 200 MPa) and a relatively high tensile strength (about 10 MPa), which minimizes crack opening and provides improved plasticity properties compared to conventional concrete. At the same time, the fibers provide plasticity both in tension and in compression. The addition of a sufficiently large amount of fibers (more than 1.5% by volume) leads to tensile strengthening [21]. In addition, ultra-high-strength fiber-reinforced concrete can significantly improve the impact resistance of panels and walls while maintaining standard thickness and appearance [22].

In [23], the impact strength of fiber-reinforced concrete slabs was studied at low-speed projectile impact. The main variables adopted in the studies were the type and volume fraction of fibers. Various types of fibers were used such as polyolefin, polyvinyl alcohol, and steel with volume fractions of 0%, 1%, and 2%, respectively. In addition, 10 square slabs 1.0 m long and 50 mm thick were cast and tested. Among various fiber types, hook-ended steel fiber concrete slabs have good energy absorption and crack resistance characteristics.

In [24], experimental and numerical studies of a fiber-reinforced concrete slab subjected to impact loading were carried out using 1%, 1.5%, and 2% steel fibers with hooks at the ends. With an increase in the volume fraction of fibers, the penetration resistance increased. Numerical analysis [25] of crater explosions in fiber-reinforced concrete using different volume fractions of steel fibers showed an increase in strength, impact strength, and fracture resistance. The best blast loading performance was obtained using concrete containing 2.0% fiber volume fraction, followed by concrete containing 1.5% and 1.0% fiber volume fraction.

In addition to fiber-reinforced concrete with steel or polymer fibers, combined fiber-reinforced concrete is of particular interest, which can be attributed to the fifth type of fiber-reinforced concrete. The combination of steel fibers with polymer fibers has led to a significant increase in the blast resistance of structures. Fiber-reinforced concrete, including the ratio of high-modulus steel fibers and relatively low-modulus polyethylene fibers, was studied in [26] to achieve a balance between deformation capacity and tensile strength required for impact and blast-resistant structures.

Combined fiber-reinforced concrete has characteristics that are superior to those of simple fiber-reinforced concrete. For example, 0.5% steel fiber with 1.5% polyethylene fiber reduced spalling and damage zones [26]. In [27], a study was made on the nature of the destruction and damage of fiber-reinforced concrete panels made of combined fiber in comparison with ordinary reinforced concrete and fiber-reinforced concrete. Composite fiber concrete panels have been shown to reduce damage and improve re-impact impact resistance, as well as improve energy absorption capacity and plasticity properties compared to reinforced concrete and fiber concrete panels [27]. In addition, steel fibers combined with polypropylene fibers in concrete have improved impact resistance compared to monofiber concrete [28]. In [29] studied four types of impact barriers, with different fiber types, including two types of synthetic fiber, nylon fiber and carbon fiber, each with 1.5% fiber volume; a combination of steel and synthetic fibers with a fiber content of 3.8% and 5%; and traditional reinforced concrete of normal density, used as a control sample. Nylon fiber-reinforced concrete with a fiber content of 1.5% showed a significant performance improvement over typical reinforced concrete. However, a mixture of steel and synthetic fibers at 3.8% and 5% showed similar improvements. Explosive testing of fiber-reinforced concrete slabs showed that the degree of destruction (particle size and velocity when exposed to an explosive load) was significantly reduced due to the addition of steel and polypropylene fibers [29].

It should be noted that the question of the best mechanical properties of fiber-reinforced concrete with steel or combined fibers in dynamic compression is still open. In connection with the foregoing, the study of the strength characteristics of fiber-reinforced concrete with different types of fibers during high-speed deformation and destruction is an urgent task. This article is devoted to the solution to this problem.

8.2 Test Method

High-speed tests under conditions of a one-dimensional compressive stress state were carried out on the experimental setup SHPB-20 [30–32], the scheme of which is shown in Fig. 8.1. The experimental setup according to the Kolsky method includes a system of measuring rods, between which a test sample is installed, a gas gun for accelerating a cylindrical impactor, strain gauges, a velocity meter, recording, and computing equipment with a software package. In one of the rods, after the impact of the impactor, a one-dimensional elastic compression wave is excited, which propagates along the rods at the speed of sound. Upon reaching the sample, this wave splits due to the difference in the acoustic stiffness of the materials of the rod and the sample, as well as the areas of their cross sections: part of it is reflected back by a tension wave, and part passes through the sample into the second rod by a compression wave. In this case, the sample undergoes elastic–plastic deformation or destruction, while the rods are deformed elastically. By registering elastic strain pulses in measuring rods with strain gauges according to the formulas proposed for the first time by Kolsky, it is possible to determine the stresses, strains, and strain rates in the sample as a function of time. The tests were carried out on samples with a diameter of 20 mm and a length of 10 mm. Measuring rods, both loading and supporting, as well as a cylindrical striker, are made of D16T duralumin alloy with an elastic modulus of 0.71×105 MPa. The amplitude of the loading wave was varied by changing the striker velocity. The loading modes were chosen in such a way that a gradual increase in the strain rate and loading rate was observed.

To fulfill the basic premise of the Kolsky method—a uniaxial stress state in the sample, the main recommendations were taken into account, namely: the ratio of the length of the sample to the diameter within 0.3–1.0; treatment of the ends of the measuring rods before the start of each experiment with Litol lubricant to reduce the effect of friction forces during the radial expansion of the sample. During dynamic tests, a FASTCAM Mini UX100 high-speed digital camera was used to visualize the deformation and fracture processes. On the basis of time-lapse photography, an analysis was made of the time and nature of the destruction of the samples.

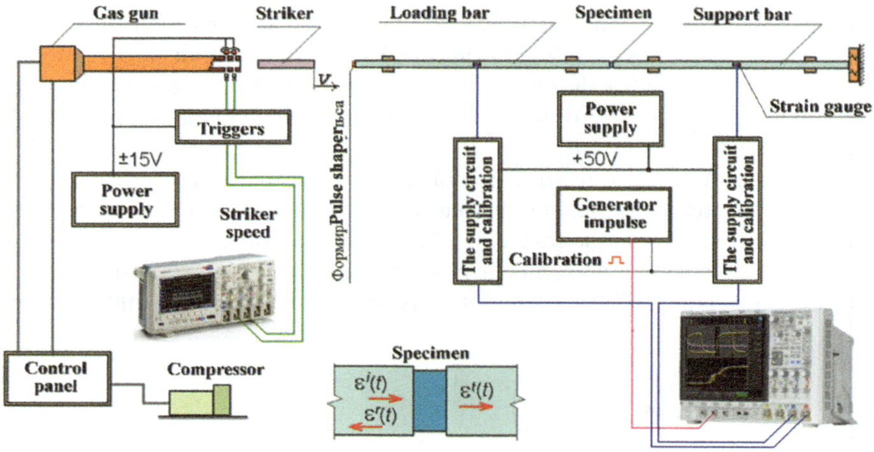

Fig. 8.1 Scheme of the installation for testing under compression in the condition of a uniaxial stress state

8.3 Characteristics of the Tested Materials

Fiber-reinforced concrete is a composite material containing reinforcing fiber and having a number of advantages over conventional concrete: increased compressive, tensile, and shear strength; impact and fatigue strength; crack resistance and fracture toughness; frost resistance, water resistance, and heat resistance (fire resistance). Improved properties of fiber-reinforced concrete provide high technical and economic efficiency. The variability of materials of this kind leads to the need for theoretical and experimental studies both to determine the dynamic characteristics of fiber-reinforced concrete and assess the influence of filler parameters (material, length, fiber shape, etc.), and to build mathematical models that describe its deformation and destruction during calculations. Taking into account the influence of dynamic loading conditions.

Three types of fiber-reinforced concrete were produced, differing in the material of the reinforcing fiber: fiber-reinforced concrete with polymer fiber (polyfiber fiber-reinforced concrete—PFRC) with a volume fraction of polypropylene fiber of 1.5%; fiber-reinforced concrete with steel fiber (steel fiber-reinforced concrete—SFRC) with a volume fraction of steel fiber of 1.5%; fiber-reinforced concrete with a combination of polymer and steel fibers (combined fiber-reinforced concrete—CFRC) with a volume fraction of polypropylene (0.75%) and steel (0.75%) fibers. Also in this section, the composition of the initial fine-grained concrete (FGC) is given for a comparative analysis of the effect of reinforcing fiber on the mechanical properties of the materials under study.

Polypropylene fiber, Fig. 8.2, manufactured by Poliarm, is a structural synthetic macrofiber 25 mm long, which is a separate rigid fiber of sinusoidal-wavy shape from oriented virgin polypropylene, treated with a special compound that improves adhesion to concrete mortar.

Polymer fiber has increased compressive and tensile strength, increased impact and fatigue strength, reduces water separation, prevents delamination of the concrete mixture, increases the fire resistance of concrete, reduces labor costs and terms of work, does not damage the mixing and supplying concrete equipment. Technical characteristics of polypropylene fiber are presented in Table 8.1.

The manufacturer of steel fiber is the Belarusian Metallurgical Plant (BMZ). Steel fiber is made from high-carbon steel wire according to GOST 9389 with a tensile strength of at least 1000 N/mm^2 (MPa). For the manufacture of steel fiber-reinforced concrete, a wave profile fiber 15 mm long and 0.3 mm in diameter was used, Fig. 8.2.

Dynamic compression tests were carried out on the RSG-20–2 unit. The compositions of fiber-reinforced concrete are presented in Table 8.2.

Fig. 8.2 Polypropylene fiber "Poliarm" and steel fiber "BMZ"

Table 8.1 Specifications of polypropylene fiber

№	Characteristic	Indicator
1	Length	25 to 55 mm
2	Material	Virgin polypropylene 100%
3	Specific gravity	0.91 kg/m^3
4	Young's modulus	3500 N/mm
5	Tensile strength	360–560 N/mm^2
6	Softening temperature	156 °C
7	Color	Transparent white
8	Chemical resistance	To acids, alkalis, and solvents

Table 8.2 Composition of fiber-reinforced concrete

№	Material	Consumption per 1 m^3 (kg)			
		Without fiber	With polymer fiber	With steel fiber	With combined fiber
1	Cement (grade D500)	480	480	480	480
2	Crushed stone (module 1–3 mm)	1250	1000	1000	1000
3	Medium sand	390	600	600	600
4	Superplasticizer	2.4	3.0	3.0	3.0
5	Water	205	200	200	200
6	Polypropylene fiber "Poliarm" 25 mm	–	14.0	–	7.0
7	Steel corrugated fiber BMZ 15 mm	–	–	120	60

All samples for dynamic testing were made by drilling with a diamond crown from pre-cast concrete blanks.

8.4 Results of Static Tests

Static tests for uniaxial compression consisted of 3–4 experiments for each material. Cylindrical samples for static tests with a length and diameter of 20 mm were drilled from concrete blanks. Static tests were carried out up to the stage of sample failure at a constant strain rate of $3\cdot10^{-5}$ s^{-1} on a Z100 Zwick-Roell testing machine. The averaged test results in the form of strain diagrams are shown in Fig. 8.3 and summarized in Table 8.6.

8.5 Results of Dynamic Tests for Uniaxial Compression

A cycle of dynamic tests of fine-grained concrete was carried out, consisting of 4 high-speed loading modes. The test parameters are shown in Table 8.7. Before testing, the samples were sorted, measured, and numbered. The strain rates were in the range from 100 to 1000 s^{-1}.

The strain pulses recorded in the measuring rods were synchronized in time. Using the formulas of the Kolsky method for each experiment, the dependences of stress on time (σ~t), strain on time (ε~t), and strain rate on time ($\dot{\varepsilon}$~t) were plotted. The obtained dependences were synchronized in time for each speed mode. After

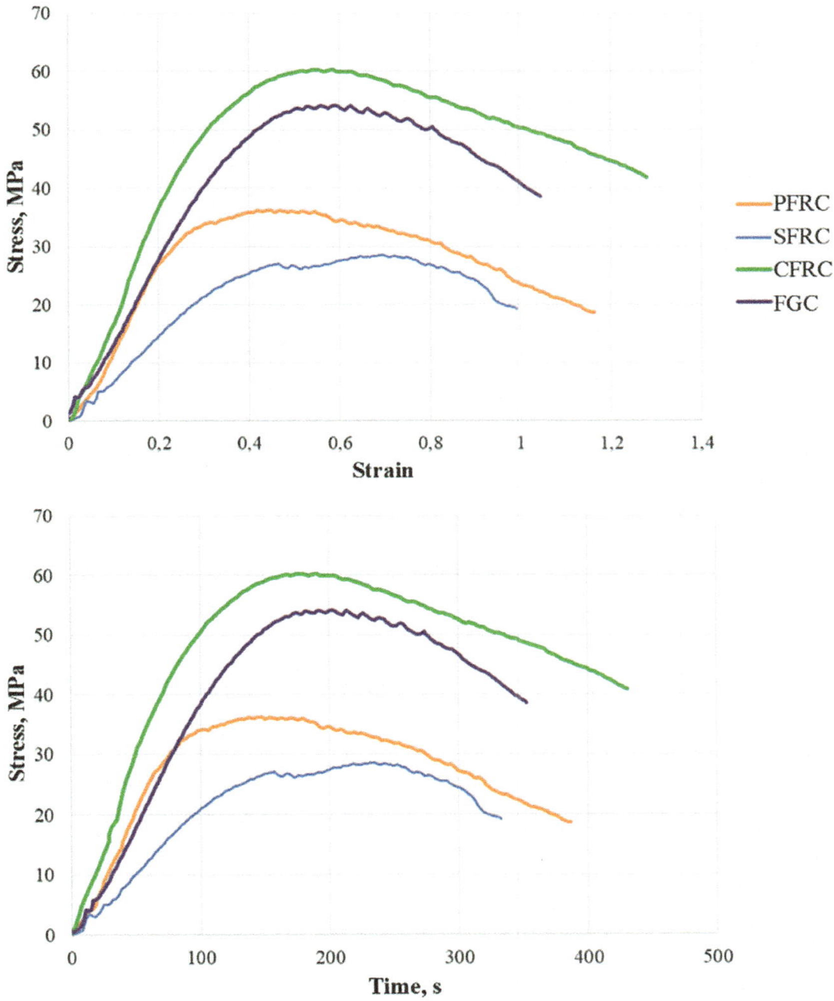

Fig. 8.3 Results of static tests

Table 8.6 Results of static tests

№	Material	Ultimate stress, MPa	Ultimate strain, %
1	Fine-grained concrete	28.65	0.7
2	Polymer fiber-reinforced concrete	36.28	0.5
3	Steel fiber-reinforced concrete	60.38	0.6
4	Combined fiber-reinforced concrete	54.14	0.6

Table 8.7 Parameters of FGC tests under dynamic compression

Mode	Experiment code	Loading parameters		Sample parameters			Experiment results					
		Pressure, bar	Striker speed m/s	D, mm	$S_{c.s.}$, mm ^{2}	L, mm	Ultimate stress, MPa	Ultimate strain, %	Strain rate, 1/s	Lifetime, ms		DIF
										Diagram	Photography	
1	c679-01	0.5	11.2	18.35	264.46	10.30	29.99	1.6	196.0	106.2	Not destroyed	1.0
	c679-02	0.5	9.8	18.45	267.35	10.40	29.37	1.1	113.0	128.0	Not destroyed	1.0
	c679-03	0.5	10.1	18.45	267.35	10.40	33.44	1.2	143.3	107.6	Not destroyed	1.2
	c679-04	0.5	9.0	18.7	274.65	10.65	35.85	0.7	115.5	109.4	Not destroyed	1.2
	c679-05	0.5	12.0	18.6	271.72	10.50	35.03	1.4	166.7	111.6	Not destroyed	1.2
2	c679-06	0.75	16.6	18.65	273.18	10.70	51.60	1.3	111.3	153.3	–	1.8
	c679-07	0.75	16.5	18.55	270.26	10.65	50.03	1.5	182.9	112.9	–	1.7
	c679-08	0.75	16.1	18.5	268.80	10.80	45.10	2.2	278.7	110.1	–	1.6
	c679-09	0.75	15.7	18.3	263.02	10.60	46.10	1.9	196.4	124.8	–	1.6
	c679-10	0.75	14.3	18.45	267.35	10.40	41.23	1.6	200.7	105.6	Not destroyed	1.4
	c679-11	0.75	16.0	18.5	268.80	10.65	44.38	1.8	222.5	108.8	109	1.5
	c679-12	0.75	14.0	18.4	265.91	10.20	45.77	1.5	152.6	126.4	125	1.6
3	c679-13	1.5	20.3	18.3	263.02	10.15	51.19	2.0	324.5	104.5	90	1.8

(continued)

Table 8.7 (continued)

Mode	Experiment code	Loading parameters		Sample parameters			Experiment results					
		Pressure, bar	Striker speed m/s	D, mm	$S_{c.s.}$, mm ^{2}	L, mm	Ultimate stress, MPa	Ultimate strain, %	Strain rate, 1/s	Lifetime, ms		DIF
										Diagram	Photography	
	c679-14	1.5	21.3	18.45	267.35	10.75	55.11	1.2	202.8	87.2	90	1.9
	c679-15	1.5	20.2	18.4	265.91	10.70	53.92	1.7	246.7	97.7	–	1.9
	c679-16	1.5	21.1	18.78	277.00	10.81	64.78	1.1	196.4	114.8	–	2.2
	c679-17	1.5	21.5	18.75	276.12	10.50	57.71	1.5	307.9	86.0	80	2.0
4	c679-19	3	27.9	18.31	263.31	10.75	66.90	1.9	778.8	40.0	40	2.3
	c679-20	3	29.9	18.76	276.41	10.65	75.82	1.5	731.2	34.8	40	2.6
	c679-21	3	30.0	18.23	261.01	10.64	61.93	2.1	961.8	43.5	40	2.1
	c679-22	3	29.4	18.78	277.00	10.51	71.08	1.7	817.8	36.2	40	2.5

the exclusion of the time parameter, a diagram of dynamic deformation (σ~ε) was constructed with the display along the additional right vertical axis of the history of changes in the strain rate ($\dot{\varepsilon}$~ε) for each experiment. Further, for loading modes of the same type, averaged diagrams of the dependence of stresses on strains (σ~ε) and strain rates on strains (~ε), as well as the dependences of stresses on time (σ~t) and strain rates on time ($\dot{\varepsilon}$~t) were plotted. Below are the test diagrams of concrete samples for each individual mode, on which solid lines show the dependence of stress on time (σ~t) or stress on strain (σ~ε), the vertical axis on the left is stress; dotted lines show the dependence of the strain rate on time ($\dot{\varepsilon}$~t) or the strain rate on strain ($\dot{\varepsilon}$~ε), the vertical axis on the right is the strain rate.

In mode No. 1, the average striker speed is 10 m/s. The average dynamic compressive strength was about 33 MPa, the average strain rate was about 150 s^{-1}. In this mode, the samples retained their integrity in the first loading cycle, which is confirmed by the elastic unloading zone on the deformation diagram. In mode No. 2, the average striker speed was 16 m/s, the average dynamic compressive strength was 46 MPa, with average strain rates of about 200 s^{-1}. In mode No. 3, the average striker speed was 21.0 m/s, the average dynamic compressive strength was 57 MPa, with average strain rates of about 250 s^{-1}. In mode No. 4, the average striker speed was 29 m/s, the average dynamic compressive strength was about 69 MPa, with average strain rates of about 820 s^{-1}. Figure 8.4 shows averaged strain diagrams with a history of strain rate changes.

In the above diagrams in the stress–strain axes (σ~ε), in the initial loading section, the growth of stress and strain occurs according to a law close to linear, and with further deformation, when the limiting stress values are reached, the concrete is intensively destroyed (in modes 2, 3, 4), which is accompanied by a decrease in stresses and an increase in deformations. Dynamic compression tests were carried out in order to determine the effect of changing the strain rate on the strain diagram and strength characteristics of the material shown below (Fig. 8.5). The dynamic increase factor (DIF), defined as the ratio of dynamic strength to static strength, ranged from 1.0 to 2.5.

The conducted cycle of dynamic tests of polyfiber-reinforced concrete consisted of 3 speed modes and a total of 17 test shots. The strain rates were in the range from 200 to 900 s^{-1}. Test parameters are given in Table 8.8.

In mode No. 1, the average striker speed is 17 m/s. The average dynamic compressive strength was 52 MPa, at average strain rates of about 220 s^{-1}. In mode No. 2, the average striker speed is 24 m/s, the average dynamic compressive strength is 57 MPa, with average strain rates of about 260 s^{-1}. In mode No. 3, the average striker speed is 33 m/s, the average dynamic compressive strength is 75 MPa, with average strain rates of about 770 s^{-1}. Figure 8.6 shows the averaged deformation diagrams with the chronology of the change in the strain rate.

The dependences of the ultimate strength and deformation characteristics on the dynamic factor (DIF) on the strain rate are shown in Fig. 8.7. The resulting time to failure decreases depending on the strain rate according to a non-linear law. Limit strains within the scatter of experimental data can be described by a straight horizontal line, which indicates that the strain rate does not affect the ultimate strain to failure.

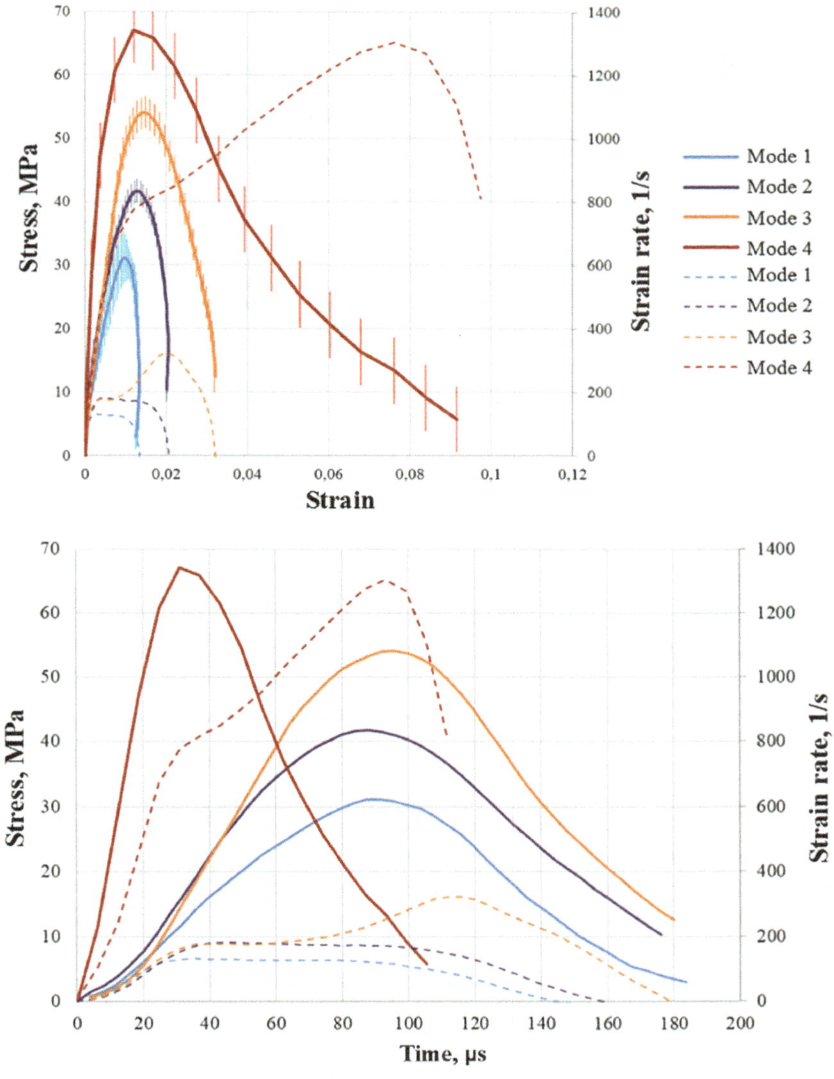

Fig. 8.4 Averaged diagrams σ(ε) and $\dot{\varepsilon}$(ε), σ(t) and $\dot{\varepsilon}$(t) under compression for all FGC modes

The conducted cycle of dynamic testing of steel fiber-reinforced concrete consisted of 3 speed modes and 14 test shots. The strain rates were in the range from 180 to 720 s^{-1}. Test parameters are shown in Table 8.9.

In mode No. 1, the average striker speed is 16 m/s. The average dynamic compressive strength was about 60 MPa, at average strain rates of about 180 s^{-1}. In mode No. 2, the average striker speed is 24 m/s, the average dynamic compressive strength is 85 MPa, with average strain rates of about 250 s^{-1}. In mode No. 3, the average striker

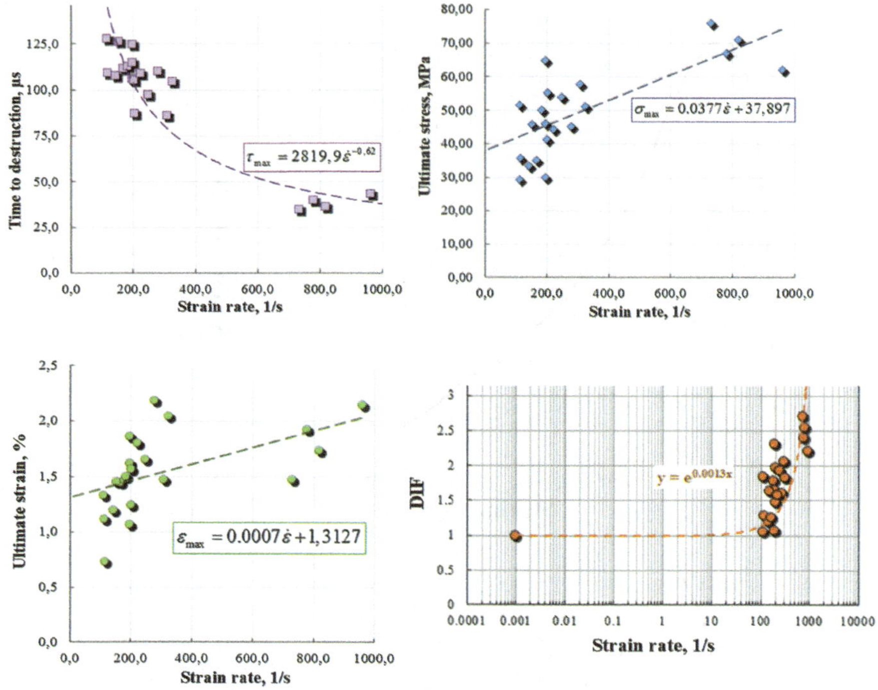

Fig. 8.5 Effect of strain rate on the mechanical properties of fine-grained concrete under dynamic compression

speed was 34 m/s, the average dynamic compressive strength was 100 MPa, with average strain rates of about 690 s^{-1}. Figure 8.8 shows the averaged deformation diagrams with the chronology of the change in the strain rate.

In the above diagrams in the stress–strain axes (σ~ε), in the initial section of loading, the growth of stress and strain occurs according to a law close to linear, and with further deformation, when the limiting stress values are reached, the concrete is intensively destroyed, which is accompanied by a decrease in stress and an increase in deformations. The effect of changing the strain rate on the strain diagram and the strength characteristics of the material is shown in Fig. 8.9.

The conducted cycle of dynamic tests of the combined fiber-reinforced concrete consisted of 3 speed modes and a total of 14 test shots. The strain rates were in the range from 140 to 740 s^{-1}. Test parameters are given in Table 8.10.

In mode No. 1, the average striker speed is 16 m/s. The average dynamic compressive strength was about 53 MPa, at average strain rates of about 160 s^{-1}. In mode No. 2, the average striker speed is 24 m/s, the average dynamic compressive strength was 68 MPa, with average strain rates of about 210 s^{-1}. In mode No. 3, the average striker speed was 34 m/s, the average dynamic compressive strength was 90 MPa, with

Table 8.8 Parameters of tests of PFC under dynamic compression

Mode	Experiment code	Loading parameters		Sample parameters			Experiment results					
		Pressure, bar	Striker speed m/s	D, mm	$S_{c.s.}$, mm ^{2}	L, mm	Ultimate stress, MPa	Ultimate strain, %	Strain rate, 1/s	Lifetime, ms		DIF
										Diagram	Photography	
1	c680-01	1	16.23	18.77	276.71	10.98	54.21	1.6	218.8	105	Not destroyed	1.5
	c680-02	1	16.67	18.67	273.77	10.97	51.86	1.7	240.3	101	Not destroyed	1.4
	c680-03	1	16.37	18.74	275.82	10.89	49.71	1.8	223.7	110	110	1.4
	c680-04	1	16.46	18.59	271.42	10.85	47.34	1.8	215.4	108	Not destroyed	1.3
	c680-05	1	17.44	18.76	276.41	10.94	55.41	1.9	215.4	112	110	1.5
2	c680-06	2	23.59	18.75	276.12	10.84	55.29	1.7	279.5	88	94	1.5
	c680-07	2	24.24	18.68	274.06	10.97	57.05	1.8	303.7	96	100	1.6
	c680-08	2	24.38	18.6	271.72	10.78	55.60	1.4	266.0	84	90	1.5
	c680-09	2	23.2	18.68	274.06	10.97	58.31	1.8	263.5	88	90	1.6
	c680-10	2	23.89	18.66	273.47	10.95	59.25	1.3	227.0	81	90	1.6
	c680-11	2	23.42	18.67	273.77	11.08	54.79	1.5	227.7	91	80	1.5
3	c680-12	4	34.04	18.76	276.41	10.82	83.08	1.5	667.1	40	40	2.3
	c680-13	4	33.87	18.57	270.84	10.98	68.18	1.8	894.7	40	40	1.9
	c680-14	4	32.91	18.71	274.94	10.99	70.83	1.6	649.8	43	50	2.0
	c680-15	4	31.02	18.58	271.13	10.52	72.15	2.0	831.3	43	50	2.0
	c680-16	4	34.35	18.75	276.12	10.75	74.76	1.7	810.7	37	40	2.1
	c680-17	4	33.71	18.7	274.65	10.64	78.09	1.5	784.5	34	30	2.2

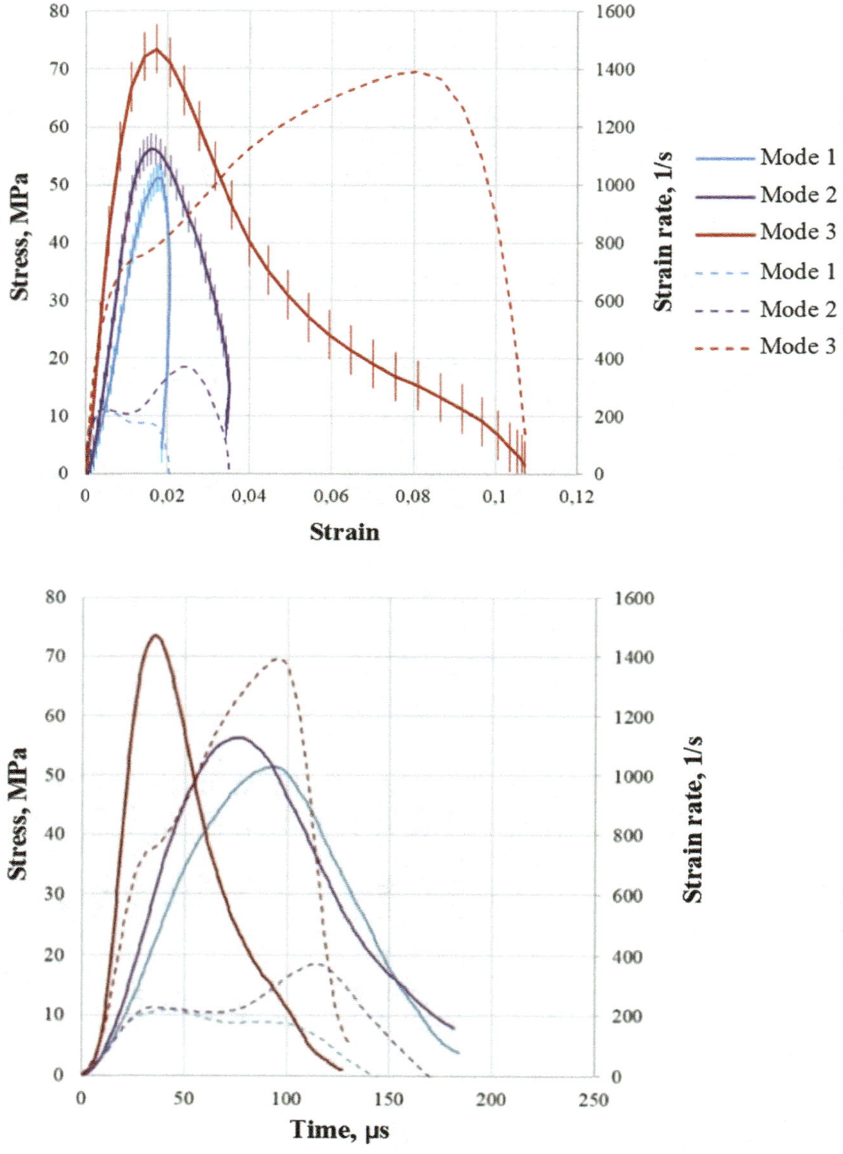

Fig. 8.6 Averaged diagrams σ(ε) and $\dot{\varepsilon}(\varepsilon)$, σ(t) and $\dot{\varepsilon}(t)$ under compression for all PFRS modes

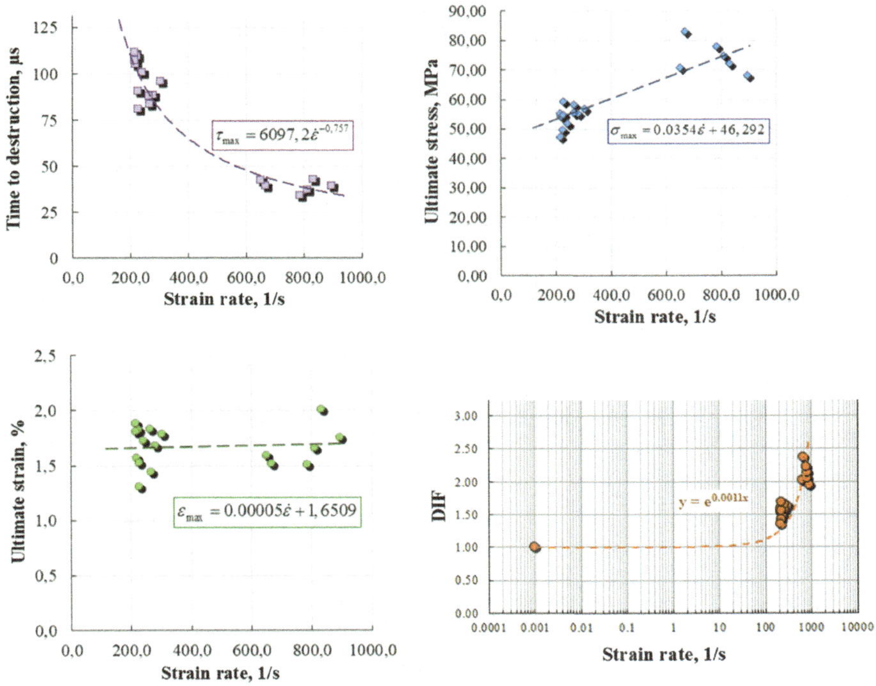

Fig. 8.7 Effect of strain rate on mechanical properties of PFRC under dynamic compression

average strain rates of about 710 s^{-1}. Figure 8.10 shows the averaged deformation diagrams with the chronology of the change in the strain rate.

Dynamic compression tests were carried out in order to determine the effect of changing the strain rate on the strain diagram and strength characteristics of the material shown below (Fig. 8.11).

For each of the resulting diagrams, characteristic points with the maximum achieved stresses were identified, after which the destruction of the samples began. For these points, the corresponding values of limiting strains and time to failure are determined. The strain rates were assumed to be maximum before the specimens began to fail, since they change during their deformation. The dependences obtained demonstrate that with an increase in the strain rate, the maximum stresses increase, the limiting strains corresponding to them also increase (according to a linear law), and the time to the onset of destruction decreases according to a power law.

Table 8.11 shows the average strength and average percent strength gain for the materials tested, relative to the composition of the original fine concrete.

Table 8.9 SFRC test parameters under dynamic compression

Mode	Experiment code	Loading parameters		Sample parameters			Experiment results					
		Pressure, bar	Striker speed m/s	D, mm	$S_{c.s.}$, mm ^{2}	L, mm	Ultimate stress, MPa	Ultimate strain, %	Strain rate, 1/s	Lifetime, ms		DIF
										Diagram	Photography	
1	c681-01	1	16.71	18.46	267.64	10.86	58.1	1.4	187.1	110.0	Not destroyed	1.0
	c681-02	1	15.53	18.45	267.35	10.81	52.0	1.3	190.9	105.2	100.0	0.9
	c681-03	1	15.54	18.54	269.97	10.97	57.6	1.2	179.9	114.0	Not destroyed	1.0
	c681-04	1	15.89	18.65	273.18	10.84	59.9	1.3	181.7	105.6	Not destroyed	1.0
	c681-05	1	16.85	18.67	273.77	10.68	62.9	1.2	181.4	109.6	Not destroyed	1.0
2	c681-06	2	23.06	18.57	270.84	10.83	85.2	1.9	251.1	109.2	100.0	1.4
	c681-07	2	24.43	18.46	267.64	10.99	82.3	1.9	268.3	113.2	112.5	1.4
	c681-08	2	23.73	18.47	267.93	10.87	79.7	1.8	228.1	116.4	112.5	1.3
	c681-09	2	24.19	18.71	274.94	10.9	91.9	1.8	272.1	118.0	112.5	1.5
3	c681-10	4	36.25	18.49	268.51	10.95	98.3	1.7	723.4	41.8	40.0	1.6
	c681-11	4	32.55	18.65	273.18	10.88	98.0	1.6	655.1	42.0	40.0	1.6
	c681-12	4	35.42	18.61	272.01	11.06	94.6	1.4	704.6	40.0	40.0	1.6
	c681-13	4	32.95	18.48	268.22	11.04	92.7	1.5	687.7	40.2	40.0	1.5
	c681-14	4	34.43	18.59	271.42	11.03	101.8	1.4	661.1	39.2	40.0	1.7

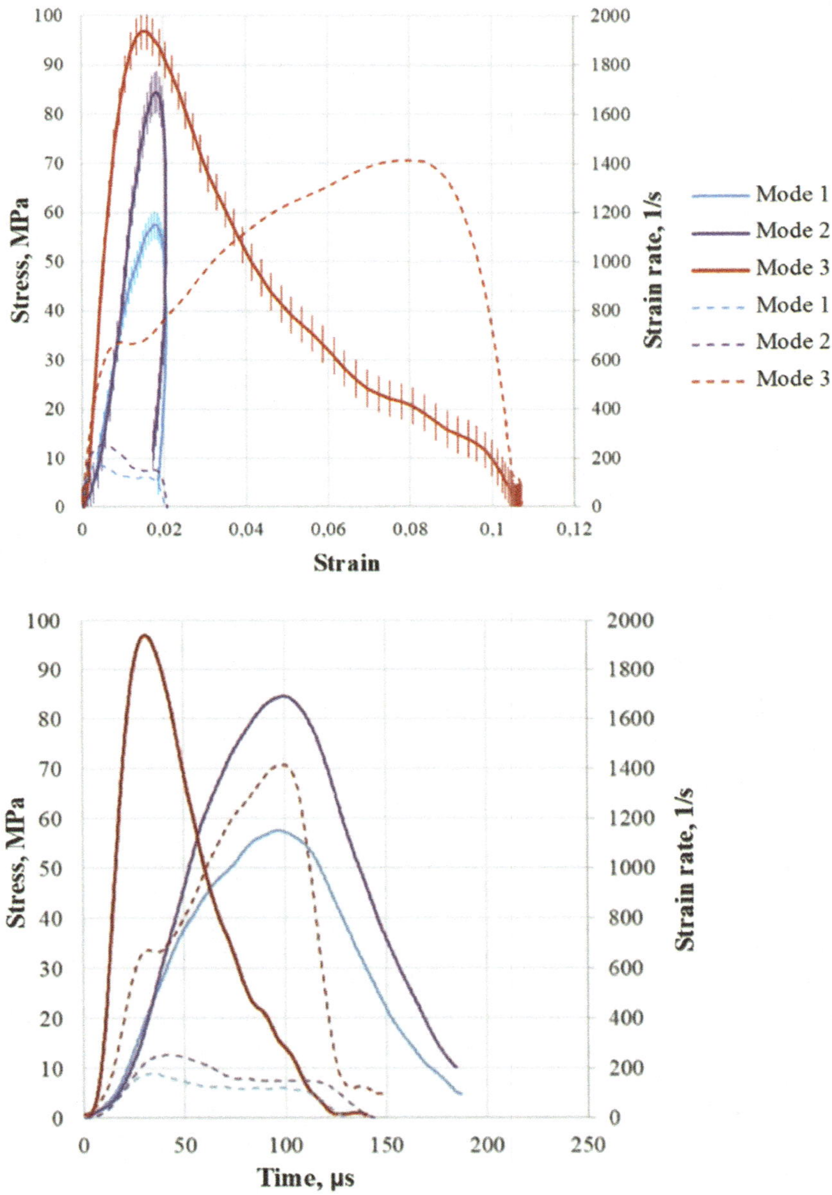

Fig. 8.8 Averaged diagrams σ(ε) and $\dot{\varepsilon}(\varepsilon)$, σ(t) and $\dot{\varepsilon}(t)$ under compression for all SFRC modes

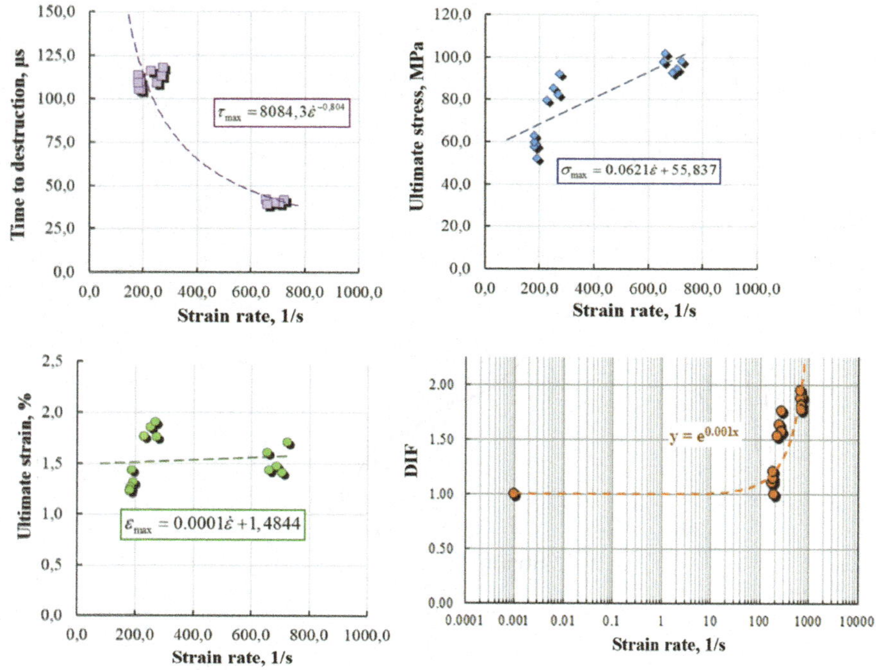

Fig. 8.9 Effect of strain rate on mechanical SFRC under dynamic compression

8.6 Conclusion

An experimental study of the dynamic properties of various types of fiber-reinforced concrete under uniaxial compression has been carried out. Diagrams of dynamic deformation at strain rates from 10^2 to 10^3 s^{-1} were obtained, based on the results of which the characteristics of strength and deformability were determined, and the dependences on various strain rates were determined.

The introduction of a reinforcing fiber into the original fine-grained concrete increased the dynamic strength of the material under uniaxial stress. The highest strength from 55 to 95 MPa under dynamic compression was shown by fiber-reinforced concrete with steel fiber. Combined fiber-reinforced concrete ranked second in terms of dynamic strengthening, and concrete with polymer fibers showed a slight increase in strength. However, the highest average DIF, equal to 1.7, was shown by the original fine-grained concrete, as well as concrete with polymer fibers, whose DIF was also equal to 1.7.

Table 8.10 CFRC test parameters under dynamic compression

Mode	Experiment code	Loading parameters		Sample parameters			Experiment results					
		Pressure, bar	Striker speed m/s	D, mm	$S_{c.s.}$, mm ^{2}	L, mm	Ultimate stress, MPa	Ultimate strain, %	Strain rate, 1/s	Lifetime, ms		DIF
										Diagram	Photography	
1	c682-01	1	15.24	18.63	272.6	11.18	55.8	0.9	142.6	103.6	100.0	1.1
	c682-02	1	15.78	18.52	269.4	11.09	51.1	1.2	145.5	104.0	100.0	1.0
	c682-03	1	17.45	18.72	275.2	11.21	61.4	0.9	147.4	100.8	–	1.2
	c682-04	1	15.77	18.57	270.8	11.17	50.4	1.0	146.1	95.2	–	1.0
	c682-05	1	16.34	18.53	269.7	11.16	46.9	1.4	201.4	98.4	–	0.9
2	c682-06	2	23.83	18.65	273.2	11.17	65.9	1.2	220.1	76.8	75.0	1.3
	c682-07	2	24.27	18.73	275.5	11.23	70.8	1.5	225.8	95.2	87.5	1.4
	c682-08	2	23.19	18.75	276.1	11.16	73.6	1.4	191.4	98.8	100.0	1.4
	c682-09	2	23.38	18.53	269.7	11.26	63.0	1.2	216.8	84.0	75.0	1.2
3	c682-10	4	34.56	18.74	275.8	11.31	97.6	1.5	654.4	44.2	50.0	1.9
	c682-11	4	32.18	18.51	269.1	11.18	86.0	1.5	712.4	43.4	50.0	1.7
	c682-12	4	34.73	18.52	269.4	11.31	91.5	1.6	725.1	43.8	50.0	1.8
	c682-13	4	32.43	18.59	271.4	11.34	89.8	1.5	694.5	44.4	50.0	1.7
	c682-14	4	34.35	18.53	269.7	11.35	85.2	1.5	744.4	43.0	50.0	1.6

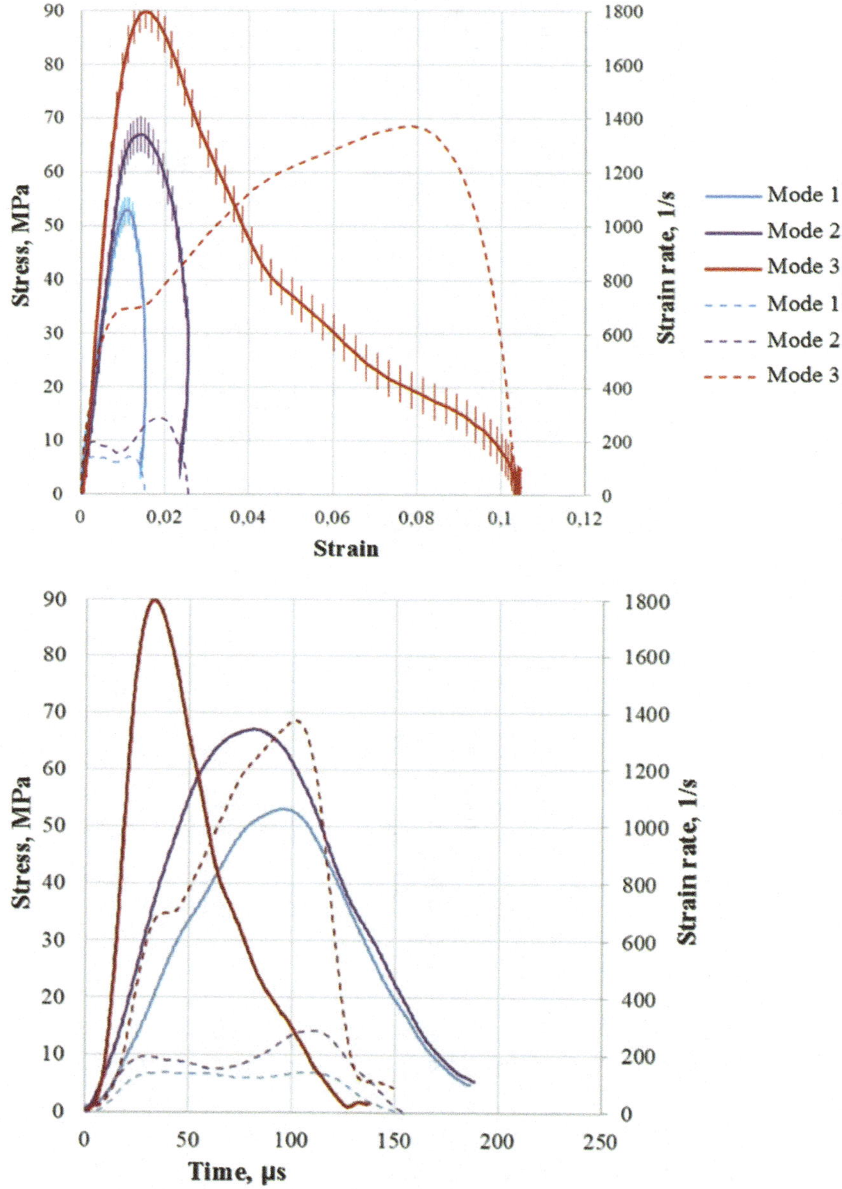

Fig. 8.10 Averaged diagrams σ(ε) and $\dot{\varepsilon}(\varepsilon)$, σ(t) and $\dot{\varepsilon}(t)$ under compression for all CFRC modes

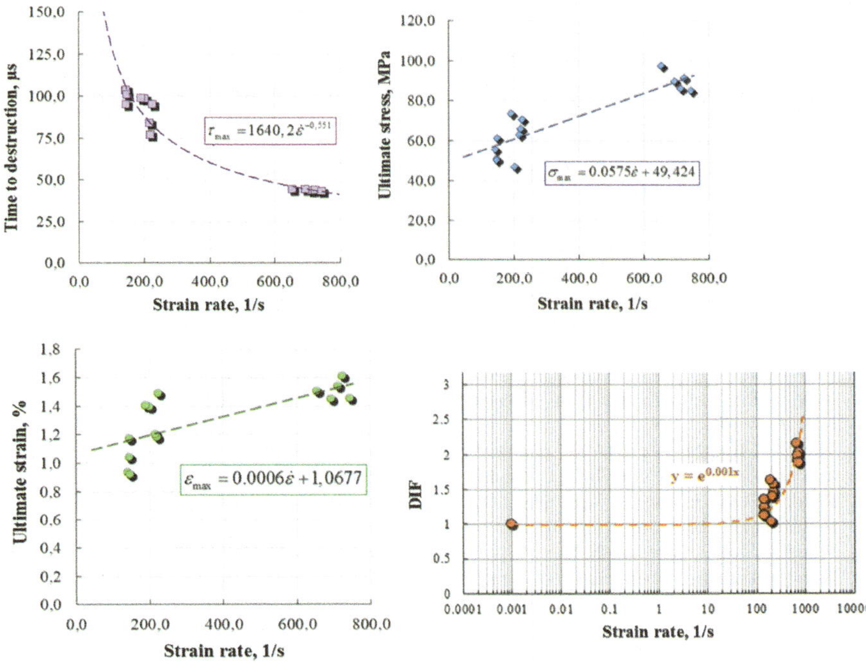

Fig. 8.11 Effect of strain rate on mechanical properties of CFRC under dynamic compression

Table 8.11 Average increase in strength of the studied materials

№	Material type	Ultimate stress, MPa			Average strength, MPa	Average increase in strength, %	Av. DIF
		Mode № 1	Mode № 2	Mode № 3			
Uniaxial dynamic compression							
1	FGC	46	57	69	57	0	1.7
2	PFRC	52	57	75	61	7	1.7
3	CFRC	53	68	90	70	18	1.3
4	SFRC	58	85	97	80	**28**	1.4

Acknowledgements The work was carried out with the financial support of the Ministry of Science and Higher Education of the Russian Federation (task 0729-2020-0054).

References

1. Yu R, van Beers L, Spiesz P, Brouwers HJH (2016) Impact resistance of a sustainable ultra-high performance fibre reinforced concrete (UHPFRC) under pendulum impact loadings. Constr Build Mater 107:203–215
2. Wang D, Shi C, Wu Z, Xiao J, Huang Z, Fang Z (2015) A review on ultra high performance concrete: part II. Hydration, Microstruct Prop, Constr Build Mater 96:368–377
3. Li PP, Yu QL (2019) Responses and post-impact properties of ultra-high performance fibre reinforced concrete under pendulum impact. Compos Struct 208:806–815
4. Li PP, Brouwers HJH, Yu Q (2020) Influence of key design parameters of ultra-high performance fibre reinforced concrete on in-service bullet resistance. Int J Impact Eng 136
5. Khosravani MR, Weinberg K (2018) A review on split Hopkinson bar experiments on the dynamic characterisation of concrete. Constr Build Mater 190:1264–1283
6. Cao YYY, Yu QL, Brouwers HJH, Chen W (2019) Predicting the rate effects on hookedend fiber pullout performance from ultra-high performance concrete (UHPC). Cem Concr Res 120:164–175
7. Yoo D-Y, Banthia N (2016) Mechanical properties of ultra-high-performance fiberreinforced concrete: a review. Cem Concr Compos 73:267–280
8. Shaikh FUA, Luhar S, Arel HŞ, Luhar I (2020) Performance evaluation of ultrahigh performance fibre reinforced concrete—a review. Constr Build Mater 232
9. Wang W, Liu J, Agostini F, Davy CA, Skoczylas F, Corvez D (2014) Durability of an ultra high performance fiber reinforced concrete (UHPFRC) under progressive aging. Cem Concr Res 55:1–13
10. Yoon Y, Lee J, Jang I, Hwang D (2013) Improved impact resistance of layered steel fiber reinforced concrte beam. In: Proceedings of the thirteenth East Asia- Pacific conference on structural engineering and construction (EASEC-13): the thirteenth East Asia-Pacific conference on structural engineering and construction (EASEC-13), p B-5-1
11. Xu T, Zhu W, Zhao G, Lin Y (2014) Dynamic spallation in fiber reinforced concrete under impact loading. In: Recent advances in material, analysis, monitoring, and evaluation in foundation and bridge engineering. ASCE, pp 17–24
12. Schenker A, Anteby I, Gal E, Kivity Y, Nizri E, Sadot O et al (2008) Full-scale field tests of concrete slabs subjected to blast loads. Int J Impact Eng 35:184–198
13. Wang F, Wan YKM, Chong OYK, Lim CH, Lim ETM (2008) Reinforced concrete slab subjected to close-in explosion. In: Proceedings of the 7th LS-DYNA Anwenderforum, Bamberg
14. Zhang X, Elazim AA, Ruiz G, Yu R (2014) Fracture behaviour of steel fibre-reinforced concrete at a wide range of loading rates. Int J Impact Eng 71:89–96
15. Xu Z, Hao H, Li H (2012) Dynamic tensile behaviour of fibre reinforced concrete with spiral fibres. Mater Des 42:72–88
16. Zhang X, Ruiz G, Elazim AMA (2015) Loading rate effect on crack velocities in steel fiber-reinforced concrete. Int J Impact Eng 76:60–66
17. Xu, Man, Hallinan B, Wille K (2016) Effect of loading rates on pullout behavior of high strength steel fibers embedded in ultra-high performance concrete (UHPC). Cement Concr Compos 70:98–109
18. Xu Z, Hao H, Li H (2012) Experimental study of dynamic compressive properties of fibre reinforced concrete material with different fibres. Mater Des 33:42–55
19. Nia AA, Hedayatian M, Nili M, Sabet VA (2012) An experimental and numerical study on how steel and polypropylene fibers affect the impact resistance in fiber-reinforced concrete. Int J Impact Eng 46:62–73
20. ACI Committee 544 R- (2008) Guide for specifying, proportioning, and production of fibre reinforced concrete. American Concrete Institute
21. Redaelli D, Muttoni A (2007) Tensile behaviour of reinforced ultra-high performance fiber reinforced concrete elements. In: FIB Symposium, Dubrovnik, pp 267–274

22. Song P, Hwang S (2004) Mechanical properties of high-strength steel fiberreinforced concrete. Constr Build Mater 18:669–673
23. Ong K, Basheerkhan M, Paramasivam P (1999) Resistance of fiber concrete slabs to low velocity projectile impact. Cement Concr Compos 21:391–401
24. Teng T-L, Chu Y-A, Chang F-A, Shen B-C, Cheng D-S (2008) Development and validation of numerical model of steel fiber reinforced concrete for highvelocity impact. Comput Mater Sci 42:90–99
25. Wang Z, Konietzky H, Huang R (2009) Elastic–plastic-hydrodynamic analysis of crater blasting in steel fiber reinforced concrete. Theoret Appl Fract Mech 52:111–116
26. Maalej M, Quek ST, Zhang J (2005) Behavior of hybrid-fiber engineered cementitious composites subjected to dynamic tensile loading and projectile impact. J Mater Civ Eng 17:143–152
27. Zhang J, Maalej M, Quek ST (2007) Performance of hybrid-fiber ECC blast/shelter panels subjected to drop weight impact. J Mater Civ Eng 19:855–863
28. Song P, Wu J, Hwang S, Sheu B (2005) Statistical analysis of impact strength and strength reliability of steel–polypropylene hybrid fiber-reinforced concrete. Constr Build Mater 19:1–9
29. Coughlin A, Musselman E, Schokker A, Linzell D (2010) Behavior of portable fiber reinforced concrete vehicle barriers subject to blasts from contact charges. Int J Impact Eng 37(521):9
30. Bragov AM, Igumnov LA, Lomunov AK (2015) High-speed deformation of fine-grained concrete and fiber-reinforced concrete. Nizhny Novgorod State University Publishing House, N. Novgorod
31. Bragov AM, Konstantinov AY, Lamzin DA, Lomunov AK, Karihaloo BL, Petrov YV, Smirnov IV (2012) High-rate deformation and fracture of fiber reinforced concrete. J Appl Mech Tech Phys 53(6):926–933
32. Bragov AM, Gonov ME, Lomunov AK (2019) Experimental study of the dynamic properties of concrete under compressive load. International conference on nonlinear solid mechanics ICoNSoM 2019, pp 403–412

Chapter 9
Unsteady Longitudinal Mechanodiffusion Vibrations of a Rectangular Plate with Inner Diffusion Flux Relaxation

Sergey A. Davydov, Anatoliy V. Vestyak, and Andrei V. Zemskov

Abstract We consider the unsteady problem of elastic diffusion deformations of a rectangular orthotropic plate considering the diffusion fluxes relaxation. External perturbations lay in the plate plane and allow us to use a two-dimensional elastic diffusion model for continuum as a mathematical model. The solution is convolutions of Green's functions with functions defining the boundary perturbations. Green's functions finding method is based on the Laplace transform and double trigonometric Fourier series. We are using residues and tables of operational calculus for transition to the originals of Green's functions. The interaction effects of mechanical and diffusion fields for a three-component material are calculated using the example of a rectangular plate under tensile forces. We have also investigated the influence of relaxation processes on mass transfer kinetics. Calculation results are in analytical and graphical forms.

Keywords Elastic diffusion · Unsteady problems · Coupled problem · Multicomponent continuum · Green's functions

9.1 Introduction

Issues related to the strength study include static/dynamic materials tests and consideration of different physical fields interaction. This is due to various modern structures and different conditions of their operation. It is often necessary to con-

S. A. Davydov · A. V. Vestyak
Moscow Aviation Institute, Moscow, Russian Federation
e-mail: xenon_93@inbox.ru

A. V. Vestyak
e-mail: kaf311@yandex.ru

A. V. Zemskov (✉)
Moscow Aviation Institute, Research Institute of Mechanics Lomonosov Moscow State University, Moscow, Russian Federation
e-mail: azemskov1975@mail.ru

H. Altenbach et al. (eds.), *Deformation and Destruction of Materials and Structures Under Quasi-static and Impulse Loading*, Advanced Structured Materials 186,
https://doi.org/10.1007/978-3-031-22093-7_9

sider the interaction between mechanical, diffusion, temperature and other fields in applied calculations.

Experimental study of the interaction between mechanical and diffusion fields began in the 20s of the twentieth century. Many scientific theories have been formed that allow describing the coupled mechanical, diffusion and other fields in a solid medium. Among the most recent works are [1–16]. Various aspects of mechanodiffusion processes modelling are considered here: beginning with initial–boundary problem formulation and ending with a description of methods for solving these problems. Here are stationary [4, 11, 12] and unsteady problems [1–3, 5–8, 13–16].

The main complexity is the Laplace inverse transform when solving mechanodiffusion problems analytically. It is often used in solving initial–boundary problems. The Durbin method [2, 3, 7, 8, 14] or similar algorithms based on the fact that the Mellin integral is expressed through the inverse Fourier transform are used here. Special quadrature formulas [1, 5, 15] allow to calculate the integral. These methods have proven themselves in calculating the originals in a certain class of functions. However, these algorithms are not suitable for Green's functions because they belong to the generalized function class, and it is difficult to use numerical integration methods in this case.

Numerical algorithms based on finite difference methods [6, 13] and the finite element methods [16] are an alternative to analytical methods. In addition, numerical methods are the only way to solve the boundary value problem in some cases. A disadvantage of numerical methods is that solution comes to a discrete set of values, which is difficult to analyse and investigate later. A rather complicated mathematical problem is the question of algorithm stability and convergence.

This article proposes an analytical method for solving the unsteady mechanodiffusion problem for a rectangular plate. The method implies the Laplace transform and Fourier series expansion in eigenfunctions of the elastic diffusion operator. This approach has been tested for one-dimensional problems of elastic diffusion and thermoelastic diffusion [17, 18].

9.2 Problem Formulation

We considered a multicomponent rectangular plate under the action of unsteady longitudinal forces. The resulting mass transfer and diffusion flux relaxations are taken into account inside the plate.

The mathematical model describing two-dimensional elastic diffusion processes for a homogeneous orthotropic medium in the rectangular Cartesian coordinate system [19, 20]:

$$
\begin{gathered}
\frac{\partial^2 u_1}{\partial x_1^2} + C_{66}\frac{\partial^2 u_1}{\partial x_2^2} + C_0\frac{\partial^2 u_2}{\partial x_1 \partial x_2} = \ddot{u}_1 + \sum_{j=1}^{N}\alpha_1^j \frac{\partial \eta_j}{\partial x_1}, \\
C_0\frac{\partial^2 u_1}{\partial x_2 \partial x_1} + C_{66}\frac{\partial^2 u_2}{\partial x_1^2} + C_{22}\frac{\partial^2 u_2}{\partial x_2^2} = \ddot{u}_2 + \sum_{j=1}^{N}\alpha_2^j \frac{\partial \eta_j}{\partial x_2},
\end{gathered}
$$

$$
\begin{gathered}
D_1^q\frac{\partial^2 \eta_q}{\partial x_1^2} + D_2^q\frac{\partial^2 \eta_q}{\partial x_2^2} = \dot{\eta}_q + \tau_q\ddot{\eta}_q + \Lambda_{11}^q\frac{\partial^3 u_1}{\partial x_1^3} + \Lambda_{21}^q\frac{\partial^3 u_1}{\partial x_2^2 \partial x_1} + \\
+\Lambda_{12}^q\frac{\partial^3 u_2}{\partial x_1^2 \partial x_2} + \Lambda_{22}^q\frac{\partial^3 u_2}{\partial x_2^3},\quad \eta_{N+1} = -\sum_{j=1}^{N}\eta_j\ \left(q = \overline{1, N}\right);
\end{gathered}
\tag{9.1}
$$

$$
\begin{gathered}
u_2|_{x_1=0} = f_{111}(x_2, \tau),\quad \eta_q\Big|_{x_1=0} = f_{q+2,11}(x_2, \tau), \\
u_1|_{x_2=0} = f_{121}(x_1, \tau),\quad \eta_q\Big|_{x_2=0} = f_{q+2,21}(x_1, \tau), \\
\left.\left(\frac{\partial u_1}{\partial x_1} + C_{12}\frac{\partial u_2}{\partial x_2} - \sum_{j=1}^{N}\alpha_1^j\eta_j\right)\right|_{x_1=0} = f_{211}(x_2, \tau), \\
\left.\left(C_{12}\frac{\partial u_1}{\partial x_1} + C_{22}\frac{\partial u_2}{\partial x_2} - \sum_{j=1}^{N}\alpha_2^j\eta_j\right)\right|_{x_2=0} = f_{221}(x_1, \tau),
\end{gathered}
\tag{9.2}
$$

$$
\begin{gathered}
u_2|_{x_1=l_1} = f_{112}(x_2, \tau),\quad \eta_q\Big|_{x_1=l_1} = f_{q+2,12}(x_2, \tau), \\
u_1|_{x_2=l_2} = f_{122}(x_1, \tau),\quad \eta_q\Big|_{x_2=l_2} = f_{q+2,22}(x_1, \tau), \\
\left.\left(\frac{\partial u_1}{\partial x_1} + C_{12}\frac{\partial u_2}{\partial x_2} - \sum_{j=1}^{N}\alpha_1^j\eta_j\right)\right|_{x_1=l_1} = f_{212}(x_2, \tau), \\
\left.\left(C_{12}\frac{\partial u_1}{\partial x_1} + C_{22}\frac{\partial u_2}{\partial x_2} - \sum_{j=1}^{N}\alpha_2^j\eta_j\right)\right|_{x_2=l_2} = f_{222}(x_1, \tau).
\end{gathered}
$$

Here, all quantities are dimensionless. Their relation to their dimensional counterparts (when written the same way, they are indicated with a dash) is defined as follows:

$$
\begin{gathered}
x_i = \frac{x_i'}{L},\ u_i = \frac{u_i'}{L},\ \tau = \frac{Ct}{L},\ C_{ij} = \frac{C_{ij}'}{C_{11}'},\ C^2 = \frac{C_{11}'}{\rho},\ \alpha_\alpha^q = \frac{\alpha_{\alpha\alpha}^{(q)}}{C_{11}'}, \\
D_\alpha^q = \frac{D_{\alpha\alpha}^{(q)}}{CL},\ \Lambda_{\alpha\beta}^q = \frac{m^{(q)}D_{\alpha\alpha}^{(qq)}\alpha_{\beta\beta}^{(q)}n_0^{(q)}}{\rho RT_0CL},\ l_i = \frac{l_i^*}{L},\ \tau_q = \frac{C\tau^{(q)}}{L},
\end{gathered}
\tag{9.3}
$$

$$
C_0 = C_{66} + C_{12},\ C_{11}' = C_{1111},\ C_{22}' = C_{2222},\ C_{66}' = C_{1212},\ C_{12}' = C_{1122},
$$

$$
f_{1kl}(x_1, \tau) = \frac{f_{1kl}^*(x_1, t)}{L},\quad f_{q+2,kl}(x_1, \tau) = f_{q+2,kl}^*(x_1, t),\quad f_{2kl}(x_1, \tau) = \frac{f_{2kl}^*(x_1, \tau)}{C_{1111}},
$$

where t is a time parametr; x_i^* is a rectangular Cartesian coordinate; u_i^* is a displacement vector component; L is the diagonal size of the plate having dimensions $l_1^* \times l_2^*$; η_q is the concentration increment of "q"th component in $N+1$-component medium; $n_0^{(q)}$ is the initial concentration of "q"th component; C_{ijkl} is the component of elastic constants tensor; ρ is a mass density; $\alpha_{ij}^{(q)}$ is a coefficient characterizing the volume change of the medium due to diffusion; $D_{ij}^{(q)}$ is a diffusion coefficient; R is the universal gas constant; T_0 is an average temperature; $m^{(q)}$ is a molar mass of "q"th component, f_{klm}^* is an external perturbation; $\tau^{(q)}$ is the relaxation time of the diffusion flux.

We assume the initial conditions are zero.

9.3 Integral Representation of the Solution

The solution to the problem is sought in the integral form as follows:

$$\begin{aligned}
\begin{Bmatrix} u_1(x_1,x_2,\tau) \\ u_2(x_1,x_2,\tau) \\ \eta_q(x_1,x_2,\tau) \end{Bmatrix} &= \sum_{k=1}^{N+2}\int_0^\tau\int_0^{l_2} \begin{Bmatrix} G_{1k1}(x_1,x_2,\zeta,t) \\ G_{2k1}(x_1,x_2,\zeta,t) \\ G_{q+2,k1}(x_1,x_2,\zeta,t) \end{Bmatrix} f_{k11}(\zeta,\tau-t)\,d\zeta dt + \\
&+\sum_{k=1}^{N+2}\int_0^\tau\int_0^{l_2} \begin{Bmatrix} -G_{1k1}(l_1-x_1,x_2,\zeta,t) \\ G_{2k1}(l_1-x_1,x_2,\zeta,t) \\ G_{q+2,k1}(l_1-x_1,x_2,\zeta,t) \end{Bmatrix} f_{k12}(\zeta,\tau-t)\,d\zeta dt + \\
&+\sum_{k=1}^{N+2}\int_0^\tau\int_0^{l_1} \begin{Bmatrix} G_{1k2}(x_1,x_2,\xi,t) \\ G_{2k2}(x_1,x_2,\xi,t) \\ G_{q+2,k2}(x_1,x_2,\xi,t) \end{Bmatrix} f_{k21}(\xi,\tau-t)\,d\xi dt + \\
&+\sum_{k=1}^{N+2}\int_0^\tau\int_0^{l_1} \begin{Bmatrix} G_{1k2}(x_1,l_2-x_2,\xi,t) \\ -G_{2k2}(x_1,l_2-x_2,\xi,t) \\ G_{q+2,k2}(x_1,l_2-x_2,\xi,t) \end{Bmatrix} f_{k22}(\xi,\tau-t)\,d\xi dt,
\end{aligned} \tag{9.4}$$

where G_{mkl} are Green's functions, which satisfy the following initial–boundary value problem (initial conditions are zero):

$$\begin{aligned}
&\frac{\partial^2 G_{1kl}}{\partial x_1^2} + C_{66}\frac{\partial^2 G_{1kl}}{\partial x_2^2} + C_0\frac{\partial^2 G_{2kl}}{\partial x_1 \partial x_2} = \ddot{G}_{1kl} + \sum_{j=1}^{N}\alpha_1^j\frac{\partial G_{j+2,kl}}{\partial x_1}, \\
&C_0\frac{\partial^2 G_{1kl}}{\partial x_2 \partial x_1} + C_{66}\frac{\partial^2 G_{2kl}}{\partial x_1^2} + C_{22}\frac{\partial^2 G_{2kl}}{\partial x_2^2} = \ddot{G}_{2kl} + \sum_{j=1}^{N}\alpha_2^j\frac{\partial G_{j+2,kl}}{\partial x_2}, \\
&D_1^q\frac{\partial^2 G_{q+2,kl}}{\partial x_1^2} + D_2^q\frac{\partial^2 G_{q+2,kl}}{\partial x_2^2} = \dot{G}_{q+2,kl} + \tau_q\ddot{G}_{q+2,kl} + \Lambda_{11}^q\frac{\partial^3 G_{1kl}}{\partial x_1^3} + \\
&+\Lambda_{21}^q\frac{\partial^3 G_{1kl}}{\partial x_2^2 \partial x_1} + \Lambda_{12}^q\frac{\partial^3 G_{2kl}}{\partial x_1^2 \partial x_2} + \Lambda_{22}^q\frac{\partial^3 G_{2kl}}{\partial x_2^3} \quad \left(q=\overline{1,N}\right);
\end{aligned} \tag{9.5}$$

$$\left(\frac{\partial G_{1kl}}{\partial x_1} + C_{12}\frac{\partial G_{2kl}}{\partial x_2} - \sum_{j=1}^{N}\alpha_1^j G_{j+2,kl}\right)\Bigg|_{x_1=0} = \delta_{1k}\delta_{1l}\delta(x_2-\zeta)\delta(\tau),$$
$$\left(C_{12}\frac{\partial G_{1kl}}{\partial x_1} + C_{22}\frac{\partial G_{2kl}}{\partial x_2} - \sum_{j=1}^{N}\alpha_2^j G_{j+2,kl}\right)\Bigg|_{x_2=0} = \delta_{2k}\delta_{2l}\delta(x_1-\xi)\delta(\tau),$$
$$\left(\frac{\partial G_{1kl}}{\partial x_1} + C_{12}\frac{\partial G_{2kl}}{\partial x_2} - \sum_{j=1}^{N}\alpha_1^j G_{j+2,kl}\right)\Bigg|_{x_1=l_1} = 0, \tag{9.6}$$
$$\left(C_{12}\frac{\partial G_{1kl}}{\partial x_1} + C_{22}\frac{\partial G_{2kl}}{\partial x_2} - \sum_{j=1}^{N}\alpha_2^j G_{j+2,kl}\right)\Bigg|_{x_2=l_2} = 0,$$
$$G_{2kl}|_{x_1=0} = \delta_{2k}\delta_{1l}\delta(x_2-\zeta)\delta(\tau),\quad G_{q+2,kl}\big|_{x_1=0} = \delta_{q+2,k}\delta_{1l}\delta(x_2-\zeta)\delta(\tau),$$
$$G_{1kl}|_{x_2=0} = \delta_{1k}\delta_{2l}\delta(x_1-\xi)\delta(\tau),\quad G_{q+2,kl}\big|_{x_2=0} = \delta_{q+2,k}\delta_{2l}\delta(x_1-\xi)\delta(\tau),$$
$$G_{2kl}|_{x_1=l_1} = 0,\quad G_{q+2,kl}\big|_{x_1=l_1} = 0,\quad G_{1kl}|_{x_2=l_2} = 0,\quad G_{q+2,kl}\big|_{x_2=l_2} = 0.$$

9.4 Algorithm for Green's Functions

To find the Green's functions, we apply the Laplace transform to (9.5) and (9.6) and then represent the unknown functions as a double Fourier series. Let's multiply the first equation in (9.5) by $\cos\mu_m x_1 \sin\lambda_n x_2$, the second by $\sin\mu_m x_1 \cos\lambda_n x_2$, and the rest by $\sin\lambda_n x_1 \sin\mu_n x_2$ ($\mu_m = \pi m/l_1$, $\lambda_n = \pi n/l_2$). Then we integrate into the rectangle $[0, l_1] \times [0, l_2]$ and obtain the following system of linear algebraic equations:

$$\begin{gathered}
k_{1mn}(s)\,G^L_{1klmn}(\xi,\zeta,s) + C_0\lambda_n\mu_m G^L_{2klmn}(\xi,\zeta,s) + \\
+\mu_m\sum_{j=1}^{N}\alpha_1^j G^L_{j+2,klmn}(\xi,\zeta,s) = F_{1mn}(\xi,\zeta,s), \\
C_0\mu_m\lambda_n G^L_{1klmn}(\xi,\zeta,s) + k_{2mn}(s)\,G^L_{2klmn}(\xi,\zeta,s) + \\
+\lambda_n\sum_{j=1}^{N}\alpha_2^j G^L_{j+2,klmn}(\xi,\zeta,s) = F_{2mn}(\xi,\zeta,s), \\
-\mu_m K_{qmn} G^L_{1klmn}(\xi,\zeta,s) - \lambda_n M_{qmn} G^L_{2klmn}(\xi,\zeta,s) + \\
+k_{q+2,mn}(s)\,G^L_{q+2,klmn}(\xi,\zeta,s) = F_{q+2,mn}(\xi,\zeta,s);
\end{gathered} \tag{9.7}$$

$$
\begin{gathered}
F_{1mn}\left(\xi,\zeta,s\right)=\frac{4}{l_1 l_2}\left[C_{66}\lambda_n\delta_{1k}\left(\delta_{1l}\cos\lambda_n\zeta+\delta_{2l}\cos\mu_m\xi\right)-\delta_{2k}\delta_{1l}\sin\lambda_n\zeta\right],\\
F_{2mn}\left(\xi,\zeta,s\right)=\frac{4}{l_1 l_2}\left[\mu_m C_{66}\delta_{1k}\left(\delta_{1l}\cos\lambda_n\zeta+\delta_{2l}\cos\mu_m\xi\right)-\delta_{2k}\delta_{2l}\sin\mu_m\xi\right],\\
F_{q+2,mn}\left(\xi,\zeta,s\right)=\frac{4}{l_1 l_2}\lambda_n\mu_m\delta_{1k}\left(\delta_{2l}\Lambda_{2q}\cos\mu_m\xi+\delta_{1l}\Lambda_{1q}\cos\lambda_n\zeta\right)+\\
+\frac{4}{l_1 l_2}\mu_m\delta_{1l}\left(D_1^q\delta_{q+2,k}-\Lambda_{11}^q\sum_{j=1}^N\alpha_1^j\delta_{j+2,k}-\Lambda_{11}^q\delta_{2k}\right)\sin\lambda_n\zeta+\\
+\frac{4}{l_1 l_2}\lambda_n\delta_{2l}\left(D_2^q\delta_{q+2,k}-\frac{\Lambda_{22}^q}{C_{22}}\sum_{j=1}^N\alpha_2^j\delta_{j+2,k}-\frac{\Lambda_{22}^q}{C_{22}}\delta_{2k}\right)\sin\mu_m\xi,\\
k_{1mn}\left(s\right)=s^2+\mu_m^2+C_{66}\lambda_n^2,\ k_{2mn}\left(s\right)=s^2+C_{66}\mu_m^2+C_{22}\lambda_n^2,\\
k_{q+2,mn}\left(s\right)=s+\tau_q s^2+D_1^q\mu_m^2+D_2^q\lambda_n^2,\ \Lambda_{1q}=\Lambda_{12}^q-\Lambda_{11}^q C_{12},\\
\Lambda_{2q}=\Lambda_{21}^q-\frac{\Lambda_{22}^q}{C_{22}}C_{12},\ K_{qmn}=\Lambda_{11}^q\mu_m^2+\Lambda_{21}^q\lambda_n^2,\ M_{qmn}=\Lambda_{12}^q\mu_m^2+\Lambda_{22}^q\lambda_n^2;
\end{gathered}
\tag{9.8}
$$

$$
G_{rkl}^L\left(x_1,x_2,\xi,\zeta,s\right)=\int_0^\infty G_{rkl}\left(x_1,x_2,\xi,\zeta,\tau\right)e^{-s\tau}d\tau,
$$

$$
\begin{gathered}
G_{1kl}^L\left(x_1,x_2,\xi,\zeta,s\right)=\sum_{m=0}^\infty\sum_{n=1}^\infty G_{1klmn}^L\left(\xi,\zeta,s\right)\cos\mu_m x_1\sin\lambda_n x_2,\\
G_{2kl}^L\left(x_1,x_2,\xi,\zeta,s\right)=\sum_{m=1}^\infty\sum_{n=0}^\infty G_{2klmn}^L\left(\xi,\zeta,s\right)\sin\mu_m x_1\cos\lambda_n x_2,\\
G_{q+2,kl}^L\left(x_1,x_2,\xi,\zeta,s\right)=\sum_{m=1}^\infty\sum_{n=1}^\infty G_{q+2,klmn}^L\left(\xi,\zeta,s\right)\sin\mu_m x_1\sin\lambda_n x_2,
\end{gathered}
\tag{9.9}
$$

$$
\begin{gathered}
G_{1klmn}^L\left(\xi,\zeta,s\right)=\frac{4}{l_1 l_2}\int_0^{l_1}\cos\mu_m x_1 dx_1\int_0^{l_2}G_{1kl}^L\left(x_1,x_2,\xi,\zeta,s\right)\sin\lambda_n x_2 dx_2,\\
G_{1kl0n}^L\left(\xi,\zeta,s\right)=\frac{2}{l_1 l_2}\int_0^{l_1}dx_1\int_0^{l_2}G_{1kl}^L\left(x_1,x_2,\xi,\zeta,s\right)\sin\lambda_n x_2 dx_2,\\
G_{2klmn}^L\left(\xi,\zeta,s\right)=\frac{4}{l_1 l_2}\int_0^{l_1}\sin\mu_m x_1 dx_1\int_0^{l_2}G_{2kl}^L\left(x_1,x_2,\xi,\zeta,s\right)\cos\lambda_n x_2 dx_2,\\
G_{2klm0}^L\left(\xi,\zeta,s\right)=\frac{2}{l_1 l_2}\int_0^{l_1}\sin\mu_m x_1 dx_1\int_0^{l_2}G_{2kl}^L\left(x_1,x_2,\xi,\zeta,s\right)dx_2,\\
G_{q+2,klmn}^L\left(\xi,\zeta,s\right)=\frac{4}{l_1 l_2}\int_0^{l_1}\sin\mu_m x_1 dx_1\int_0^{l_2}G_{q+2,kl}^L\left(x_1,x_2,\xi,\zeta,s\right)\sin\lambda_n x_2 dx_2.
\end{gathered}
$$

Below are the systems of equations for the zero harmonics:

$$
\begin{aligned}
k_{10n}(s)\,G^L_{1kl0n}(\xi,\zeta,s) &= F_{10n}(\xi,\zeta,s), \\
k_{2m0}(s)\,G^L_{2klm0}(\xi,\zeta,s) &= F_{2m0}(\xi,\zeta,s), \\
F_{10n}(\xi,\zeta,s) &= \frac{2}{l_1 l_2}\left[C_{66}\lambda_n\delta_{1k}\left(\delta_{1l}\cos\lambda_n\zeta+\delta_{2l}\right)-\delta_{2k}\delta_{1l}\sin\lambda_n\zeta\right], \\
F_{2m0}(\xi,\zeta,s) &= \frac{2}{l_1 l_2}\left[\mu_m C_{66}\delta_{1k}\left(\delta_{1l}+\delta_{2l}\cos\mu_m\xi\right]-\delta_{2k}\delta_{2l}\sin\mu_m\xi\right].
\end{aligned}
\tag{9.10}
$$

The solution of the systems (9.7) and (9.10) has the following form:

$$
\begin{aligned}
G^L_{1kl0n}(\xi,\zeta,s) &= \frac{2}{l_1 l_2}\frac{C_{66}\lambda_n\delta_{1k}(\delta_{1l}\cos\lambda_n\zeta+\delta_{2l})-\delta_{2k}\delta_{1l}\sin\lambda_n\zeta}{k_1(0,\lambda_n,s)}, \\
G^L_{2klm0}(\xi,\zeta,s) &= \frac{2}{l_1 l_2}\frac{\mu_m C_{66}\delta_{1k}(\delta_{1l}+\delta_{2l}\cos\mu_m\xi)-\delta_{2k}\delta_{2l}\sin\mu_m\xi}{k_2(\mu_m,0,s)};
\end{aligned}
\tag{9.11}
$$

$$
\begin{aligned}
G^L_{iklmn}(\xi,\zeta,s) &= \frac{P_{iklmn}(\xi,\zeta,s)}{P_{mn}(s)}\quad(i=1,2), \\
G^L_{q+2,klmn}(\xi,\zeta,s) &= \hat{G}^L_{q+2,klmn}(\xi,\zeta,s)+\frac{P_{q+2,klmn}(\xi,\zeta,s)}{Q_{qmn}(s)};
\end{aligned}
\tag{9.12}
$$

$$
\begin{aligned}
\hat{G}^L_{q+2,21mn}(\xi,\zeta,s) &= -\frac{4}{l_1 l_2}\frac{\Lambda^q_{11}\mu_m}{k_{q+2,mn}(s)}\sin\lambda_n\zeta, \\
\hat{G}^L_{q+2,12mn}(\xi,\zeta,s) &= \frac{4}{l_1 l_2}\frac{\Lambda_{2q}\lambda_n\mu_m}{k_{q+2,mn}(s)}\cos\mu_m\xi, \\
\hat{G}^L_{q+2,11mn}(\xi,\zeta,s) &= \frac{4}{l_1 l_2}\frac{\Lambda_{1q}\lambda_n\mu_m}{k_{q+2,mn}(s)}\cos\lambda_n\zeta, \\
\hat{G}^L_{q+2,22mn}(\xi,\zeta,s) &= -\frac{4}{l_1 l_2 C_{22}}\frac{\Lambda^q_{22}\lambda_n}{k_{q+2,mn}(s)}\sin\mu_m\xi, \\
\hat{G}^L_{q+2,p+2,1mn}(\xi,\zeta,s) &= \frac{4D_{1jq}\mu_m}{l_1 l_2}\frac{\sin\lambda_n\zeta}{k_{q+2,mn}(s)}, \\
\hat{G}^L_{q+2,p+2,2mn}(\xi,\zeta,s) &= \frac{4D_{2qp}\lambda_n}{l_1 l_2}\frac{\sin\mu_m\xi}{k_{q+2,mn}(s)},
\end{aligned}
\tag{9.13}
$$

where

$$
\begin{aligned}
P_{mn}(s) &= \left[k_{1mn}(s)\,k_{2mn}(s)-\mu_m^2\lambda_n^2C_0^2\right]\Pi_{mn}(s)- \\
&-\lambda_n^2\sum_{j=1}^{N}S_{1mn}(s)\,\Pi_{jmn}(s)\,M_{jmn}-\mu_m^2\sum_{j=1}^{N}S_{2mn}(s)\,\Pi_{jmn}(s)\,K_{jmn}+ \\
&+\mu_m^2\lambda_n^2\sum_{i=1}^{N}\sum_{j=1}^{N}\mathrm{A}_{ij}K_{imn}M_{jmn}\Pi_{ijmn}(s), \\
Q_{qmn}(s) &= k_{q+2,mn}(s)\,P_{mn}(s),
\end{aligned}
\tag{9.14}
$$

$$P_{121mn}(\xi,\zeta,s) = -\frac{4}{l_1 l_2} k_{2mn}(s)\,\Pi_{mn}(s)\sin\lambda_n\zeta -$$
$$-\frac{4}{l_1 l_2}\sum_{j=1}^{N}\left[\mu_m^2\Lambda_{11}^j S_{2mn}(s) + \alpha_2^j\lambda_n^2 M_{jmn}\right]\Pi_{jmn}(s)\sin\lambda_n\zeta +$$
$$+\frac{4}{l_1 l_2}\mu_m^2\lambda_n^2\sum_{j=1}^{N}\Lambda_{11}^j\sum_{i=1}^{N}\mathrm{A}_{ji}M_{imn}\Pi_{ijmn}(s)\sin\lambda_n\zeta,$$

$$P_{112mn}(\xi,\zeta,s) = \frac{4}{l_1 l_2} C_{66}\lambda_n\left(k_{2mn}(s) - \mu_m^2 C_0\right)\Pi_{mn}(s)\cos\mu_m\xi +$$
$$+\frac{4}{l_1 l_2}\lambda_n\sum_{j=1}^{N}\left[C_{66}B_{jmn}M_{jmn} + \mu_m^2\Lambda_{2j}S_{2mn}(s)\right]\Pi_{jmn}(s)\cos\mu_m\xi -$$
$$-\frac{4}{l_1 l_2}\lambda_n^3\mu_m^2\sum_{j=1}^{N}\Lambda_{2j}\sum_{i=1}^{N}\mathrm{A}_{ji}M_{imn}\Pi_{ijmn}(s)\cos\mu_m\xi,$$

$$P_{111mn}(\xi,\zeta,s) = \frac{4}{l_1 l_2} C_{66}\lambda_n\left(k_{2mn}(s) - C_0\mu_m^2\right)\Pi_{mn}(s)\cos\lambda_n\zeta +$$
$$+\frac{4}{l_1 l_2}\lambda_n\sum_{j=1}^{N}\left[C_{66}B_{jmn}M_{jmn} + \mu_m^2\Lambda_{1j}S_{2mn}(s)\right]\Pi_{jmn}(s)\cos\lambda_n\zeta -$$
$$-\frac{4}{l_1 l_2}\lambda_n^3\mu_m^2\sum_{j=1}^{N}\Lambda_{1j}\sum_{i=1}^{N}\mathrm{A}_{ji}M_{imn}\Pi_{ijmn}(s)\cos\lambda_n\zeta,$$

$$P_{122mn}(\xi,\zeta,s) = \frac{4}{l_1 l_2}\lambda_n\mu_m C_0\Pi_{mn}(s)\sin\mu_m\xi +$$
$$+\frac{4}{l_1 l_2}\mu_m\lambda_n\sum_{j=1}^{N}\left[\alpha_1^j M_{jmn}(s) - \frac{\Lambda_{22}^j}{C_{22}}S_{2mn}(s)\right]\Pi_{jmn}(s)\sin\mu_m\xi +$$
$$+\frac{4}{l_1 l_2}\mu_m\lambda_n^3\sum_{j=1}^{N}\frac{\Lambda_{22}^j}{C_{22}}\sum_{i=1}^{N}\mathrm{A}_{ji}M_{imn}\Pi_{ijmn}(s)\sin\mu_m\xi,$$

$$P_{221mn}(\xi,\zeta,s) = \frac{4}{l_1 l_2}\mu_m\lambda_n C_0\Pi_{mn}(s)\sin\mu_m\zeta +$$
$$+\frac{4}{l_1 l_2}\mu_m\lambda_n\sum_{j=1}^{N}\left[\alpha_2^j K_{jmn} - \Lambda_{11}^j S_{1mn}(s)\right]\Pi_{jmn}(s)\sin\lambda_n\zeta +$$
$$+\frac{4}{l_1 l_2}\mu_m^3\lambda_n\sum_{j=1}^{N}\Lambda_{11}^j\sum_{i=1}^{N}\mathrm{A}_{ji}K_{imn}\Pi_{ijmn}(s)\sin\lambda_n\zeta,$$

$$P_{212mn}(\xi,\zeta,s)=\frac{4}{l_1l_2}C_{66}\mu_m\left(k_{1mn}-C_0\lambda_n^2\right)\Pi_{mn}(s)\cos\mu_m\xi+$$
$$+\frac{4}{l_1l_2}\mu_m\sum_{j=1}^{N}\left[-C_{66}B_{jmn}K_{jmn}+\lambda_n^2\Lambda_{2j}S_{1mn}(s)\right]\Pi_{jmn}(s)\cos\mu_m\xi-$$
$$-\frac{4}{l_1l_2}\lambda_n^2\mu_m^3\sum_{j=1}^{N}\Lambda_{2j}\sum_{i=1}^{N}\mathrm{A}_{ji}K_{imn}\Pi_{ijmn}(s)\cos\mu_m\xi,$$

$$P_{211mn}(\xi,\zeta,s)=\frac{4}{l_1l_2}C_{66}\mu_m\left(k_{1mn}(s)-C_0\lambda_n^2\right)\Pi_{mn}(s)\cos\lambda_n\zeta+$$
$$+\frac{4}{l_1l_2}\mu_m\sum_{j=1}^{N}\left[-C_{66}B_{jmn}K_{jmn}+\lambda_n^2\Lambda_{1j}S_{1mn}(s)\right]\Pi_{jmn}(s)\cos\lambda_n\zeta-$$
$$-\frac{4}{l_1l_2}\lambda_n^2\mu_m^3\sum_{j=1}^{N}\Lambda_{1j}\sum_{i=1}^{N}\mathrm{A}_{ji}K_{imn}\Pi_{ijmn}(s)\cos\lambda_n\zeta,$$

$$P_{222mn}(\xi,\zeta,s)=-\frac{4}{l_1l_2}k_{1mn}(s)\Pi_{mn}(s)\sin\lambda_n\xi-$$
$$-\frac{4}{l_1l_2}\sum_{j=1}^{N}\left[\frac{\Lambda_{22}^j}{C_{22}}\lambda_n^2S_{1mn}(s)+\mu_m^2\alpha_1^jK_{jmn}\right]\Pi_{jmn}(s)\sin\mu_m\xi+$$
$$+\frac{4}{l_1l_2}\mu_m^2\lambda_n^2\sum_{j=1}^{N}\frac{\Lambda_{22}^j}{C_{22}}\sum_{i=1}^{N}\mathrm{A}_{ji}K_{imn}\Pi_{ijmn}(s)\sin\mu_m\xi,$$

$$P_{1,q+2,1mn}(\xi,\zeta,s)=\frac{4}{l_1l_2}\mu_m^2\sum_{j=1}^{N}D_{1jq}S_{2mn}(s)\Pi_{jmn}(s)\sin\lambda_n\zeta-$$
$$-\frac{4}{l_1l_2}\mu_m^2\lambda_n^2\sum_{j=1}^{N}D_{1jq}\sum_{r=1}^{N}\mathrm{A}_{jr}M_{rmn}\Pi_{rjmn}(s)\sin\lambda_n\zeta,$$
$$P_{1,q+2,2mn}(\xi,\zeta,s)=\frac{4}{l_1l_2}\mu_m\lambda_n\sum_{j=1}^{N}D_{2jq}S_{2mn}(s)\Pi_{jmn}(s)\sin\mu_m\xi-$$
$$-\frac{4}{l_1l_2}\mu_m\lambda_n^3\sum_{j=1}^{N}D_{2jq}\sum_{r=1}^{N}\mathrm{A}_{jr}M_{rmn}\Pi_{rjmn}(s)\sin\mu_m\xi,$$

$$P_{2,q+2,1mn}(\xi,\zeta,s)=\frac{4}{l_1l_2}\lambda_n\mu_m\sum_{j=1}^{N}D_{1jq}S_{1mn}(s)\Pi_{jmn}(s)\sin\lambda_n\zeta-$$
$$-\frac{4}{l_1l_2}\mu_m^3\lambda_n\sum_{j=1}^{N}D_{1jq}\sum_{r=1}^{N}\mathrm{A}_{jr}K_{rmn}\Pi_{rjmn}(s)\sin\lambda_n\zeta,$$

$$P_{2,q+2,2mn}(\xi,\zeta,s)=\frac{4}{l_1l_2}\lambda_n^2\sum_{j=1}^{N}D_{2jq}S_{1mn}(s)\,\Pi_{jmn}(s)\sin\mu_m\xi-$$
$$-\frac{4}{l_1l_2}\mu_m^2\lambda_n^2\sum_{j=1}^{N}D_{2jq}\sum_{r=1}^{N}\mathrm{A}_{jr}K_{rmn}\Pi_{rjmn}(s)\sin\mu_m\xi,$$

$$P_{q+2,klmn}(\xi,\zeta,s)=\mu_mK_{qmn}P_{1klmn}(\xi,\zeta,s)+\lambda_nM_{qmn}P_{2klmn}(\xi,\zeta,s),$$

$$\Pi_{mn}(s)=\prod_{j=1}^{N}k_{j+2,mn}(s),\ \ \Pi_{qmn}(s)=\prod_{j=1,j\neq q}^{N}k_{j+2,mn}(s),\ \ \Pi_{qpmn}(s)=\prod_{j=1,j\neq q,p}^{N}k_{j+2,mn}(s),$$
$$S_{1mn}(s)=\alpha_1^jC_0\mu_m^2-\alpha_2^jk_{1mn}(s),\ \ S_{2mn}(s)=\alpha_2^jC_0\lambda_n^2-\alpha_1^jk_{2mn}(s),$$
$$\mathrm{A}_{ji}=\alpha_1^j\alpha_2^i-\alpha_2^j\alpha_1^i,\ \ B_{jmn}=\alpha_2^j\lambda_n^2-\mu_m^2\alpha_1^j,$$
$$D_{1jq}=D_1^j\delta_{qj}-\Lambda_{11}^j\alpha_1^q,\ \ D_{2jq}=D_2^j\delta_{qj}-\frac{\Lambda_{22}^j}{C_{22}}\alpha_2^q.$$

The transition to the original domain is done by residues and tables of operational calculus as follows [21]:

$$G_{1kl0n}(\xi,\zeta,\tau)=\frac{2\sqrt{C_{66}}}{l_1l_2}\left[\delta_{1k}(\delta_{1l}+\delta_{2l})-\frac{\delta_{2k}\delta_{1l}\sin\lambda_n\zeta}{\lambda_nC_{66}}\right]\sin\left(\lambda_n\sqrt{C_{66}}\tau\right),$$
$$G_{2klm0}^{sc}(\xi,\zeta,\tau)=\frac{2\sqrt{C_{66}}}{l_1l_2}\left[\delta_{1k}(\delta_{1l}+\delta_{2l})-\frac{\delta_{2k}\delta_{2l}\sin\mu_m\xi}{\mu_mC_{66}}\right]\sin\left(\mu_m\sqrt{C_{66}}\tau\right);\tag{9.15}$$

$$G_{iklmn}(\xi,\zeta,\tau)=\sum_{j=1}^{2N+4}A_{iklmn}^{(j)}(\xi,\zeta)\,e^{s_{jmn}\tau}\ \ (i=1,2),$$
$$G_{q+2,klmn}(\xi,\zeta,\tau)=\hat{G}_{q+2,klmn}(\xi,\zeta,\tau)+\sum_{j=1}^{2N+6}A_{q+2,klmn}^{(j)}(\xi,\zeta)e^{s_{jmn}\tau},\tag{9.16}$$
$$A_{iklmn}^{(j)}(\xi,\zeta)=\frac{P_{iklmn}(\xi,\zeta,s_{jmn})}{P'_{mn}(s_{jmn})},\ \ A_{q+2,klmn}^{(j)}(\xi,\zeta)=\frac{P_{q+2,klmn}(\xi,\zeta,s_{jmn})}{Q'_{qmn}(s_{jmn})},$$

$$\hat{G}_{q+2,21mn}(\xi,\zeta,\tau)=-\frac{4}{l_1l_2}\Lambda_{11}^q\mu_m\sin\lambda_n\zeta\sum_{j=1}^{2}\frac{e^{\chi_{jmn}\tau}}{k'_{q+2,mn}(\chi_{jmn})},$$
$$\hat{G}_{q+2,12mn}(\xi,\zeta,\tau)=\frac{4}{l_1l_2}\Lambda_{2q}\lambda_n\mu_m\cos\mu_m\xi\sum_{j=1}^{2}\frac{e^{\chi_{jmn}\tau}}{k'_{q+2,mn}(\chi_{jmn})},$$
$$\hat{G}_{q+2,11mn}(\xi,\zeta,\tau)=\frac{4}{l_1l_2}\Lambda_{1q}\lambda_n\mu_m\cos\lambda_n\zeta\sum_{j=1}^{2}\frac{e^{\chi_{jmn}\tau}}{k'_{q+2,mn}(\chi_{jmn})},$$
$$\hat{G}_{q+2,22mn}(\xi,\zeta,\tau)=-\frac{4}{l_1l_2}\frac{\Lambda_{22}^q\lambda_n}{C_{22}}\sin\mu_m\xi\sum_{j=1}^{2}\frac{e^{\chi_{jmn}\tau}}{k'_{q+2,mn}(\chi_{jmn})}.$$

9.5 Calculation Example

Let us take a three-component material ($N = 2$, independent components, zinc and copper, diffusing into aluminium) with the following characteristics [22]:

$$C'_{12} = 5.11 \cdot 10^{10} \frac{\text{N}}{\text{m}^2}, \; C'_{66} = 2.63 \cdot 10^{10} \frac{\text{N}}{\text{m}^2}, \; T_0 = 700\ K, \; \rho = 2700 \frac{\text{kg}}{\text{m}^3},$$
$$\alpha_{11}^{(1)} = \alpha_{22}^{(1)} = 6.55 \cdot 10^7 \frac{\text{J}}{\text{kg}}, \; \alpha_{11}^{(2)} = \alpha_{22}^{(2)} = 6.14 \cdot 10^7 \frac{\text{J}}{\text{kg}}, \; l = 10^{-2}\ \text{m},$$

$$D_{11}^{(1)} = D_{22}^{(1)} = 2.62 \cdot 10^{-12} \frac{\text{m}^2}{\text{s}}, \; D_{11}^{(2)} = D_{22}^{(2)} = 2.89 \cdot 10^{-15} \frac{\text{m}^2}{\text{s}},$$
$$n_0^{(1)} = 0.01, \; n_0^{(2)} = 0.045, \; m^{(1)} = 0.065 \frac{\text{kg}}{\text{mol}}, \; m^{(2)} = 0.064 \frac{\text{kg}}{\text{mol}}.$$

Assume that the plate is under the action of tensile forces applied to the boundaries $x_1 = 0$ and $x_1 = l_1$:

$$f_{211}(x_2, \tau) = f_{212}(x_2, \tau) = H(\tau) \sin \frac{\pi x_2}{l_2},$$

where $H(\tau)$ is the Heaviside function.

All other load parameters in the boundary conditions (9.2) are assumed to be zero. Calculating the convolutions (9.4), we obtain

$$u_1(x_1, x_2, \tau) = \int_0^{l_2} G_{121}(x_1, x_2, \zeta, \tau) * f_{211}(\zeta, \tau)\, d\zeta -$$
$$- \int_0^{l_2} G_{121}(l_1 - x_1, x_2, \zeta, \tau) * f_{212}(\zeta, \tau)\, d\zeta =$$
$$= l_2 \sin \lambda_1 x_2 \sum_{m=1}^{\infty} (-1)^{m+1} \sin\left[\mu_m \left(\frac{l_1}{2} - x_1\right)\right] \sum_{j=1}^{2N+4} \tilde{A}_{121m1}^{(j)} \frac{e^{s_{jm1}\tau} - 1}{s_{jm1}},$$

$$u_2(x_1, x_2, \tau) = \int_0^{l_2} G_{221}(x_1, x_2, \zeta, \tau) * f_{211}(\zeta, \tau)\, d\zeta +$$
$$+ \int_0^{l_2} G_{221}(l_1 - x_1, x_2, \zeta, \tau) * f_{212}(\zeta, \tau)\, d\zeta =$$
$$= l_2 \cos \lambda_1 x_2 \sum_{m=1}^{\infty} (-1)^{m+1} \cos\left[\mu_m \left(\frac{l_1}{2} - x_1\right)\right] \sum_{j=1}^{2N+4} \tilde{A}_{221m1}^{(j)} \frac{e^{s_{jm1}\tau} - 1}{s_{jm1}},$$

$$\eta_q(x_1, x_2, \tau) = \int_0^{l_2} G_{q+2,21}(x_1, x_2, \zeta, \tau) * f_{211}(\zeta, \tau)\, d\zeta +$$
$$+ \int_0^{l_2} G_{q+2,21}(l_1 - x_1, x_2, \zeta, \tau) * f_{212}(\zeta, \tau)\, d\zeta =$$
$$= \frac{4}{l_1} \sin \lambda_1 x_2 \sum_{m=1}^{\infty} (-1)^{m+1} \cos\left[\mu_m \left(\frac{l_1}{2} - x_1\right)\right] \sum_{j=1}^{2} \frac{\Lambda_{1q} \lambda_1 \mu_m}{k'_{q+2,m1}(\chi_{jm1})} \frac{e^{\chi_{jm1}\tau} - 1}{\chi_{jm1}} +$$
$$+ l_2 \sin \lambda_1 x_2 \sum_{m=1}^{\infty} (-1)^{m+1} \cos\left[\mu_m \left(\frac{l_1}{2} - x_1\right)\right] \sum_{j=1}^{2N+6} \tilde{A}_{q+2,21m1}^{(j)} \frac{e^{s_{jm1}\tau} - 1}{s_{jm1}},$$

Fig. 9.1 Displacement $u_1(x_1, x_2, \tau)$ at $x_2 = l_2/2$

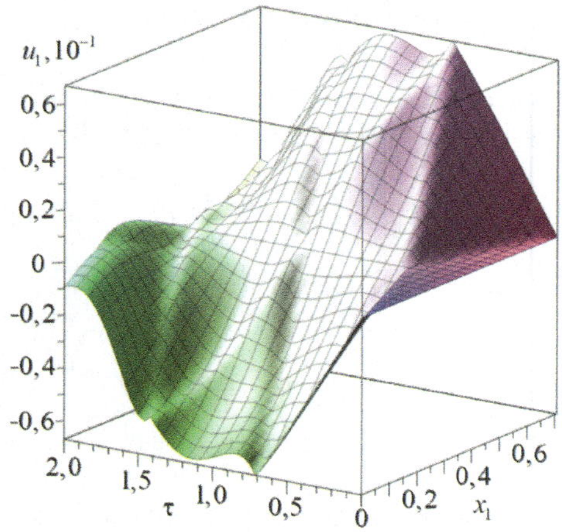

Fig. 9.2 Displacement $u_1(x_1, x_2, \tau)$ at $x_1 = l_1/4$

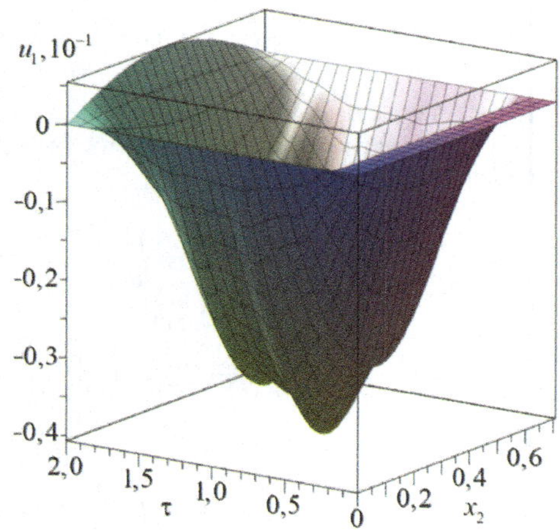

$$\tilde{A}^{(j)}_{k21mn} = \frac{A^{(j)}_{k21mn}(\zeta)}{\sin \lambda_1 \zeta}.$$

The calculation results are shown in the Figs. 9.1, 9.2, 9.3, 9.4, 9.5, 9.6, 9.7 and 9.8. Figures 9.1, 9.2, 9.3 and 9.4 show dependences of longitudinal displacements u_i on time and coordinates. Calculations show that diffusion processes at the initial stages of deformation do not affect the displacement field. At initial times, the elastic and elastic diffusion displacements coincide.

Figures 9.5 and 9.6 show the zinc and copper concentration increments, respectively, caused by longitudinal deformations.

The influence of relaxation effects on the mass transfer kinetics is shown in Figs. 9.7 and 9.8. Here, different lines show zinc concentration increments for models with a finite and infinite speed of diffusion fluxes. The relaxation effects manifest themselves at some finite interval of time and then disappear. Thus, in Fig. 9.8, corresponding to the time $\tau = 10^9$, both curves already coincide.

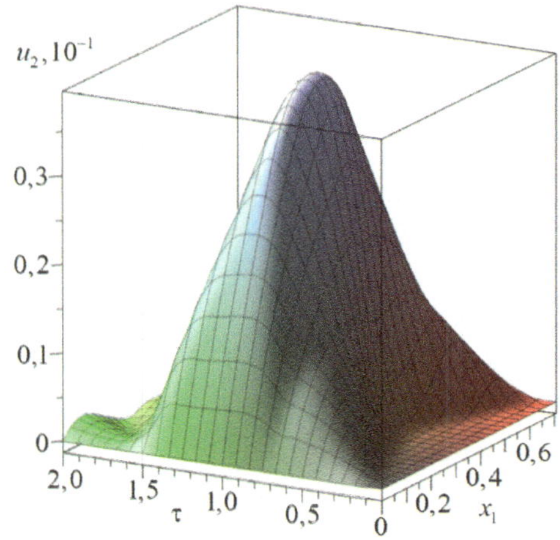

Fig. 9.3 Displacement $u_2(x_1, x_2, \tau)$ at $x_2 = l_2/4$

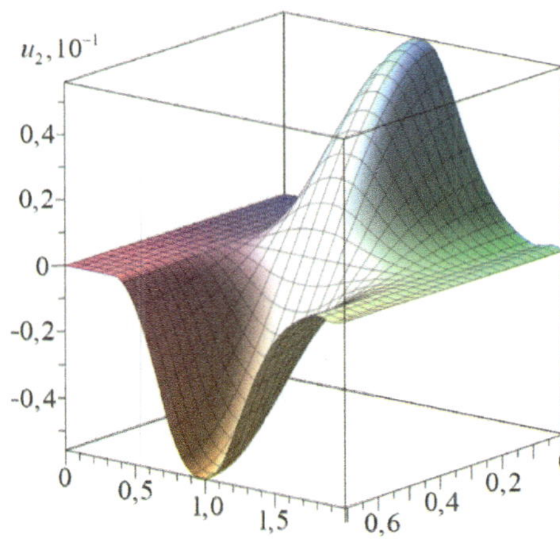

Fig. 9.4 Displacements $u_2(x_1, x_2, \tau)$ at $x_1 = l_1/2$

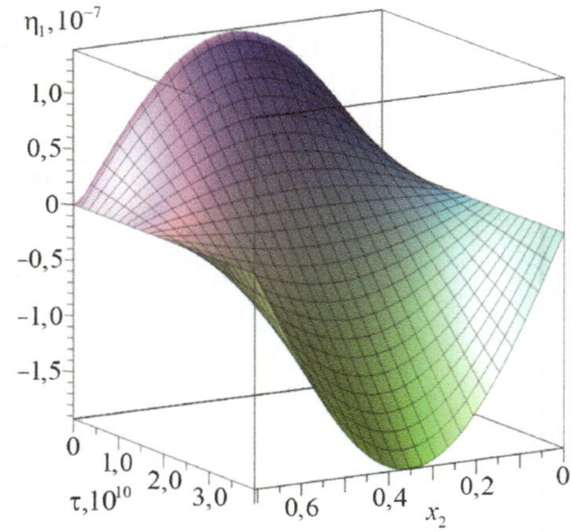

Fig. 9.5 Zinc concentration increment $\eta_1\ (x_1, x_2, \tau)$ at $x_1 = l_1/2$

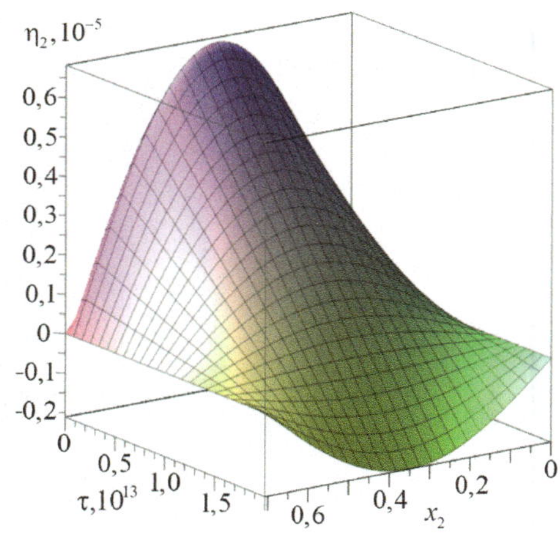

Fig. 9.6 Copper concentration increment $\eta_2\ (x_1, x_2, \tau)$ at $x_1 = l_1/2$

Fig. 9.7 Zinc concentration increment $\eta_1(x_1, x_2, \tau)$. The solid line corresponds to the time $\tau^{(q)} = 200\,\mathrm{s}$, the dotted line to $\tau^{(q)} = 0$

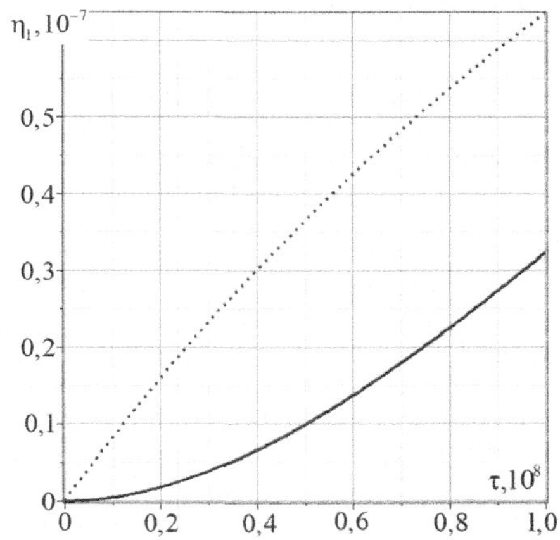

Fig. 9.8 Zinc concentration increment $\eta_1(x_1, x_2, \tau)$. The solid line corresponds to the time $\tau^{(q)} = 200\,\mathrm{s}$, the dotted line to $\tau^{(q)} = 0$

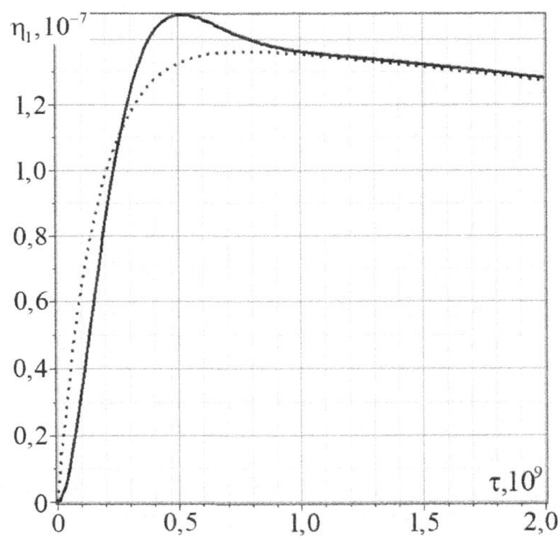

9.6 Conclusions

We propose a mathematical model describing the effects of the interaction between mechanical and diffusion fields during a rectangular orthotropic plate deformation. The algorithm is developed for finding Green's functions based on Laplace transform and decomposition into trigonometric Fourier series. It allows us to reduce the general problem of Laplace transform inversion to the problem of rational function inversion. Originals of Green's functions are found using residues and tables of operational calculus. As a result, it is possible to obtain the solution in analytical form, which provides ample opportunities for various kinds of numerical experiments.

The effect of interaction between the mechanical and diffusion fields is demonstrated by the example of a rectangular plate under the action of tensile forces. Unsteady loads initiate the process of mass transfer. At the same time, the relaxation diffusion processes reduce with time. The kinetics of mass transfer at significant times can be described by classical mechanodiffusion with the infinite speed of diffusion fluxes. All results are presented in analytical and graphical forms.

Acknowledgements This work was supported by the Russian Science Foundation (project No 20-19-00217).

References

1. Abbas AI (2015) Eigenvalue approach on fractional order theory of thermoelastic diffusion problem for an infinite elastic medium with a spherical cavity. Appl Math Model 39(20):6196–6206
2. Afram AY, Khader SE (2014) 2D problem for a half-space under the theory of fractional thermoelastic diffusion. Am J Sci Ind Res 6(3):47–57
3. Aouadi M (2005) Variable electrical and thermal conductivity in the theory of generalized thermoelastic diffusion. Z Angew Math Phys 57(2):350–366
4. Bachher M, Sarkar N (2016) Fractional order magneto-thermoelasticity in a rotating media with one relaxation time. Math Model Eng 2(1):57–68
5. Bhattacharya D, Kanoria M (2014) The influence of two temperature generalized thermoelastic diffusion inside a spherical shell. Int J Eng Tech Res (IJETR) 2(5):151–159
6. Deswal S, Kalkal KK, Sheoran SS (2016) Axi-symmetric generalized thermoelastic diffusion problem with two-temperature and initial stress under fractional order heat conduction. Phys B 496:57–68
7. Elhagary MA (2013) A two-dimensional generalized thermoelastic diffusion problem for a half-space subjected to harmonically varying heating. Acta Mech 224:3057–3069
8. El-Sayed AM (2016) A two-dimensional generalized thermoelastic diffusion problem for a half-space. Math Mech Solids 21(9):1045–1060
9. Freidin A, Morozov N, Petrenko S, Vilchevskaya E (2016) Chemical reactions in spherically symmetric problems of mechanochemistry. Acta Mech 227(1):43–56
10. Indeitsev DA, Semenov BN, Sterlin MD (2014) The phenomenon of localization of diffusion process in a dynamically deformed solid. Dokl Phys 11:299–319
11. Kumar R, Ahuja S, Garg SK (2014) Latin Am J Solids Struct Mater Phys Mech 11:299–319
12. Kumar R, Chawla V (2012) A study of Green's functions for two-dimensional problem in orthotropic magnetothermoelastic media with mass diffusion. Mater Phys Mech 15:78–95
13. Parfenova ES, Knyazeva AG (2021) The influence of chemical reaction parameters on the interaction of thermal, diffusion and mechanical waves in the condition of surface treatment by particles beam. Comput Contin Mech 14(1):77–90
14. Sherief HH, El-Maghraby NM (2009) A thick plate problem in the theory of generalized thermoelastic diffusion. Int J Thermophys 30:2044–2057

15. Tripathi JJ, Kedar GD, Deshmukh KC (2015) Two-dimensional generalized thermoelastic diffusion in a half-space under axisymmetric distributions. Acta Mech 226:3263–3274
16. Zhang J, Li Y (2014) A two-dimensional generalized electromagnetothermoelastic diffusion problem for a rotating half-space. Math Probl Eng 2014(964218): 12
17. Davydov SA, Zemskov AV, Tarlakovskii DV (2015) An elastic half-space under the action of one-dimensional time-dependent diffusion perturbations. Lobachevskii J Math 36(4):503–509
18. Davydov SA, Zemskov AV (2022) Thermoelastic diffusion phase-lag model for a layer with internal heat and mass sources. Int J Heat Mass Transf 183(C):122213
19. Zemskov AV, Tarlakovskii DV (2019) Bulk Green's functions in two-dimensional coupled unsteady problems of elastic diffusion for orthotropic continuum. Lobachevskii J Math 40(3):375–383
20. Igumnov LA, Tarlakovskii DV, Zemskov AV (2017) A two-dimensional nonstationary problem of elastic diffusion for an orthotropic one-component layer. Lobachevskii J Math 38(5):808–817
21. Ditkin VA, Prudnikov AP (1965) Handbook on operational calculus. Vysshaya Shkola, Moscow [In Russian]
22. Grigoriev IS, Meylikhov IZ (eds) (1991) Fizicheskiye velichiny: Spravochnik. Energoatomizdat, Moscow [In Russian]

Chapter 10
The Influence of the Detailed Model of the Structure on the Stress–Strain State of the Soil Base in the Calculations of Seismic Resistance

Nadezhda S. Dyukina

Abstract The research is aimed at selecting an adequate design model of the structure in the tasks of assessing the strength of underground pipelines adjacent to the structure. In a series of computational experiments, the influence of the structure on the soil base under static and dynamic influences was studied. To analyze the influence of the structure's detail on the seismic response and dynamic deformation of the foundation, three design models of a structure are considered. Invariants of the calculation experiments were the mass and the center of mass of the structure, the shape of the footprint and the overall dimensions of structures, the dynamic influence, and the elastic constants for all materials. It is established that the simplest model of a structure—a homogeneous array—shows the greatest differences in the static effect on the soil base during settlement and the dynamic effect from compression and shear waves. It is shown, some degree of detail of the structure should be considered when calculating seismic impacts of underground pipelines adjacent to the structure.

Keywords Numerical experiment · SSE · Deformation of the foundation · Seismic impacts

10.1 Introduction

The calculation of structures and adjacent pipelines for seismic impacts is usually carried out either with a simplified model of the structure [1, 2] or with a simplified model of the soil base [3]. Simplification of the construction model is allowed in [4] for nuclear power plants that differ from civil structures by a significant thickness of foundation plates, whereas calculations of civil structures show maximum stresses in the foundation plate [3]. The use of a simplified model of the soil foundation implies the calculation of the Winkler coefficient of the foundation [5] under the assumption

N. S. Dyukina (✉)
National Research Lobachevsky State University of Nizhny Novgorod, 23/6, Gagarina Av., Nizhny Novgorod 603022, Russia
e-mail: ndyukina@inbox.ru

H. Altenbach et al. (eds.), *Deformation and Destruction of Materials and Structures Under Quasi-static and Impulse Loading*, Advanced Structured Materials 186,
https://doi.org/10.1007/978-3-031-22093-7_10

that the spot of the projected structure transmits to the soil a load of a certain constant intensity applied at the level of the sole of the foundation. The limitations of this method of accounting for the soil base are associated with replacing the nonlinear dependence of the subsidence of the soil base on the applied load with a linear one and accepting one of the most likely load values—its own weight, long-term loads, or a long part of short-term loads. These limitations do not allow us to apply simplified models of the soil base in calculations for seismic impacts, since taking into account wave effects in the base is extremely important for determining the dynamic stiffness of the soil base [6–9]. Seismic waves passing under the foundation of the building change the parameters of the stress–strain state of the "building-soil" system. Since the soil base has malleability and inertia, waves of the initial seismic impact and waves associated with the presence and movement of the foundation of the structure propagate in it. The phenomenon of wave damping, when mechanical energy does not pass into heat but is carried away by waves from a moving die, is combined with traditional damping associated with internal friction characteristic of soils, and often turns out to be much more important for the overall reaction of the structure. It is also necessary to take into account the possibility of local stress concentrations in the soil base, leading to the transition of the soil medium into a loose and elastic–plastic state and, as a consequence, a change in the bearing capacity of the base.

It is required to substantiate the permissible limits of the idealization of the structure and the soil foundation in the tasks of calculating seismic resistance. It is interesting to study the influence of the details of the structure model on the stress–strain state of the soil foundation and the dynamic deformation of the foundation in the calculations of seismic resistance.

10.2 The Mathematical Model and Numerical Experiment

A rectangular soil massif of 2000 × 2000 × 200 m is considered. The soil is modeled by a continuous ideally elastic medium with characteristics E = 2.4 GPa, $\nu = 0.4$, $\rho = 2000$ kg/m^3. On the upper soil surface in the center there is a structure with mechanical characteristics E = 20 GPa, $\nu = 0.25$, and mass of M = 8000 tons. To analyze the influence of the structure's detail on the seismic response and dynamic deformation of the foundation, several design models of a structure of a given mass are considered (Table 10.1). When the density of the construction material changed, the new volume V^{new} was determined from the ratio:

$$V^{\text{new}} = \frac{M}{\rho^{\text{new}}}$$

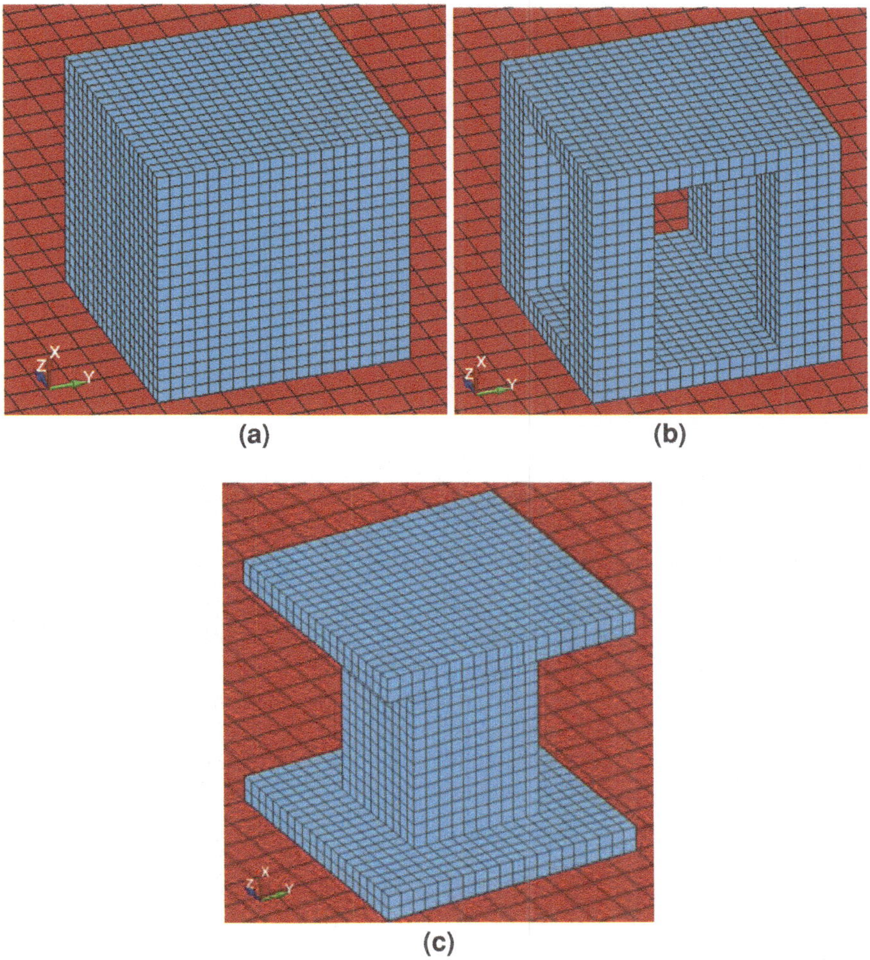

Fig. 10.1 Structures **a**—variant 1 in Table 10.1, **b**—variant 2 in Table 10.1, **c**—variant 3 in Table 10.1

Table 10.1 The description of the calculation experiments

Variant of calculation	Density of material for construction ρ, kg/m^3	Geometric parameters of the structure
1	$\rho = \frac{M}{V} = 1000$	Cube 20 × 20 × 20 m (Fig. 10.1a)
2	$\rho = 2500$	Two plates 2 × 20 × 20 m, connected by 4 columns 16 × 5 × 5 m (Fig. 10.1b)
3	$\rho = 2500$	Two plates 2 × 20 × 20 m, connected by 1 column 16 × 10 × 10 m (Fig. 10.1c)

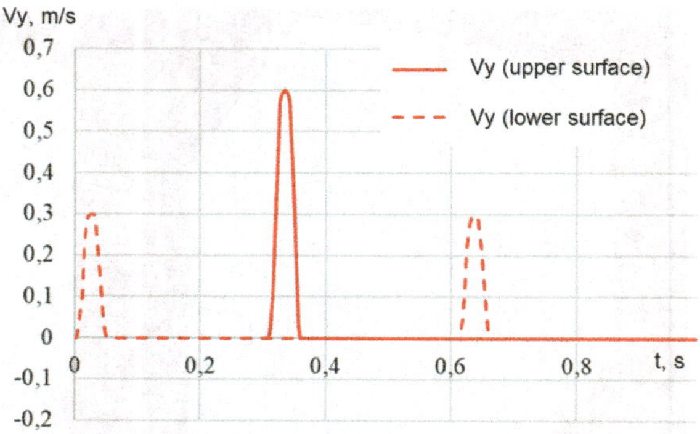

Fig. 10.2. Shear wave at the lower boundary of the soil

Invariants of the calculation experiments were:

- The mass and the center of mass of the structure;
- The shape of the footprint and the overall dimensions of structures;
- The dynamic influence;
- The elastic constants E, ν for all materials.

Numerical simulation is performed in LS-Dyna. The length of the edge of cubic finite elements for the soil was assumed to be 2.5 m, for the structure—1 m. To simulate the effect of gravity, a constant compressive stress is applied to the computational domain. Kinematic boundary conditions were set at the lower boundary of the soil massif, simulating a shear wave propagating to the day surface of the soil (Fig. 10.2). The determination of the kinematic effect at the lower boundary of the soil was carried out in accordance with [10].

10.3 Numerical Results

Calculations made it possible to detect inhomogeneities in the stress fields and displacement fields in complex structures (Fig. 10.3).

Qualitative differences in the process of the influence of settlement and dynamics of the structures under consideration on the soil base at a given impact are shown in Fig. 10.4. In accordance with the structural features of buildings, inhomogeneous stress fields were obtained on the lower plate of the structure and in the near-surface layer of the soil from the settlement (t = 0 s) and with a dynamic reaction to a shear pulse (t = 0.33 s).

Figures 10.5, 10.6, 10.7, 10.8 and 10.9 demonstrate the distribution of stresses and velocity components V_x, V_y in the soil horizontal profile buried at 0.5 m and lying on the plane of symmetry of structures with a normal {0; 0; 1}. An increase in the difference in stresses is typically at the stage of the settlement of buildings (t = 0 s—solid lines in Figs. 10.5 and 10.6) and at the stage of the dynamic impact of

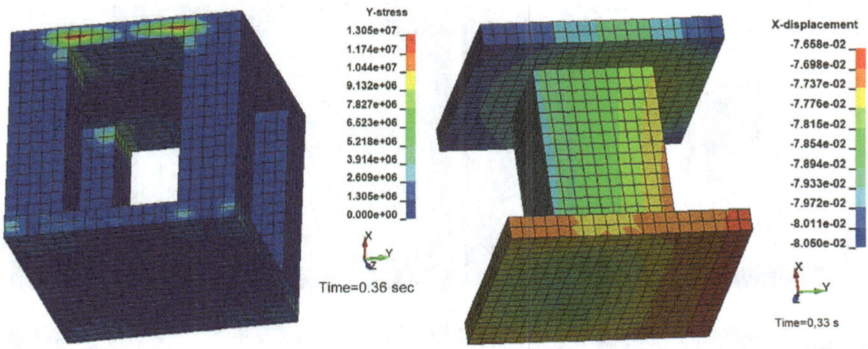

(a) (b)

Fig. 10.3 Inhomogeneities in the stress fields (**a**) and displacement fields (**b**)

the shear wave on buildings (t = 0.33 s—dotted lines in Figs. 10.5 and 10.6, lines with markers in Fig. 10.7). Here the interval of 40–60 m of the profile corresponds to the foundation of the structure. The most significant difference obtained for the components stress tensor σ_{xx}, σ_{yy}.

Figures 10.8 and 10.9 demonstrate as shear wave head to the structure (t < 0.35 s) and reflected from it (t > 0.35 s). The velocity profiles show differences in shear velocities up to 17% for different design variants (Fig. 10.8). The difference in the vertical component of the velocity V_x, which characterizes the rocking of the structure, is more significant (Fig. 10.9). The worst scenario is implemented for a homogeneous version of the design.

Similar calculations were carried out for the short pulse influence of the longitudinal compression-stretching wave. The characteristic differences for this type of impact are the occurrence of long–term vibrations in complex structures and the subsequent imposition of these vibrations (Fig. 10.10).

The response of all types of structures to the impact of a compression-stretching wave was more significant than for the impact of a shear wave (Figs. 10.11, 10.12, 10.13). The stress profiles (Fig. 10.11) at the stage of static impact from the gravity field and at the stage of dynamic impact differ by more than 4 times for all design variants. The difference in the shapes of the profiles is more significant than when exposed to a shear wave.

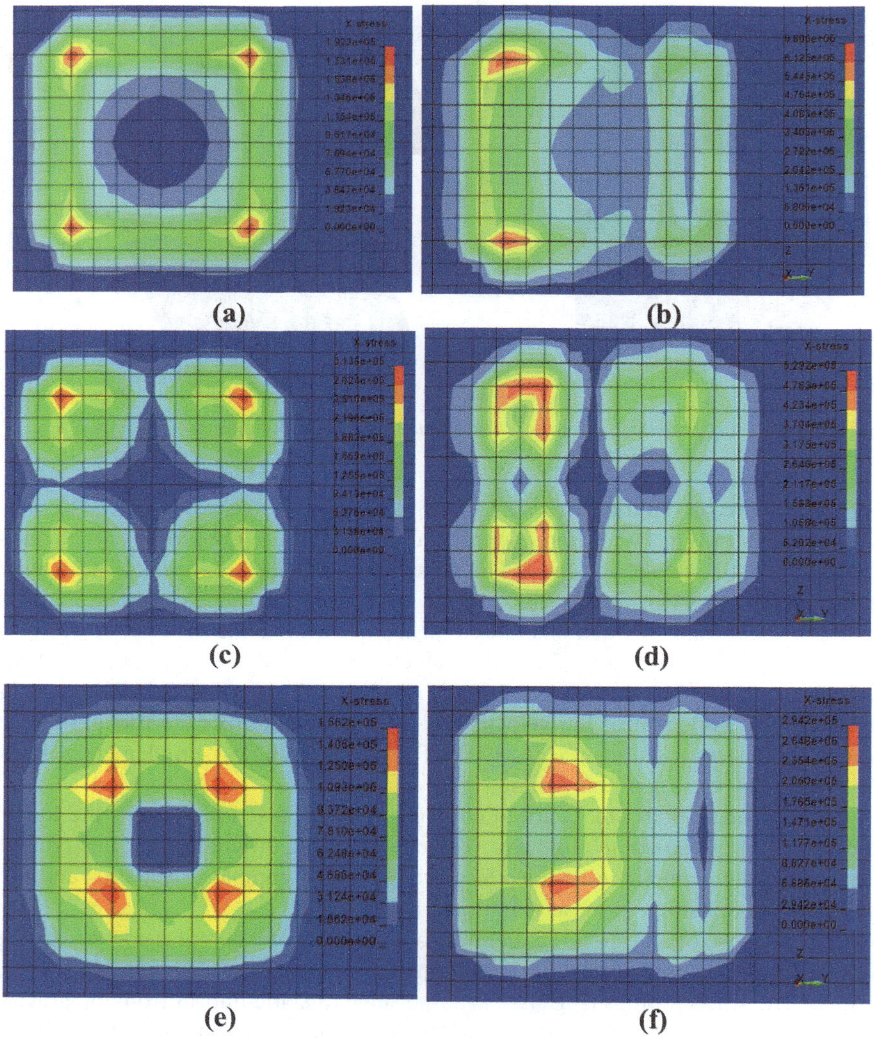

Fig. 10.4 Stresses σ_{xx} in the soil under the foundation plate after settlement (**a**, **c**, **e**) and at the time of the approach of the shear wave to the building (**b**, **d**, **f**): **a**, **b**—structure N1, **c**, **d**—N2, **e**, **f**—N3

10.4 Conclusion

In a series of computational experiments, the influence of the structure on the ground base under static and dynamic influences was studied. The research is aimed, first, at selecting an adequate design model of the structure in the tasks of assessing the strength of underground pipelines adjacent to the structure.

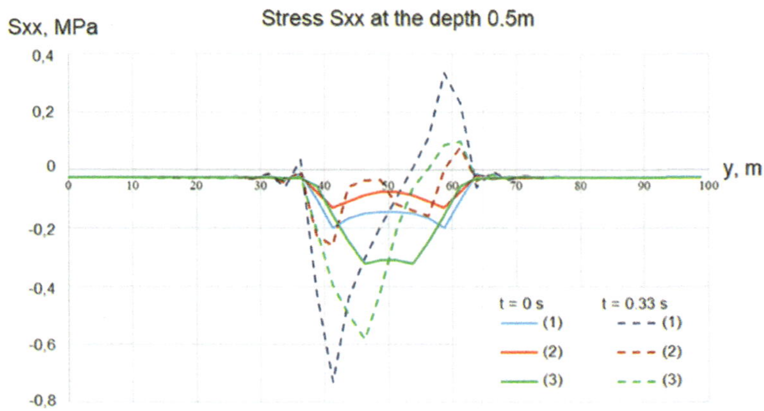

Fig. 10.5. Stress σ_{xx} at the depth 0.5 m

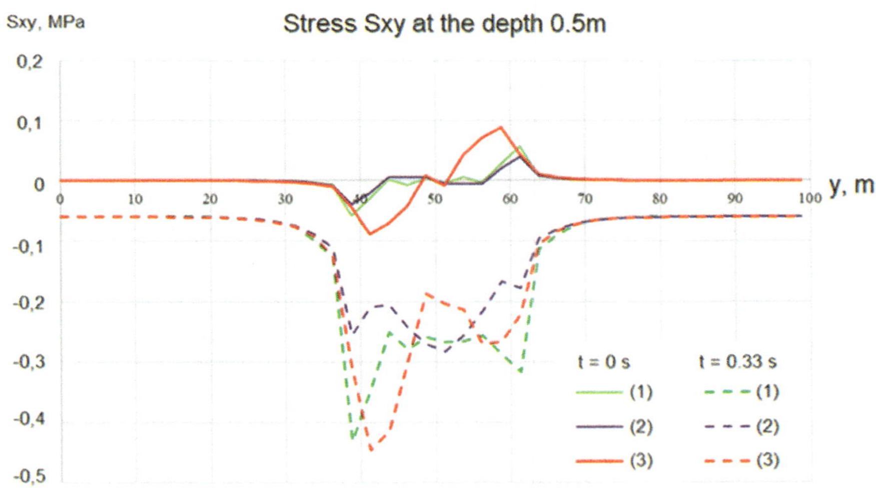

Fig. 10.6. Stress σ_{xy} at the depth 0.5 m

In the study, the worst-case scenario is implemented for the crudest model of the structure, which, on the one hand, allows us to guarantee more favorable scenarios for the reaction of the structure and the soil base for real, more detailed models. At the same time, this can lead to overestimated load estimates when assessing the seismic resistance of the structure and, as a result, the cost of construction increases.

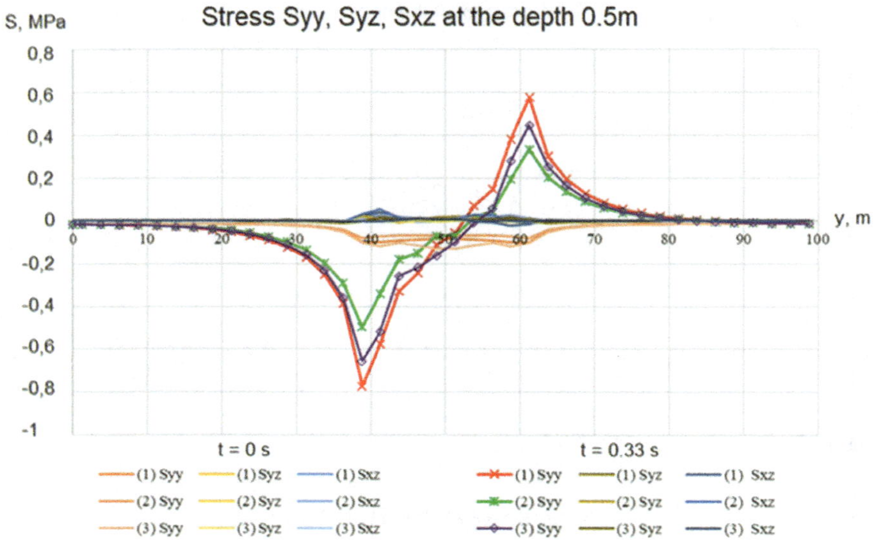

Fig. 10.7. Stress σ_{yy}, σ_{yz}, σ_{xz} at the depth 0.5 m

The choice of the construction model affects the results of numerical modeling of the behavior of the structure under dynamic loads, as well as the change in the stress–strain state of the soil base. Therefore, some degree of detail of the structure should be taken into account when calculating seismic impacts.

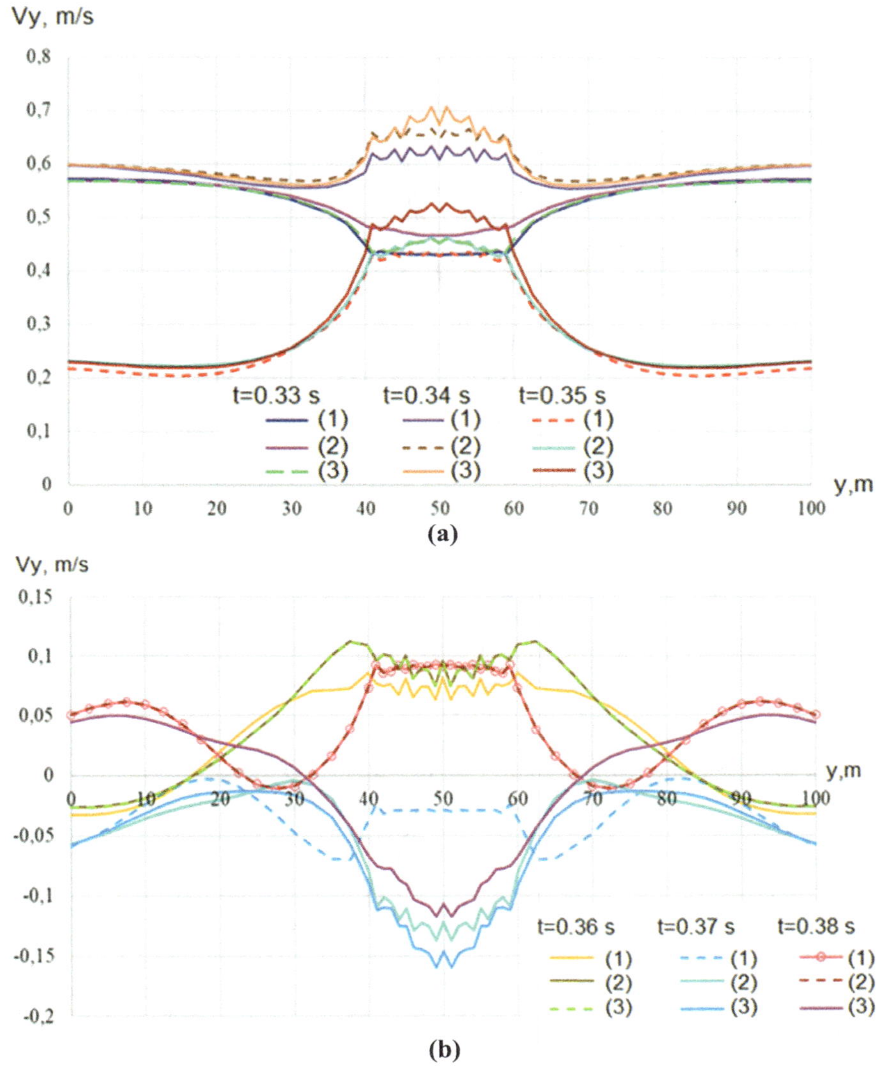

Fig. 10.8. Velocity profiles V_y in the soil at the depth 0.5 m under the structure: **a**—the beginning of the dynamic impact, **b**—dynamic response of the structures

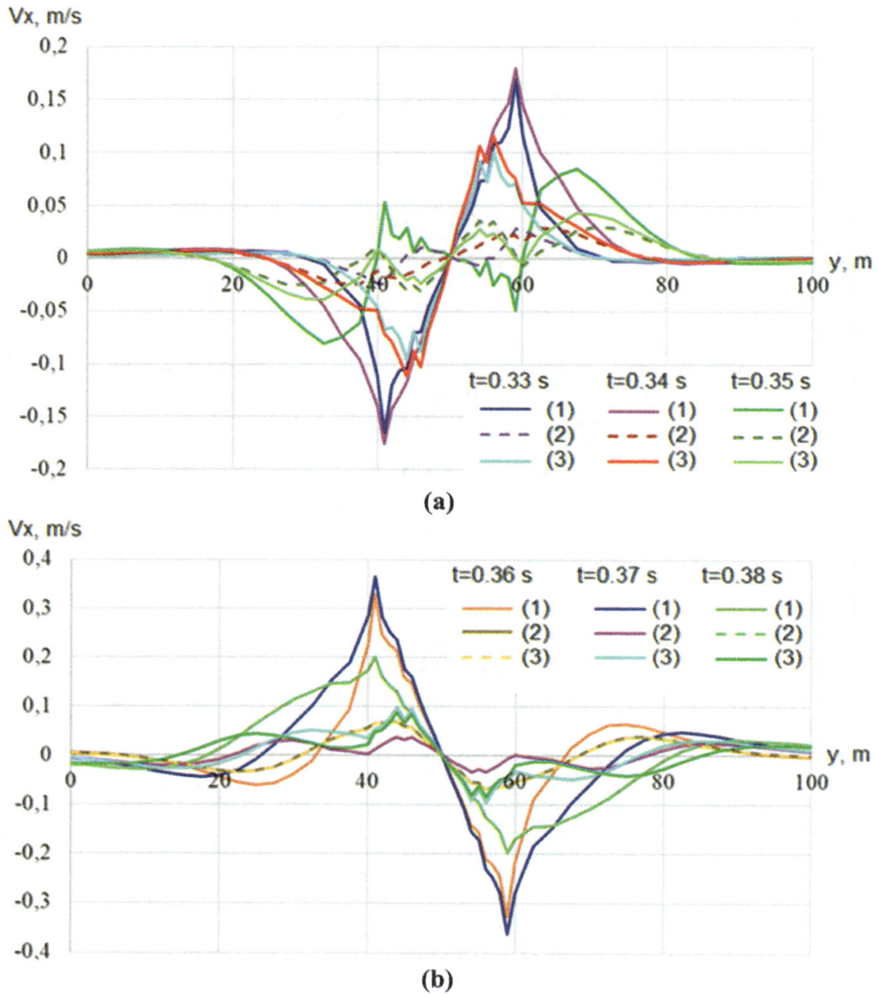

Fig. 10.9. Velocity profiles V_x in the soil at the depth 0.5 m under the structure: **a**—the beginning of the dynamic impact, **b**—dynamic response of the structures

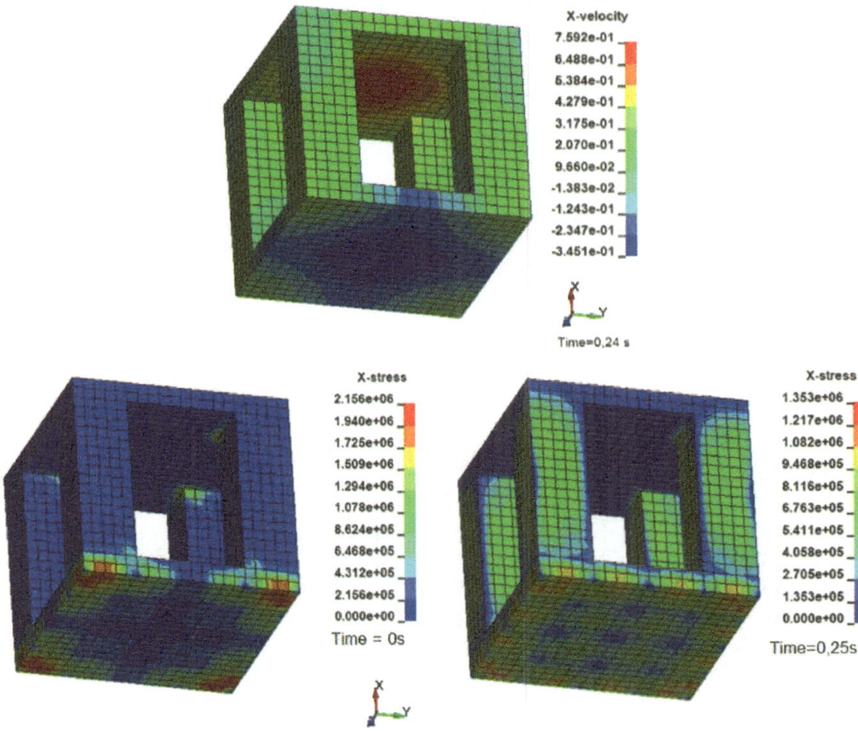

Fig. 10.10. Velocity V_x (**a**) and stress σ_{xx} (**b**, **c**) distribution, caused by the settlement (**b**) and compression-stretching wave influence alone the X-axis (**a**, **c**)

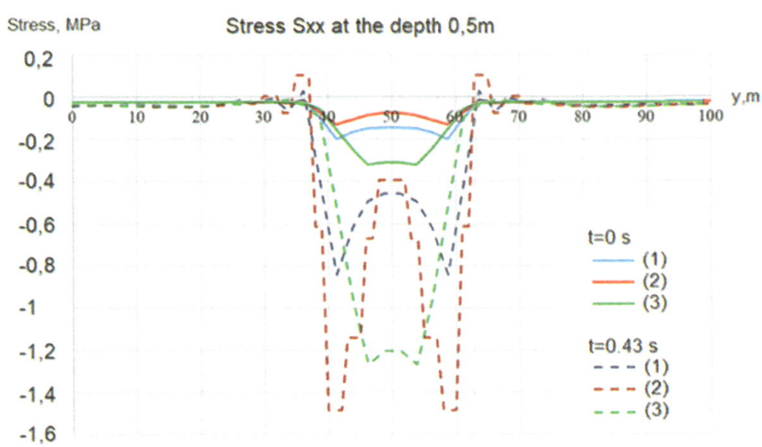

Fig. 10.11 Stress profile σ_{xx} in the soil at the depth 0.5 m

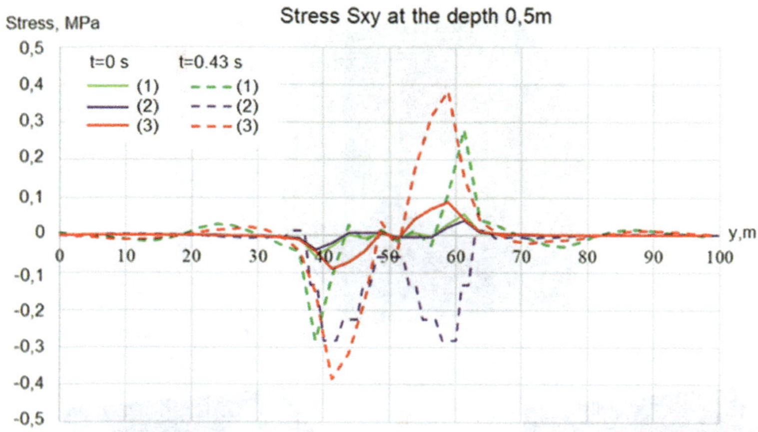

Fig. 10.12. Stress profile σ_{xy} in the soil at the depth 0.5 m

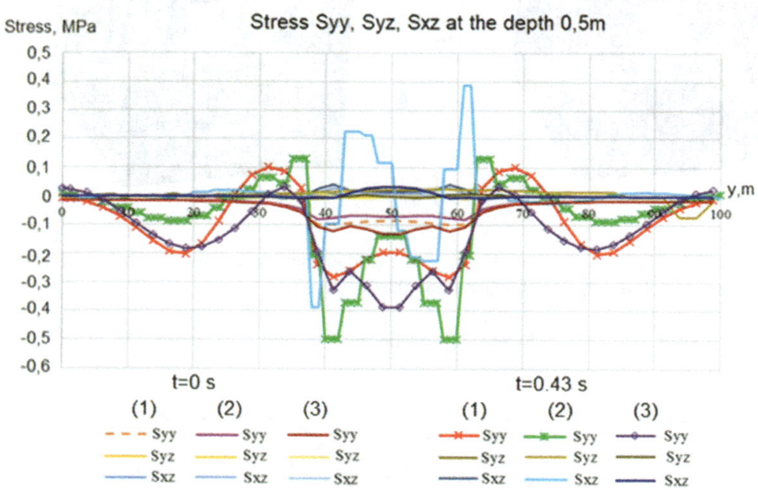

Fig. 10.13. Stress profile σ_{yy}, σ_{yz}, σ_{xz} in the soil at the depth 0.5 m

Acknowledgements The work was carried out with the financial support of the Ministry of Science and Higher Education of the Russian Federation (task 0729-2020-0054).

References

1. Tyapin AG (2013) Features of calculation of rigid foundations on "pillows" arranged under the foundation in a rigid foundation using the SASSI program Earthquake-resistant construction. Safety of structures, vol 3, pp 17–23
2. Dyukina NS, Kotov VL, Dyanov DY, Borlyaev VV (2021) Analysis of the capabilities of the Logos software package for calculating the seismic impact on a structure. Problems of strength and plasticity, pp 160–169
3. Erofees VI, Satanov AA, Hazov PA, Shishova MA (2020) Frequencies and forms of natural oscillations of a high-rise building in Vladivostok when calculating the maximum calculated earthquake. Survivability and Structural Materials Science (ZHIVKOM—2020), pp 237–241
4. Seismic Analysis of Safety-Related Nuclear Structures and Commentary ASCE4–98 (1999) Reston, Virginia, USA
5. Fedorovskii VG, Bezvolev SG (2000) Forecast of precipitation of shallow foundations and selection of the base model for the calculation of slabs. Foundations, basements and soil mechanics, vol 4, pp 10–18
6. Nazarov YP, Pozdnyak EV (2017)Calculations of building structures for wave seismic effects in the STARKON software package. Digital means of engineering analysis production. Collection of materials of the first all-Russian conference with international participation, pp 239–245
7. Nazarov YP, Poznyak EV, Filimonov AV (2015) A brief theory and computing of seismic ground rotations for structural analyses. Soil dynamics and earthquake engineering, vol 71, pp 31–41
8. Tyapin AG (2013) Calculation of structures for seismic impacts, taking into account the interaction with the ground base. ACB, Moscow
9. Tyapin AG (2017) Degradation of soil properties for longitudinal wave velocities under seismic impacts. Earthquake-resistant construction. Safety of structures, vol 5, pp 16–19
10. Dyukina NS, Bazhenov VG (2009) Numerical solution of two-dimensional and three-dimensional problems of interaction of structures with the ground base under seismic influences. Bulletin of the V.A. Kucherenko Central Research Institute "Research on the theory of structures", vol 1, pp 172–178

Chapter 11
Generalized Linear Model of Dynamics of Elastic Moment Shells

Quoc Chien Mai, Marina Yu. Ryazantseva, and Dmitry V. Tarlakovskii

Abstract The initial and boundary value problem describing the nonstationary dynamics of a homogeneous thin anisotropic elastic moment shell is constructed. The model takes into account the material anisotropy, free rotation, independent normal unit vector rotation, and its transverse normal strain. The functional and the physical law for the shell are constructed using the Hamilton functional, physical relations for a three-dimensional body described by the Cosserat model, the hypothesis of direct normal for the displacement field, and an analogous supposition for the rotation vector. The boundary value problem for the elastic moment shell is obtained as a necessary condition for an extremum of this functional. Various versions of natural boundary conditions are considered. It is pointed out that transition to an isotropic material only simplifies the physical relations without reducing the number of unknowns. The latter can be realized only at the expense of introducing some additional hypotheses concerning the displacement and rotation fields or considering special geometries of the middle surface.

Keywords Elastic moment shells · Unsteady problems · Cosserat model · Hamilton's principle · Anisotropic shell

---

Q. C. Mai · D. V. Tarlakovskii
Moscow Aviation Institute, Moscow, Russian Federation
e-mail: chienvk23@gmail.com

D. V. Tarlakovskii
e-mail: tdvhome@mail.ru

M. Yu. Ryazantseva (✉) · D. V. Tarlakovskii
Research Institute of Mechanics, Lomonosov Moscow State University, Moscow, Russian Federation
e-mail: marina-ryazantseva@yandex.ru

D. V. Tarlakovskii
N.I. Lobachevsky Institute of Mathematics and Mechanics, Kazan Federal University, Kazan, Russian Federation

H. Altenbach et al. (eds.), *Deformation and Destruction of Materials and Structures Under Quasi-static and Impulse Loading*, Advanced Structured Materials 186,
https://doi.org/10.1007/978-3-031-22093-7_11

11.1 Introduction

At present, the theory of elasticity of three-dimensional moment media has been developed in sufficient detail and there have been many studies concerning this issue. We will restrict ourselves only to the reference to publications [1–5] used in this study.

The studies concerned with elastic moment shells are considerably fewer in number. We note only certain publications pertaining to the linear models, namely, [6–8]. In particular, the authors of [7] use the Hamilton functional and expansions into power series in the normal coordinate which are then replaced by partial sums.

Different questions associated with the Saint-Venant hypothesis, the special features of boundary conditions, the presence of holes, and the temperature field effect for cylindrical shells were considered in [9–12]. In [13], an elastic sloping normal moment shell under the action of a given nonstationary temperature field was studied. The material behavior was modeled by the isotropic Cosserat medium. Under the suppositions of Timoshenko type, the standard for theory of shells, the equations of motion are constructed. The examples of the statics of a hinge-supported rectangular-in-plan spherical shell and the dynamics of a freely supported square plate, as well as the stability of a circular cylindrical shell under the action of uniformly distributed temperature, were considered.

By virtue of the known reasons, the behavior of elastic moment plates and beams is studied most simply. These questions are considered in [14–21]. In particular, in [17], the equilibrium equations were constructed using certain simplifying hypotheses. In [18–21], the problem of the motion of a plate of constant thickness was formulated using the Cosserat model. The asymptotic equations were derived, which break down into relations governing the stress–strain state and the boundary effect.

In this study, we develop initial and boundary value problems for elastic moment shells, which, in our opinion, are most general within the framework of the direct normal hypothesis and the Cosserat model. They take into account the medium anisotropy, its free rotation, the independent normal unit vector rotation, and its transverse strain.

11.2 Problem Formulation

We will consider a homogeneous anisotropic elastic moment shell of constant thickness h, whose material possesses the symmetry about its smooth oriented middle surface Π (closed or having a boundary $\Gamma = \partial \Pi$) with the unit normal vector $\boldsymbol{n}$ to the outer side. The geometric domain $G \subset R^3$ occupied by the shell is bounded by the surface [22, 24] $\partial G = \Pi_- \cup \Pi_+ \cup \Pi_b$, where Π_- and Π_+ are the inner and outer surfaces and Π_b is the lateral surface.

It is assumed that the shell is thin, that is, there holds the inequality $h/\lambda = \delta \ll 1$, where λ is the characteristic dimension of its middle surface. The middle surface and its boundary Γ, if any, are preassigned as follows:

$$\Pi : \boldsymbol{r} = \boldsymbol{r}_0\left(\xi^1, \xi^2\right), \left(\xi^1, \xi^2\right) \in D \subset R^2, \quad \Gamma : \ \boldsymbol{r} = \boldsymbol{r}_0\left(\xi^1, \xi^2\right), \left(\xi^1, \xi^2\right) \in \partial D.$$

Then the outer, inner, and lateral surfaces are described, respectively, by the following equations:

$$\Pi_{\pm} : \boldsymbol{r} = \boldsymbol{r}_{\pm}\left(\xi^1, \xi^2\right) = \boldsymbol{r}_0\left(\xi^1, \xi^2\right) \pm \frac{h}{2}\boldsymbol{n}, \left(\xi^1, \xi^2\right) \in D,$$
$$\Pi_b\colon \boldsymbol{r} = \boldsymbol{r}_0\left(\xi^1, \xi^2\right) + z\boldsymbol{n}, \ \left(\xi^1, \xi^2\right) \in \partial D, \ -\frac{h}{2} \le z \le \frac{h}{2}.$$

The basis of the space tangent to the middle surface (tangent plane) $P = \partial \Pi$ is determined as follows:

$$\boldsymbol{\pi}_j = \frac{\partial \boldsymbol{r}_0}{\partial \xi^j}. \tag{11.1}$$

Here and in what follows, unless otherwise specified, the Roman indices take the values 1 and 2. The summation over the Greek indices is not made.

In domain G, the curvilinear coordinates ξ^1, ξ^2, z are introduced, as follows:

$$\boldsymbol{r} = \boldsymbol{r}_0\left(\xi^1, \xi^2\right) + z\boldsymbol{n}\left(\xi^1, \xi^2\right), \left(\xi^1, \xi^2\right) \in D, \ -\frac{h}{2} \le z \le \frac{h}{2}. \tag{11.2}$$

In accordance with Eqs. (11.1) and (11.2), the spatial basis $\boldsymbol{e}_1, \boldsymbol{e}_2, \boldsymbol{e}_3$ is determined as follows:

$$\boldsymbol{e}_j = \frac{\partial \boldsymbol{r}}{\partial \xi^j} = \boldsymbol{\pi}_j + z\frac{\partial \boldsymbol{n}}{\partial \xi^j}, \ \boldsymbol{e}_3 = \frac{\partial \boldsymbol{r}}{\partial z} = \boldsymbol{n}.$$

We note also that, for the components $\hat{g}^{ij}$ of the metric tensor in the basis $\boldsymbol{\pi}_1, \boldsymbol{\pi}_2, \boldsymbol{n}$, there hold the equalities [5, 25] $\hat{g}^{ij} = g^{ij}, \ \hat{g}^{13} = \hat{g}^{23} = 0, \ \hat{g}^{33} = 1$, where g_{ij} are the components of the metric tensor of surface Π.

Then for the vectors $\boldsymbol{u}$ and the second-rank tensors $\boldsymbol{d}$ we will use the following notation:

$$\boldsymbol{u} = \tilde{u}^i\boldsymbol{e}_i + \tilde{u}^3\boldsymbol{n} = \tilde{u}_i\boldsymbol{e}^i + u_3\boldsymbol{n} = \hat{u}^j\boldsymbol{\pi}_j + \hat{u}^3\boldsymbol{n} = \hat{u}_j\boldsymbol{\pi}^j + u_3\boldsymbol{n}, \ \hat{u}^3 = \tilde{u}^3 = u_3,$$
$$\boldsymbol{d} = \tilde{d}^{ij}\boldsymbol{e}_i\boldsymbol{e}_j + \tilde{d}^{i3}\boldsymbol{e}_i\boldsymbol{n} + \tilde{d}^{3j}\boldsymbol{n}\boldsymbol{e}_j + \tilde{d}^{33}\boldsymbol{n}\boldsymbol{n} = \tilde{d}_{ij}\boldsymbol{e}^i\boldsymbol{e}^j + \tilde{d}_{i3}\boldsymbol{e}^i\boldsymbol{n} + \tilde{d}_{3j}\boldsymbol{n}\boldsymbol{e}^j + \tilde{d}_{33}\boldsymbol{n}\boldsymbol{n} =$$
$$= \hat{d}^{ij}\boldsymbol{\pi}_i\boldsymbol{\pi}_j + \hat{d}^{i3}\boldsymbol{\pi}_i\boldsymbol{n} + \hat{d}^{3j}\boldsymbol{n}\boldsymbol{\pi}_j + \hat{d}^{33}\boldsymbol{n}\boldsymbol{n} = \hat{d}_{ij}\boldsymbol{\pi}^i\boldsymbol{\pi}^j + \hat{d}_{i3}\boldsymbol{\pi}^i\boldsymbol{n} + \hat{d}_{3j}\boldsymbol{n}\boldsymbol{\pi}^j + \hat{d}_{33}\boldsymbol{n}\boldsymbol{n}.$$

In constructing initial and boundary value problems for shells, we will use the Hamilton functional for three-dimensional elastic moment shells, which can be written in different forms [1–4, 25]. Here, we will write it as follows:

$$H(\boldsymbol{u},\boldsymbol{\omega}) = \int\limits_{t_1}^{t_2} (I - E)\,dt,\quad I(\boldsymbol{u},\boldsymbol{\omega}) = W - A \ (t_2 > t_1)\,, \tag{11.3}$$

where

$$W = \frac{1}{2}\int\limits_G \left(\sigma^{ij}\gamma_{ij} + \mu^{ij}\chi_{ij}\right)d\boldsymbol{y},\ \ E = \frac{1}{2}\int\limits_G \left(\rho v^2 + J\boldsymbol{\Omega}^2\right)d\boldsymbol{y},\ \ \boldsymbol{v} = \dot{\boldsymbol{u}},\ \ \boldsymbol{\Omega} = \dot{\boldsymbol{\omega}},$$

$$A = \int\limits_G \rho\left[(\boldsymbol{F},\boldsymbol{u}) + (\boldsymbol{M},\boldsymbol{\omega})\right]d\boldsymbol{y} + \iint\limits_{\partial G}\left[(\boldsymbol{p}^{\nu},\boldsymbol{u}) + (\boldsymbol{m}^{\nu},\boldsymbol{\omega})\right]dS.$$

Here, E and W are the kinetic and potential energies of the body, respectively, A is the work of external forces, ρ is the density, J is the mass measure of medium inertia under rotation, $\boldsymbol{u} = u^i\boldsymbol{e}_i$ and $\boldsymbol{\omega} = \omega^i\boldsymbol{e}_i$ are the displacement and rotation angle vectors, respectively, t is time, $\boldsymbol{\gamma} = \gamma^{ij}\boldsymbol{e}_i\boldsymbol{e}_j$ and $\boldsymbol{\chi} = \chi^{ij}\boldsymbol{e}_i\boldsymbol{e}_j$ are the strain and bending/torsion tensors, respectively, $\boldsymbol{\sigma} = \sigma^{ij}\boldsymbol{e}_i\boldsymbol{e}_j$ and $\boldsymbol{\mu} = \mu^{ij}\boldsymbol{e}_i\boldsymbol{e}_j$ are the stress and moment stress tensors, respectively, $\boldsymbol{F} = F^j\boldsymbol{e}_j$ and $\boldsymbol{M} = M^j\boldsymbol{e}_j$ are the bulk force and moment, respectively, and $\boldsymbol{p}^{\nu} = \sigma^{ij}\nu_i\boldsymbol{e}_j$ and $\boldsymbol{m}^{\nu} = \mu^{ij}\nu_i\boldsymbol{e}_j$ are the stress and moment stress vectors, respectively, on an area with the unit outward normal vector $\boldsymbol{\nu} = \nu^i\boldsymbol{e}_i$; the dots refer to time derivatives. In these qualities, and below in this section, the Roman indices take the values 1, 2, and 3.

The components of the strain and bending/torsion tensors are related with the coordinates of the displacement and rotation angle vectors by the equations $\gamma_{ij} = \nabla_i u_j - \varepsilon_{kij}\omega^k$, $\chi_{ij} = \nabla_i\omega_j$, where ε_{kij} are the components of the Levi-Civita pseudotensor [25].

The general form of the physical law is constructed using the linearization of the free energy and takes the form:

$$\sigma^{ij} = C^{ijkl}\gamma_{kl} + D^{ijkl}\chi_{kl},\ \ \mu^{ij} = B^{ijkl}\chi_{kl} + D^{ijkl}\gamma_{kl}. \tag{11.4}$$

Here, C^{ijkl}, B^{ijkl}, D^{ijkl} are the physical constants of the medium or the components of fourth-rank tensors having the following symmetry:

$$C^{ijkl} = C^{klij},\quad B^{ijkl} = B^{klij},\quad D^{ijkl} = D^{klij}.$$

For the isotropic Cosserat medium, these tensors and the physical law can be written in the form:

$$\begin{gathered}
C^{ijkl} = (\mu+\alpha)\,g^{ik}g^{jl} + (\mu-\alpha)\,g^{jk}g^{il} + \lambda g^{ij}g^{kl},\\
B^{ijkl} = (\gamma+\varepsilon)\,g^{ik}g^{jl} + (\gamma-\varepsilon)\,g^{jk}g^{il} + \beta g^{ij}g^{kl},\quad D^{ijkl} = 0,\\
\sigma_{ij} = (\mu+\alpha)\,\gamma_{ij} + (\mu-\alpha)\,\gamma_{ji} + \lambda g_{ij}\mathrm{div}\boldsymbol{u},\\
\mu_{ij} = (\gamma+\varepsilon)\,\chi_{ij} + (\gamma-\varepsilon)\,\chi_{ji} + \beta g_{ij}\mathrm{div}\boldsymbol{\omega}.
\end{gathered} \tag{11.5}$$

Here, $\lambda,\ \mu$ are the elastic Lamé constants, while $\alpha,\ \beta, \gamma,\ \varepsilon$ are some additional physical parameters of the medium in the presence of moment effects. The last four parameters satisfy the following restrictions [3, 4]:

$$\alpha, \gamma, \varepsilon \geq 0,\ \ 3\beta + 2\gamma \geq 0,\ \ \mu\,(2\gamma + \beta) = (\alpha + \mu)\,(\gamma + \varepsilon)\,.$$

Using the methods of calculus of variations, the following assertion can be fairly easily demonstrated.

Assertion 1 The necessary conditions for an extremum of functional (11.3), when the conditions $\delta \boldsymbol{u}|_{t=t_1,t_2} = \mathbf{0}$, $\delta \boldsymbol{\omega}|_{t=t_1,t_2} = \mathbf{0}$ are fulfilled (the symbol δ refers to a variation), are the equations of motion:

$$\rho \frac{\partial^2 u^j}{\partial t^2} = \nabla_i \sigma^{ij} + \rho F^j,\ \ J \frac{\partial^2 \omega^j}{\partial t^2} = \nabla_i \mu^{ij} + \varepsilon^{jkl} \sigma_{kl} + \rho M^j$$

and the boundary conditions

$$u_i|_{\Pi_u} = u_{i0},\ \ \omega_i|_{\Pi_\omega} = \omega_{i0},\ \ \sigma^{ji} \nu_j\big|_{\Pi_\sigma} = P^i,\ \ \mu^{ji} \nu_j\big|_{\Pi_\mu} = K^i.$$

Here, $u_{i0}, \omega_{i0}, P^i, K^i$ are given functions and $\partial G = \Pi_u \cup \Pi_\sigma = \Pi_\omega \cup \Pi_\mu$ where the surfaces Π_u and Π_σ as well as Π_ω and Π_μ, intersect in a null-measure set.

11.3 Hamiltonian for an Elastic Moment Shell

Analogously to [22–24], we will use the direct normal hypothesis for the displacement field thus arriving at the following equalities:

$$\begin{aligned} \hat{u}_i\left(\xi^1, \xi^2, z\right) &= u_i\left(\xi^1, \xi^2\right) + \psi_i\left(\xi^1, \xi^2\right) z, \\ u_3\left(\xi^1, \xi^2, z\right) &= w\left(\xi^1, \xi^2\right) + \psi_3\left(\xi^1, \xi^2\right) z; \end{aligned} \tag{11.6}$$

$$\hat{\nabla}_i \hat{u}_j = \alpha_{ij} + z\beta_{ij},\ \ \hat{\nabla}_i u_3 = -\vartheta_i + \left(b_i^k \theta_k + \nabla_i \psi_3\right) z,\ \ \hat{\nabla}_3 \hat{u}_i = \psi_i,\ \hat{\nabla}_3 u_3 = \psi_3, \tag{11.7}$$

where

$$\begin{aligned} \alpha_{ij} = \nabla_i u_j - b_{ij} w,\ \ -\vartheta_i &= \nabla_i w + b_i^k u_k,\ \ \theta_k = \psi_k - \vartheta_k, \\ \beta_{ij} &= \nabla_i \psi_j - b_{ij} \psi_3 + b_i^k \nabla_k u_j - c_{ij} w = \\ &= \nabla_i \psi_j - b_{ij} \psi_3 + b_i^k \alpha_{kj},\ \ c_{ij} = b_{ik} b_j^k. \end{aligned} \tag{11.8}$$

Here, ∇_i is the operator of covariant differentiation on the surface Π; b_{ij} are the components of the tensor of the middle surface curvature; u_i and w are the tangential and normal displacements, respectively; and the quantities ψ_i determine the angle between the vector $\boldsymbol{n}$ and the directing normal unit vector in the deformed state.

As it is agreed in the theory of shells, here and in what follows, the terms nonlinear in the coordinate z in these equalities are discarded and approximate equalities are replaced by exact ones.

We assume that, for the rotation angle vector, there exist relations analogous to Eqs. (11.6)–(11.8):

$$\begin{aligned}\hat{\omega}_i\left(\xi^1,\xi^2,z\right)&=\omega_i\left(\xi^1,\xi^2\right)+\varphi_i\left(\xi^1,\xi^2\right)z,\\ \omega_3\left(\xi^1,\xi^2,z\right)&=\omega\left(\xi^1,\xi^2\right)+\varphi_3\left(\xi^1,\xi^2\right)z;\end{aligned}\tag{11.9}$$

$$\hat{\nabla}_i\hat{\omega}_j=\eta_{ij}+\lambda_{ij}z,\ \ \hat{\nabla}_i\omega_3=\eta_{i3}+\lambda_{i3}z,\ \ \hat{\nabla}_3\hat{\omega}_i=\varphi_i,\ \ \hat{\nabla}_3\omega_3=\varphi_3,\tag{11.10}$$

where

$$\begin{aligned}\eta_{ij}&=\nabla_i\omega_j-b_{ij}\omega,\ \ \eta_{i3}=\nabla_i\omega+b_i^k\omega_k,\ \ \lambda_{i3}=b_i^k\left(\varphi_k+\eta_{k3}\right)+\nabla_i\varphi_3,\\ \lambda_{ij}&=\nabla_i\varphi_j-b_{ij}\varphi_3+b_i^k\nabla_k\omega_j-c_{ij}\omega=\nabla_i\varphi_j-b_{ij}\varphi_3+b_i^k\eta_{kj}.\end{aligned}\tag{11.11}$$

Then the components of the strain and bending/torsion tensors take the form:

$$\hat{\chi}_{ij}=\eta_{ij}+\lambda_{ij}z,\ \ \hat{\chi}_{i3}=\eta_{i3}+\lambda_{i3}z,\ \ \hat{\chi}_{3i}=\varphi_i,\ \ \hat{\chi}_{33}=\varphi_3;\tag{11.12}$$

$$\hat{\gamma}_{ij}=\xi_{ij}+\zeta_{ij}z,\ \ \hat{\gamma}_{i3}=\xi_{i3}+\zeta_{i3}z,\ \ \hat{\gamma}_{3i}=\xi_{3i}+\zeta_{3i}z,\ \ \hat{\gamma}_{33}=\psi_3,\tag{11.13}$$

where

$$\begin{aligned}\xi_{ij}&=\alpha_{ij}-\kappa_{ij}\omega=\nabla_i u_j-b_{ij}w-\kappa_{ij}\omega,\\ \xi_{i3}&=-\vartheta_i-\kappa_{ki}\omega^k=\nabla_i w+b_i^k u_k-\kappa_{ki}\omega^k,\ \ \xi_{3i}=\psi_i+\kappa_{ki}\omega^k,\\ \zeta_{ij}&=\beta_{ij}-\kappa_{ij}\varphi_3=\nabla_i\psi_j-b_{ij}\psi_3+b_i^k\nabla_k u_j-c_{ij}w-\kappa_{ij}\varphi_3,\\ \zeta_{i3}&=b_i^k\theta_k+\nabla_i\psi_3-\kappa_{ki}\varphi^k=b_i^k\left(\psi_k+\nabla_k w\right)+c_i^k u_k+\nabla_i\psi_3-\kappa_{ki}\varphi^k,\\ \zeta_{3i}&=\kappa_{ki}\varphi^k.\end{aligned}\tag{11.14}$$

Here, $\kappa_{jk}=\varepsilon_{jk3}$ are the components of a discriminant pseudotensor on the surface Π.

Then, using Eqs. (11.6)–(11.14) and the procedure of the passage from the triple over G integrals and the surface integrals over $\Pi_\pm$ to the integrals over the middle surface Π which is standard in theory of shells, and the passage from the integral Π_b to the curvilinear integral over the curve $\Gamma=\partial\Pi$, we can write down the components of the functional (11.3) as follows:

$$\begin{aligned}W&=W_0\left(\boldsymbol{u},\boldsymbol{\psi},\boldsymbol{\omega},\boldsymbol{\varphi},w,\psi_3,\omega,\varphi_3\right)=\\ &=\frac{1}{2}\iint\limits_{\Pi}\Big(\hat{T}^{ij}\xi_{ij}+\hat{T}^{i3}\xi_{i3}+\hat{T}^{3i}\xi_{3i}+\hat{T}^{33}\psi_3+\hat{M}^{ij}\zeta_{ij}+\hat{M}^{i3}\zeta_{i3}+\\ &+\hat{M}^{3i}\zeta_{3i}+\hat{R}^{ij}\eta_{ij}+\hat{R}^{i3}\eta_{i3}+\hat{R}^{3i}\varphi_i+\hat{R}^{33}\varphi_3+S^{ij}\lambda_{ij}+S^{i3}\lambda_{i3}\Big)dS;\end{aligned}\tag{11.15}$$

$$
\begin{aligned}
&E = E_0\left(\boldsymbol{u}, \boldsymbol{\psi}, \boldsymbol{\omega}, \boldsymbol{\varphi}, w, \psi_3, \omega, \varphi_3\right) = \\
&= \frac{\rho}{2} \iint\limits_{\Pi} \left[h\left(\dot{u}_i \dot{u}^i + \dot{w}^2\right) + I\left(\dot{\psi}_i \dot{\psi}^i + \partial \dot{\psi}_3^2\right)\right] dS + \\
&+ \frac{J}{2} \iint\limits_{\Pi} \left[h\left(\dot{\omega}_i \dot{\omega}^i + \partial \dot{\omega}^2\right) + I\left(\partial \dot{\varphi}_i \partial \dot{\varphi}^i + \partial \dot{\varphi}_3^2\right)\right] dS, \quad I = h^3/12;
\end{aligned}
\tag{11.16}
$$

$$
\begin{aligned}
&A = A_0\left(\boldsymbol{u}, \boldsymbol{\psi}, \boldsymbol{\omega}, \boldsymbol{\varphi}, w, \psi_3, \omega, \varphi_3\right) = \\
&= \iint\limits_{\Pi} \left[(\boldsymbol{q}, \boldsymbol{u}) + (\boldsymbol{m}, \boldsymbol{\psi}) + q w + m \psi_3 + (\boldsymbol{m}_M, \boldsymbol{\omega}) + \right. \\
&\left. + (\boldsymbol{m}_{2M}, \boldsymbol{\varphi}) + \tilde{m}_M \omega + \tilde{m}_{2M} \varphi_3\right] dS + \\
&+ \int\limits_{\Gamma} \left[\left(\boldsymbol{T}_{(0)}, \boldsymbol{u}\right) + \left(\boldsymbol{M}_{(0)}, \boldsymbol{\psi}\right) + T_{(0)} w + M_{(0)} \psi_3 + \right. \\
&\left. + \left(\boldsymbol{R}_{(0)}, \boldsymbol{\omega}\right) + \left(\boldsymbol{S}_{(0)}, \boldsymbol{\varphi}\right) + R_{(0)} \omega + S_{(0)} \varphi_3\right] ds,
\end{aligned}
\tag{11.17}
$$

where

$$
\begin{gathered}
\hat{T}^{ij} = \int\limits_{-h/2}^{h/2} \hat{\sigma}^{ij} dz, \;\; M^{ij} = \int\limits_{-h/2}^{h/2} z \hat{\sigma}^{ij} dz, \;\; \hat{T}^{i3} = \int\limits_{-h/2}^{h/2} \hat{\sigma}^{i3} dz, \;\; \hat{T}^{3i} = \int\limits_{-h/2}^{h/2} \hat{\sigma}^{3i} dz, \\
M^{i3} = \int\limits_{-h/2}^{h/2} z \hat{\sigma}^{i3} dz, \;\; M^{3i} = \int\limits_{-h/2}^{h/2} z \hat{\sigma}^{3i} dz, \;\; \hat{T}^{33} = \int\limits_{-h/2}^{h/2} \hat{\sigma}^{33} dz, \\
S^{i3} = \int\limits_{-h/2}^{h/2} z \hat{\mu}^{i3} dz, \;\; \hat{R}^{ij} = \int\limits_{-h/2}^{h/2} \hat{\mu}^{ij} dz, \;\; S^{ij} = \int\limits_{-h/2}^{h/2} z \hat{\mu}^{ij} dz, \\
\hat{R}^{i3} = \int\limits_{-h/2}^{h/2} \hat{\mu}^{i3} dz, \;\; \hat{R}^{3i} = \int\limits_{-h/2}^{h/2} \hat{\mu}^{3i} dz, \;\; \hat{R}^{33} = \int\limits_{-h/2}^{h/2} \hat{\mu}^{33} dz;
\end{gathered}
\tag{11.18}
$$

$$
\begin{gathered}
\boldsymbol{q} = q^i \boldsymbol{\pi}_i, \;\; \boldsymbol{m} = m^i \boldsymbol{\pi}_i, \;\; \boldsymbol{m}_M = \tilde{m}_M^i \boldsymbol{\pi}_i, \;\; \boldsymbol{m}_{2M} = \tilde{m}_{2M}^i \boldsymbol{\pi}_i, \\
q^i = q_F^i + q_+^i + q_-^i, \;\; m^i = m_F^i + m_+^i + m_-^i, \;\; q = q_F + q_+ + q_-, \;\; m = m_F + m_+ + m_-, \\
\tilde{m}_M^i = m_M^i + m_{M\pm}^i, \;\; \tilde{m}_{2M}^i = m_{2M}^i + m_{2M\pm}^i, \;\; \tilde{m}_M = m_M + m_{M\pm}, \;\; \tilde{m}_{2M} = m_{2M} + m_{2M\pm}, \\
q_\pm^i = \left.\hat{P}^i\right|_{z=\pm h/2}, \;\; q_\pm = \left.\hat{P}^3\right|_{z=\pm h/2}, \;\; m_\pm^i = \pm \frac{h}{2} q_\pm^i, \;\; m_\pm = \pm \frac{h}{2} q_\pm \\
m_{M\pm}^i = \left.\hat{K}^i\right|_{z=\pm h/2}, \;\; m_{M\pm} = \left.\hat{K}^3\right|_{z=\pm h/2}, \;\; m_{2M\pm}^i = \pm \frac{h}{2} q_{M\pm}^i, \;\; m_{2M\pm} = \pm \frac{h}{2} q_{M\pm},
\end{gathered}
$$

$$q_F^i = \rho \int\limits_{-h/2}^{h/2} \hat{F}^i dz,\; q_F = \rho \int\limits_{-h/2}^{h/2} F^3 dz,\; m_F^i = \rho \int\limits_{-h/2}^{h/2} z\hat{F}^i dz,\; m_F = \rho \int\limits_{-h/2}^{h/2} zF^3 dz,$$
$$m_M^i = \rho \int\limits_{-h/2}^{h/2} \hat{M}^i dz,\; m_M = \rho \int\limits_{-h/2}^{h/2} M^3 dz,\; m_{2M}^i = \rho \int\limits_{-h/2}^{h/2} z\hat{M}^{ii} dz,\; m_{2M} = \rho \int\limits_{-h/2}^{h/2} zM^3 dz,$$
$$T_{(0)}^i = \int\limits_{-h/2}^{h/2} \hat{P}^i dz,\; M_{(0)}^i = \int\limits_{-h/2}^{h/2} z\hat{P}^i dz,\; T_{(0)} = \int\limits_{-h/2}^{h/2} \hat{P}^3 dz,\; M_{(0)} = \int\limits_{-h/2}^{h/2} z\hat{P}^3 dz,$$
$$R_{(0)}^i = \int\limits_{-h/2}^{h/2} \hat{K}^i dz,\; S_{(0)}^i = \int\limits_{-h/2}^{h/2} z\hat{K}^i dz,\; R_{(0)} = \int\limits_{-h/2}^{h/2} \hat{K}^3 dz,\; S_{(0)} = \int\limits_{-h/2}^{h/2} z\hat{K}^3 dz.$$

Here, $\hat{\boldsymbol{T}} = \hat{T}^{ij}\boldsymbol{\pi}_i\boldsymbol{\pi}_j$ and $\boldsymbol{M} = M^{ij}\boldsymbol{\pi}_i\boldsymbol{\pi}_j$ are the tensors of tangential forces and moments, respectively, $\hat{T}^{i3}\boldsymbol{\pi}_i$, $\hat{T}^{3i}\boldsymbol{\pi}_i$ and $M^{i3}\boldsymbol{\pi}_i$, $M^{3i}\boldsymbol{\pi}_i$ are the vectors of shear stresses and additional moments, respectively, and $\hat{T}^{33}$ is the normal force. We retain the analogous names for the quantities $\hat{\boldsymbol{R}} = \hat{R}^{ij}\boldsymbol{\pi}_i\boldsymbol{\pi}_j$ and $\boldsymbol{S} = S^{ij}\boldsymbol{\pi}_i\boldsymbol{\pi}_j$, $\hat{R}^{i3}\boldsymbol{\pi}_i$, $\hat{R}^{3i}\boldsymbol{\pi}_i$ and $S^{i3}\boldsymbol{\pi}_i$ and $\hat{R}^{33}$ supplementing them with the adjective "moment". The vectors $\boldsymbol{q} + q\boldsymbol{n}$ and $\boldsymbol{m} + m\boldsymbol{n}$, $\boldsymbol{m}_M + \tilde{m}_M\boldsymbol{n}$ have the meaning of the surface pressure and the moments per unit area, while $\boldsymbol{m}_{2M} + \tilde{m}_{2M}\boldsymbol{n}$ is the analogous second-order moment. The vectors $\boldsymbol{u}$, $\boldsymbol{\psi}$, $\boldsymbol{\omega}$ and $\boldsymbol{\varphi}$ are understood to mean the quantities $\boldsymbol{u} = u_i\boldsymbol{\pi}^i$, $\boldsymbol{\psi} = \psi_i\boldsymbol{\pi}^i$, $\boldsymbol{\omega} = \omega_i\boldsymbol{\pi}^i$, and $\boldsymbol{\varphi} = \varphi_i\boldsymbol{\pi}^i$ on the surface Π. Thus, the Hamiltonian for an elastic moment shell has the following form:

$$H_0(\boldsymbol{u}, \boldsymbol{\psi}, \boldsymbol{\omega}, \boldsymbol{\varphi}, w, \psi_3, \omega, \varphi_3) = \int\limits_{t_1}^{t_2} (I_0 - E_0)\, dt,\;\; I_0 = W_0 - A_0\,(t_2 > t_1)\,, \tag{11.19}$$

where W_0, E_0, and A_0 are determined by Eqs. (11.15)–(11.17).

11.4 Physical Law for an Elastic Moment Shell

We assume that the shell material possesses the symmetry about the middle surface, which is equivalent to the following equalities for the components of tensors of physical constants (here, the sign pointing to their correspondence with the basis $\boldsymbol{\pi}_1, \boldsymbol{\pi}_2, \boldsymbol{n}$ is omitted):

$$C^{ijk3} = C^{ij3k} = C^{i333} = C^{3i33} = 0,\;\; B^{ijk3} = B^{ij3k} = B^{i333} = B^{3i33} = 0,$$
$$D^{ijk3} = D^{ij3k} = D^{i333} = D^{3i33} = 0.$$

Rewriting Eqs. (11.4) in terms of Eqs. (11.12) and (11.13), we obtain the laws of stress variation throughout the shell thickness

$$\begin{aligned}
\hat{\sigma}^{ij} &= C^{ijkl}\left(\xi_{kl} + z\zeta_{kl}\right) + C^{ij33}\psi_3 + D^{ijkl}\left(\eta_{kl} + \lambda_{kl}z\right) + D^{ij33}\varphi_3,\\
\hat{\sigma}^{i3} &= C^{i33l}\left(\xi_{3l} + z\zeta_{3l}\right) + C^{i3k3}\left(\xi_{k3} + z\zeta_{k3}\right) + D^{i3k3}\left(\eta_{k3} + \lambda_{k3}z\right) + D^{i33k}\varphi_k,\\
\hat{\sigma}^{3i} &= C^{3i3l}\left(\xi_{3l} + z\zeta_{3l}\right) + C^{3ik3}\left(\xi_{k3} + z\zeta_{k3}\right) + D^{3ik3}\left(\eta_{k3} + \lambda_{k3}z\right) + D^{3i3k}\varphi_k,\\
\hat{\sigma}^{33} &= C^{33kl}\left(\xi_{kl} + z\zeta_{kl}\right) + C^{3333}\psi_3 + D^{33kl}\left(\eta_{kl} + \lambda_{kl}z\right) + D^{3333}\varphi_3,\\
\hat{\mu}^{ij} &= D^{ijkl}\left(\xi_{kl} + z\zeta_{kl}\right) + D^{ij33}\psi_3 + B^{ijkl}\left(\eta_{kl} + \lambda_{kl}z\right) + B^{ij33}\varphi_3,\\
\hat{\mu}^{i3} &= D^{i33l}\left(\xi_{3l} + z\zeta_{3l}\right) + D^{i3k3}\left(\xi_{k3} + z\zeta_{k3}\right) + B^{i33l}\varphi_l + B^{i3k3}\left(\eta_{k3} + \lambda_{k3}z\right),\\
\hat{\mu}^{3j} &= D^{3j3l}\left(\xi_{3l} + z\zeta_{3l}\right) + D^{3jk3}\left(\xi_{k3} + z\zeta_{k3}\right) + B^{3j3l}\varphi_l + B^{3jk3}\left(\eta_{k3} + \lambda_{k3}z\right),\\
\hat{\mu}^{33} &= D^{33kl}\left(\xi_{kl} + z\zeta_{kl}\right) + D^{3333}\psi_3 + B^{33kl}\left(\eta_{kl} + \lambda_{kl}z\right) + B^{3333}\varphi_3.
\end{aligned}$$

Considering these equalities in combination with Eqs. (11.18) leads to a relation between the internal force factors and the kinematic parameters of the shell, that is, to the physical law:

$$\begin{aligned}
\hat{T}^{ij} &= h\left(C^{ijkl}\xi_{kl} + C^{ij33}\psi_3 + D^{ijkl}\eta_{kl} + D^{ij33}\varphi_3\right),\\
M^{ij} &= I\left(C^{ijkl}\zeta_{kl} + D^{ijkl}\lambda_{kl}\right),\\
\hat{T}^{i3} &= h\left(C^{i33l}\xi_{3l} + C^{i3k3}\xi_{k3} + D^{i33k}\varphi_k + D^{i3k3}\eta_{k3}\right),\\
M^{i3} &= I\left(C^{i33l}\zeta_{3l} + C^{i3k3}\zeta_{k3} + D^{i3k3}\lambda_{k3}\right),\\
\hat{T}^{3i} &= h\left(C^{3i3l}\xi_{3l} + C^{3ik3}\xi_{k3} + D^{3i3k}\varphi_k + D^{3ik3}\eta_{k3}\right),\\
M^{3i} &= I\left(C^{3i3l}\zeta_{3l} + C^{3ik3}\zeta_{k3} + D^{3ik3}\lambda_{k3}\right),\\
\hat{T}^{33} &= h\left(C^{33kl}\xi_{kl} + C^{3333}\psi_3 + D^{33kl}\eta_{kl} + D^{3333}\varphi_3\right);
\end{aligned} \tag{11.20}$$

$$\begin{aligned}
\hat{R}^{ij} &= h\left(D^{ijkl}\xi_{kl} + D^{ij33}\psi_3 + B^{ijkl}\eta_{kl} + B^{ij33}\varphi_3\right),\\
S^{ij} &= I\left(D^{ijkl}\zeta_{kl} + B^{ijkl}\lambda_{kl}\right),\\
\hat{R}^{i3} &= h\left(D^{i33l}\xi_{3l} + D^{i3k3}\xi_{k3} + B^{i33k}\varphi_k + B^{i3k3}\eta_{k3}\right),\\
S^{i3} &= I\left(D^{i33l}\zeta_{3l} + D^{i3k3}\zeta_{k3} + B^{i3k3}\lambda_{k3}\right),\\
\hat{R}^{3i} &= h\left(D^{3i3l}\xi_{3l} + D^{3ik3}\xi_{k3} + B^{3i3l}\varphi_l + B^{3ik3}\eta_{kl}\right),\\
\hat{R}^{33} &= h\left(D^{33kl}\xi_{kl} + D^{3333}\psi_3 + B^{33kl}\eta_{kl} + B^{3333}\varphi_3\right).
\end{aligned} \tag{11.21}$$

We note that the tensors of the internal force factors having the components $\hat{T}^{ij}$, M^{ij}, $\hat{R}^{ij}$, S^{ij} are nonsymmetric. Letting the rotation angle vector in Eq. (11.20) be zero and assuming that the quantities C^{ijkl} $(i, j, k, l = 1, 2, 3)$ are symmetric with respect to the first and second pairs of indices we arrive, accurate to certain designations, to the formulas for the internal force factors of an anisotropic elastic shell derived in [23].

In the case of an isotropic material, the physical law is simplified, when Eq. (11.5) is used:

$$\begin{aligned}
\hat{T}^{ij} &= h\left[(\mu + \alpha)\,\xi^{ij} + (\mu - \alpha)\,\xi^{ji} + \lambda g^{ij}\left(\varepsilon_s + \psi_3\right)\right],\\
M^{ij} &= I\left[(\mu + \alpha)\,\zeta^{ij} + (\mu - \alpha)\,\zeta^{ji} + \lambda g^{ij}\kappa_s\right],\\
\hat{T}^{33} &= h\left[\lambda\varepsilon_s + (\lambda + 2\mu)\,\psi_3\right],\\
\hat{T}^{i3} &= hg^{ik}\left[(\mu - \alpha)\,\xi_{3k} + (\mu + \alpha)\,\xi_{k3}\right],\\
M^{i3} &= Ig^{ik}\left[(\mu - \alpha)\,\zeta_{3k} + (\mu + \alpha)\,\zeta_{k3}\right],\\
\hat{T}^{3i} &= hg^{ik}\left[(\mu + \alpha)\,\xi_{3k} + (\mu - \alpha)\,\xi_{k3}\right],\\
M^{3i} &= Ig^{ik}\left[(\mu + \alpha)\,\zeta_{3k} + (\mu - \alpha)\,\zeta_{k3}\right];
\end{aligned} \tag{11.22}$$

$$
\begin{aligned}
\hat{R}^{ij} &= h\left[(\gamma+\varepsilon)\,\eta^{ij} + (\gamma-\varepsilon)\,\eta^{ji} + \beta g^{ij}\,(\eta_s+\varphi_3)\right],\\
S^{ij} &= I\left[(\gamma+\varepsilon)\,\lambda^{ij} + (\gamma-\varepsilon)\,\lambda^{ji} + \beta g^{ij}\lambda_s\right],\\
\hat{R}^{i3} &= hg^{ik}\left[(\gamma+\varepsilon)\,\eta_{k3} + (\gamma-\varepsilon)\,\varphi_k\right],\; S^{i3} = I\,(\gamma+\varepsilon)\,g^{ik}\lambda_{k3},\\
\hat{R}^{3i} &= hg^{ik}\left[(\gamma+\varepsilon)\,\varphi_k + (\gamma-\varepsilon)\,\eta_{k3}\right],\; \hat{R}^{33} = h\left[\beta\eta_s + (\beta+2\gamma)\,\varphi_3\right].
\end{aligned}
\tag{11.23}
$$

Here, the following equalities are used:

$$
\begin{aligned}
\varepsilon_s = g^{kl}\xi_{kl} = g^{kl}\alpha_{kl} &= \mathrm{div}\boldsymbol{u} - 2Hw,\;\; \eta_s = g^{kl}\eta_{kl} = \mathrm{div}\boldsymbol{\omega} - 2H\omega,\;\; g^{kl}\pi_{kl} = 0,\\
\kappa_s = g^{kl}\zeta_{kl} = g^{kl}\beta_{kl} &= \nabla_k\psi^k - b_{kl}g^{kl}\psi_3 + b^{kl}\nabla_k u_l - c_{kl}g^{kl}w =\\
&= \mathrm{div}\boldsymbol{\psi} - 2H\psi_3 + b^{kl}\nabla_k u_l - 2\left(2H^2-K\right)w,\\
\lambda_s = g^{kl}\lambda_{kl} &= \nabla_k\varphi^k - b_{kl}g^{kl}\varphi_3 + b^{kl}\nabla_k\omega_l - c_{kl}g^{kl}\omega =\\
&= \mathrm{div}\boldsymbol{\varphi} - 2H\varphi_3 + b^{kl}\nabla_k\omega_l - 2\left(2H^2-K\right)\omega,
\end{aligned}
$$

where $H = b_{ij}g^{ij}/2$ and $K = \det\left(b^i_j\right)$ are the mean and total (Gaussian) curvatures of the surface Π, respectively.

11.5 Initial and Boundary Value Problems for an Anisotropic Shell

Assertion 2 A necessary condition for an extremum of functional (11.19), when the following conditions are fulfilled:

$$
\delta\boldsymbol{u}|_{t=t_1,t_2} = \boldsymbol{0},\;\; \delta\boldsymbol{\psi}|_{t=t_1,t_2} = \boldsymbol{0},\;\; \delta\boldsymbol{\omega}|_{t=t_1,t_2} = \boldsymbol{0},\;\; \delta\boldsymbol{\varphi}|_{t=t_1,t_2} = \boldsymbol{0},
$$

$$
\delta w|_{t=t_1,t_2} = 0,\;\; \delta\psi_3|_{t=t_1,t_2} = 0,\;\; \delta\omega|_{t=t_1,t_2} = 0,\;\; \delta\varphi_3|_{t=t_1,t_2} = 0
$$

are the equations of motion

$$
\begin{aligned}
\rho h\frac{\partial^2 u^i}{\partial t^2} &= \nabla_j T^{ji} - b^i_k T^{k3} + q^i,\;\; \rho h\frac{\partial^2 w}{\partial t} = \nabla_j T^{j3} + b_{ij}T^{ij} + q,\\
\rho I\frac{\partial^2 \psi^i}{\partial t^2} &= \nabla_j M^{ji} - T^{3i} + m^i,\;\; \rho I\frac{\partial^2 \psi_3}{\partial t^2} = \nabla_j M^{j3} - N + m,\\
hJ\frac{\partial^2 \omega_i}{\partial t^2} &= g_{ik}\nabla_j R^{jk} - \pi_{ik}\left(T^{3k} - T^{k3}\right) - b_{ik}R^{k3} + \tilde{m}_{Mi},\\
hJ\frac{\partial^2 \omega}{\partial t} &= \nabla_j R^{j3} + \pi_{ik}\hat{T}^{ik} + b_{ij}R^{ij} + \tilde{m}_M,\\
IJ\frac{\partial^2 \varphi_i}{\partial t^2} &= g_{ik}\left(\nabla_j S^{jk} - R^{3k}\right) - \pi_{ik}\left(M^{3k} - M^{k3}\right) + \tilde{m}_{2Mi},\\
hJ\frac{\partial^2 \omega}{\partial t} &= \nabla_j R^{j3} + \pi_{ik}\hat{T}^{ik} + b_{ij}R^{ij} + \tilde{m}_M,\\
IJ\frac{\partial^2 \varphi_3}{\partial t^2} &= \nabla_j S^{j3} - N_\omega + \pi_{ij}M^{ij} + \tilde{m}_{2M},
\end{aligned}
\tag{11.24}
$$

where

$$T^{ij} = \hat{T}^{ij} + b^i_k M^{kj},\ T^{i3} = \hat{T}^{i3} + b^i_j M^{j3},\ T^{3i} = \hat{T}^{3i} + b^i_k M^{k3},\ N = \hat{T}^{33} - b_{ij} M^{ij},$$
$$R^{ij} = \hat{R}^{ij} + b^i_k S^{kj},\ R^{i3} = \hat{R}^{i3} + b^i_j S^{j3},\ R^{3i} = \hat{R}^{3i} + b^i_k S^{k3},\ N_\omega = \hat{R}^{33} - b_{ij} S^{ij}$$

and the boundary conditions

$$\begin{gathered} u_i|_{\Gamma_u} = u_{i0},\ \ w|_{\Gamma_u} = w_0,\ \ \psi_i|_{\Gamma_u} = \psi_{i0},\ \ \psi_3|_{\Gamma_u} = \psi_{30}, \\ \omega_i|_{\Gamma_\omega} = \omega_{i0},\ \ \omega|_{\Gamma_\omega} = \omega_0,\ \ \varphi_i|_{\Gamma_\omega} = \varphi_{i0},\ \ \varphi_3|_{\Gamma_\omega} = \varphi_{30}, \\ T^{ji}\nu_j\big|_{\Gamma_p} = T^i_{(0)},\ \ T^{j3}\nu_j\big|_{\Gamma_p} = T_{(0)},\ \ M^{ji}\nu_j\big|_{\Gamma_p} = M^i_{(0)}, \\ M^{j3}\nu_j\big|_{\Gamma_p} = M_{(0)},\ \ R^{ji}\nu_j\big|_{\Gamma_m} = R^i_{(0)}, \\ R^{j3}\nu_j\big|_{\Gamma_m} = R_{(0)},\ \ S^{ji}\nu_j\big|_{\Gamma_m} = S^i_{(0)},\ \ S^{j3}\nu_j\big|_{\Gamma_m} = S_{(0)}. \end{gathered} \tag{11.25}$$

Here, $u_{i0}, w_0, \psi_{i0}, \psi_{30}, \omega_{i0}, \omega_0, \varphi_{i0}, \varphi_{30}$ and $T^i_{(0)}, T_{(0)}, M^i_{(0)}, M_{(0)}, R^i_{(0)}, R_{(0)}, S^i_{(0)}$, $S_{(0)}$ are some given functions; $\Gamma = \Gamma_u \cup \Gamma_p = \Gamma_\omega \cup \Gamma_m$, where the curve pairs Γ_u, Γ_p and Γ_ω, Γ_m can have only points in common; and $\boldsymbol{\nu}$ is the unit vector of the outward normal to the lateral surface at $z = 0$ (at the line of intersection $\Pi_b \cap \Pi$)

Proof A necessary condition of an extremum of functional (11.19) is the variational equation

$$\delta H_0(\boldsymbol{u}, \boldsymbol{\psi}, \boldsymbol{\omega}, \boldsymbol{\varphi}, w, \psi_3, \omega, \varphi_3) = \int_{t_1}^{t_2} (\delta W_0 - \delta A_0)\, dt - \int_{t_1}^{t_2} \delta E_0 dt = 0. \tag{11.26}$$

□

In calculating the variation δW_0 we take into account that the integrand in Eq. (11.15), when Eqs. (11.22) and (11.23) are taken into account, is the linear form with the coefficients given by the following linear forms:

$$f(\boldsymbol{x}) = \sum_{i=1}^{n} g_i(\boldsymbol{x})\, x_i,\ \boldsymbol{x} = (x_1, x_2, \ldots, x_n),\ \ g_i(\boldsymbol{x}) = \sum_{j=1}^{n} a_{ij} x_j,\ \ (a_{ij} = \text{const}).$$

For these functions, we can easily prove the equality

$$\delta f(\boldsymbol{x}) = 2 \sum_{i=1}^{n} g_i(\boldsymbol{x})\, \delta x_i.$$

Using this equality, together with Eq. (11.14) and the generalized Ostrogradskii-Gauss formula for a surface [25], we obtain

$$
\begin{aligned}
\delta W_0 = \iint\limits_{\Pi} \{ & p - \left(b_k^i T^{k3} - \nabla_j T^{ji}\right) \delta u_i - \left(b_{ij} T^{ij} + \nabla_j T^{j3}\right) \delta w + \\
& + \left(T^{3i} - \nabla_j M^{ji}\right) \delta \psi_i + \left(N - \nabla_j M^{j3}\right) \delta \psi_3 + \\
& + \left[\pi_{ik} \left(T^{3k} - T^{k3}\right) + b_{ik} R^{k3} - g_{ik} \nabla_j R^{jk}\right] \delta \omega^i - \\
& - \left(\pi_{ik} \hat{T}^{ik} + b_{ij} R^{ij} + \nabla_j R^{j3}\right) \delta \omega + \\
& + \left[g_{ij} R^{3j} + \pi_{ik} \left(M^{3k} - M^{k3}\right) - g_{ik} \nabla_j S^{jk}\right] \delta \varphi^i + \\
& + \left(N_\omega - \pi_{ij} M^{ij} - \nabla_j S^{j3}\right) \delta \varphi_3 \} dS + \\
& + \int\limits_{\Gamma} \nu_j \left[T^{ji} \delta u_i + T^{j3} \delta w + M^{ji} \delta \psi_i + M^{j3} \delta \psi_3 + + \right. \\
& \left. + R^{ji} \delta \omega_i + R^{j3} \delta \omega + S^{ji} \delta \varphi_i + S^{j3} \delta \varphi_3 \right] ds.
\end{aligned}
\tag{11.27}
$$

Here, $\boldsymbol{\nu} = \nu^i \boldsymbol{\pi}_i = [\boldsymbol{\tau}, \boldsymbol{n}]$, where $\boldsymbol{\tau} = \tau^i \boldsymbol{\pi}_i$ is the unit tangent vector to the curve Γ. As shown in [23], it coincides with the unit vector of the outward normal to the lateral surface at $z = 0$.

The variation of the work of external forces is determined as follows:

$$
\begin{aligned}
\delta A_0 = \iint\limits_{\Pi} & \left[q^i \delta u_i + m^i \delta \psi_i + q \delta w + m \delta \psi_3 + \tilde{m}_M^i \delta \omega_i + \tilde{m}_{2M}^i \delta \varphi_i + \right. \\
& \left. + \tilde{m}_M \delta \omega + \tilde{m}_{2M} \delta \varphi_3 \right] dS + \int\limits_{\Gamma} \left[T_{(0)}^i \delta u_i + M_{(0)}^i \delta \psi_i + \right. \\
& \left. + T_{(0)} \delta w + M_{(0)} \psi_3 + R_{(0)}^i \delta \omega_i + S_{(0)}^i \delta \varphi_i + R_{(0)} \delta \omega + S_{(0)} \delta \varphi_3 \right] ds.
\end{aligned}
\tag{11.28}
$$

In transforming the last integral in Eq. (11.26), we assume that the order of integrations with respect to time and over the surface can be changed and also apply the formula of integration by parts. As a result, we obtain

$$
\begin{aligned}
\int\limits_{t_1}^{t_2} \delta E_0 dt = & -\rho \int\limits_{t_1}^{t_2} dt \iint\limits_{\Pi} \left[h \left(\ddot{u}^i \delta u_i + \ddot{w} \delta w\right) + I \left(\ddot{\psi}^i \delta \psi_i + \ddot{\psi}_3 \delta \psi_3\right)\right] dS - \\
& - J \int\limits_{t_1}^{t_2} dt \iint\limits_{\Pi} \left[h \left(\ddot{\omega}^i \delta \omega_i + \ddot{\omega} \delta \omega\right) + I \left(\ddot{\varphi}^i \delta \varphi_i + \ddot{\varphi}_3 \delta \varphi_3\right)\right] dS.
\end{aligned}
\tag{11.29}
$$

Substituting then Eqs. (11.27)–(11.29) into Eq. (11.26) and taking into account the arbitrariness of variations of the functional arguments, we arrive at the assertion formulated.

The corresponding initial and boundary value problem is formed by Eqs. (11.24) supplemented with the physical (11.20), (11.21) and kinematic (11.11), (11.14) relations, the boundary conditions (11.25), and the initial conditions. The assumption on the material isotropy only simplifies the physical relations without reducing the number of unknowns. The latter can be achieved only by means of introducing certain additional hypotheses concerning the displacement and rotation fields or considering special geometries of the middle surface.

11.6 Conclusions

The most general initial and boundary value problems for elastic moment shells are constructed within the framework of the direct normal hypothesis and the Cosserat model. They take into account the medium anisotropy, free rotation, independent rotation of a normal fiber, and its transverse normal strain. Using the results obtained it is possible to pass in the future to different simplified versions of the equations, in which the material is assumed to be isotropic, the normal strain and the perpendicularity of the normal fiber to the middle surface are neglected, the problem dimension is reduced (the case of plates and beams), etc.

Acknowledgements This paper has been supported by the Kazan Federal University Strategic Academic Leadership Program ("PRIORITY-2030", Sects. 11.1–11.3) and Russian Science Foundation (grant No 20-19-00217, Sects. 11.4–11.6).

References

1. Novacky V (1975) Teorija uprugosti (Theory of elasticity). Mir, Moscow [in Russian]
2. Sedov LI. Mexanika sploshnoj sredy. Tom 1. Continuum mechanics, vol 1. Nauka, Moscow [in Russian]
3. Erofeyev VI (1999) Volnovye processy v tverdyx telax s mikrostrukturoj (Wave processes in solid bodies with microstructure). Publishing House of Moscow State University, Moscow [in Russian]
4. Gerasimov SI, Erofeev VI, Soldatov IN (2012) Volnovye processy v sploshnyx sredax (Wave processes in continuous media). RFNC-VNIIEF Publishing House, Sarov [in Russian]
5. Bagdoev AG, Erofeyev VI, Shekoyan AV (2016) Wave dynamics of generalized continua. Advanced structured materials, vol 24. Springer, Heidelberg
6. Green AE, Naghdi PM (1968) The linear elastic Cosserat surface and shell theory. Int J Solid Struct 4:585–592
7. Babich DV (1966) Osnovnye uravneniya dvizheniya obolochki s uchetom nesimmetrichnosti tenzora napryazhenij (The basic equations of motion of the shell with allowance for the asymmetry of the stress tensor). Prikladnaya mexanika 2(12):41–48
8. Vanin GA (2004) Couple-stress mechanics of thin shells. Mech Solid 39(4):92–101
9. Birsan M (2004) The solution of Saint-Venant's problem in the theory of Cosserat shells. J Elast 74:185–214
10. Birsan M (2006) Several results in the dynamic theory of thermoelastic Cosserat shells with voids. Mech Res Commun 33(2):157–176
11. Birsan M (2007) On the theory of loaded general cylindrical Cosserat elastic shells. Int J Solids Struct 44:7399–7419
12. Birsan M (2009) Thermal stresses in cylindrical Cosserat elastic shells. Eur J Mech A 28(1):94–101
13. Ambartsumyan SA (2002) The problem of asymmetric thermoelasticity of a very gently sloping shell. Izv AN RA Mech 3:20–33

14. Palmov WW, Altenbach H (1982) Über eine Cosseratsche Theorie für elastische Platten. Tech Mech 3(3):5–9
15. Altenbach H, Eremeyev V (2009) On the linear theory of micropolar plates. Z Angew Math Mech (ZAMM) 89(4):242–256
16. Hoffman O (1964) On the bending of thin elastic plates in the presence of moment stresses. ASME J Appl Mech 31(4):706–707
17. Ambartsumyan SA (2000) Temperature problem of a micropolar plate. Izv. Universities. North Caucasian region. Ser: Nat Sci 3:17–20
18. Sargsyan SH (2004) On some interior and boundary effects in thin plates based on the asymmetric theory of elasticity. In: Kienzler R, Ott I, Altenbach H (eds) Theories of plates and shells. Lecture notes in applied and computational mechanics, vol 16. Springer, Berlin, Heidelberg
19. Sargsyan SO (2008) General models of micropolar elastic thin plates. Vestn. Perm GTU. Math Model Syst Process 16:111–120
20. Sargsyan SO (2008) Boundary value problems of thin plates in asymmetric elasticity theory. PMM 72(1):129–147
21. Sargsyan SO, Sargsyan AA (2011) General dynamic theory of micropolar elastic thin plates with free rotation and features of their free vibrations. Acoustic J 4:461–469
22. Vestyak VA, Tarlakovskii DV (2019) The model of thin electromagnetoelastic shells dynamics. In: Proceedings of the second international conference on theoretical, applied and experimental mechanics. Structural integrity. Springer Nature Switzerland AG, pp 254–258
23. Mihajlova EY, Tarlakovskii DV, Fedotenkov GV (2018) A generalized linear model of dynamics of thin elastic shells. Uchenye Zapiski Kazanskogo Universiteta, Seriya Fiziko-Matematicheskie Nauki 160(3):561–577
24. Okonechnikov AS, Ryazantseva MY, Tarlakovskii DV (2021) Variational principle and equations of dynamics of electromagnetoelastic shells. In: Altenbach H, Eremeyev VA, Igumnov LA (eds) Multiscale solid mechanics. Advanced structured materials, vol 141. Springer, Cham
25. Gorshkov AG, Rabinskii LN, Tarlakovskii DV (2000) Osnovy tenzornogo analiza i mekhanika sploshnojj sredy (Fundamentals of tensor analysis and continuum mechanics). Nauka, Moscow [in Russian]

Chapter 12
Transient Interaction of a Rigid Indenter with a Membrane Accounting for Adhesive Forces

Anatoly. S. Okonechnikov, Grigory. V. Fedotenkov, and Elena. S. Feoktistova

Abstract This paper is devoted to the study of adhesive force influence on transient deformation of a membrane interacted with rigid indenter. The research is carried out by splitting the process of interaction into two phases. The phases are analyzed sequentially. The first phase of interaction includes time period without mechanical contact. After the mechanical contact takes place, the second phase begins: it includes the influence of contact pressure as adhesive forces. In this study, we considered both phases of transient interaction for an infinite membrane with a rigid indenter accounting for adhesive forces. The original numerical–analytical algorithm for determining the contact stresses under the indenter is proposed. The graphical results for the first phase of interaction are shown.

Keywords Adhesive force · Transient contact · Rigid indenter · Influence function · Membrane · Integral equations · Influence function method

12.1 Introduction

Adhesion (from Latin adhaesio—adhesion, sticking), in this work, we will understand as attraction to each other of the surfaces of interacting solids, based on the action of intermolecular forces. At present, the issue of the influence of adhesive forces on the contact interaction of bodies is most widely covered in a stationary formulation. In [1], the essential role of adhesion forces at the nanoscale level is

Anatoly. S. Okonechnikov (✉) · Elena. S. Feoktistova
Moscow Aviation Institute, Moscow, Russia
e-mail: oas.mai@mail.ru

Elena. S. Feoktistova
e-mail: feoktistova0206@mail.ru

Grigory. V. Fedotenkov
Institute of Mechanics, Moscow Aviation Institute, Lomonosov Moscow State University, Moscow, USSR
e-mail: greghome@mail.ru

H. Altenbach et al. (eds.), *Deformation and Destruction of Materials and Structures Under Quasi-static and Impulse Loading*, Advanced Structured Materials 186,
https://doi.org/10.1007/978-3-031-22093-7_12

shown, and the influence of adhesion on the contact of bodies is assessed. Lyashenko and Popov in [2] investigated the destruction of the adhesive contact between the indenter and the elastic half-space. The influence of the friction force on the contact interaction is shown, and the possible ways of changing the friction force as the condition of contact between two bodies violated are investigated. In [3], the interaction of several indenters with an elastic half-space was studied taking into account adhesion in a transient formulation. The adhesive force was considered, in one case, as the force of intermolecular attraction according to the Maugis model [4]. In the other case, adhesive force considered as an interaction that takes into account the liquid minisk located between the surfaces of the bodies. Goryacheva and Makhovskaya [5] studied the attraction of two flat half-spaces under the action of the force of adhesive attraction, taking into account the roughness of the surfaces. The influence of the parameters of the adhesive force and the geometry of surfaces on gap between the studied bodies was analyzed. In [6], Makhovskaya investigated the contact of a rigid surface with protrusions with an elastic half-space under the forces of intermolecular attraction and capillary adhesion. An example of calculating the actual contact area and the surface area occupied by the liquid, depending on the external nominal pressure at different relative humidities of the ambient air and different distances between the protrusions, is given. In [7], the interaction of an indenter with an elastic half-plane was studied taking into account adhesion in an axisymmetric formulation. A study of the influence of microgeometry on the nature of the contact interaction of an indenter with an elastic half-plane has been carried out. Tkachuk in [8] proposes a method for solving axisymmetric problems for elastic bodies, taking into account adhesion. A variational principle was proposed that allows one to construct an approximate solution. A discretization is constructed and nonlinear equations are obtained to determine the unknown radius of the circular contact spot and the nodal values of the contact pressure.

A number of papers are devoted to the study of indenter motion along the boundary of a deformable body. Thus, in [9], the contact problem of sliding at a constant speed of rigid indenter's doubly periodic wavy surface over a viscoelastic layer lying on a rigid base is considered. The adhesive force also takes place between the surfaces. In this paper, the influence of the friction force on the deformation of the layer is estimated for various parameters of the layer viscosity and adhesive force. Meshcheryakova in [10] studied the effect of adhesive forces on the rolling of a spherical body on a viscoelastic layer on a rigid base. In the course of the study, a method was used that makes it possible to approximately represent the contact area in the form of a union of non-intersecting bands. The adhesive force is described by the Lennard-Jones potential. The contribution of adhesive forces to the rolling resistance of a spherical body over a viscoelastic layer is shown. In [11], Aleksandrov et al. considered the problem of sliding at a constant speed of a smooth indenter along the boundary of a viscoelastic half-space. A numerical–analytical determination method for the contact pressure distribution and its dependence on the movement speed is proposed. The problem was solved by constructing Green's function for a viscoelastic half-space with a concentrated force sliding along it at a constant speed.

As the presented review of studies shows, there are very few studies in the field of transient contact interaction taking into account adhesive forces, which adds to the relevance of this work devoted to the study of transient processes taking into account adhesive forces. It is caused by the complexity of the mathematical description and approaches to the study of transient processes. References [12–18] reflect the main approaches and methods for studying transient processes in the mechanics of a deformable solid body. These approaches are mainly based on the integral transformations, the influence function method, and the theory of generalized functions.

There are quite a large number of models describing adhesive interactions. However, the most widely used model of the interaction of solids, proposed in the 20s of the twentieth century by the English theoretical physicist, a member of the English Royal Society—Lennard-Jones. According to his theory, the force of interaction $p_a(h)$ between two surfaces depends on the gap h between the bodies:

$$p_a(h) = -\frac{8w_a}{3h_0}\left(\left(\frac{h_0}{h}\right)^3 - \left(\frac{h_0}{h}\right)^3\right), \tag{12.1}$$

where w_a is the specific work of adhesion, h_0 is the equilibrium distance between the surfaces of the bodies, at which the adhesive force is absent.

In addition to the potential of Lennard-Jones, no less important is the model proposed at the end of the twentieth century. It gives a simpler idea of the adhesive interaction. According to this theory, an adhesive attraction of constant intensity p_0 arises between the surfaces of bodies if the gap $h(x, t)$ is less than some known critical value of the gap h_{max}. In the area of mechanical contact, as well as when the gap between the bodies exceeds the critical value, we assume the absence of adhesive attraction. This model is called the Maugis–Dagdale theory [4] or the Maugis model.

$$p_a(x, t) = \begin{cases} p_0, 0 < h \leq h_{\max}, \\ 0, h > h_{\max} \\ 0, \ x \in \Omega(x, t), \end{cases} \tag{12.2}$$

where $\Omega(x, t)$ is the area of mechanical contact between two bodies. A graphic representation of the effect of adhesive forces on the deformation of the body according to the Lennard-Jones and Maugis models is shown in Fig. 12.1.

12.2 Problem Statement

The transient interaction of an absolutely rigid indenter and an infinite membrane is considered with the presence of a gap between the contacting bodies at the initial time moment. We introduce a fixed rectangular Cartesian coordinate system Oxz, associated with the plane of the membrane in the undeformed state, as well as a moving coordinate system $O_1x_1z_1$, associated with the centroid of the indenter. The

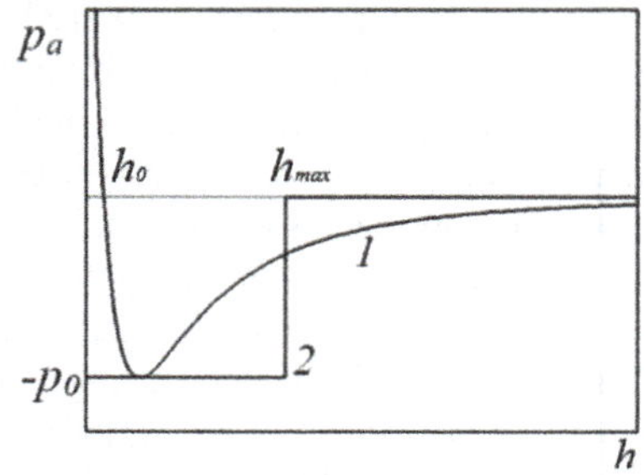

Fig. 12.1 Curve 1—potential of Lennard-Jones, Curve 2—Maugis model

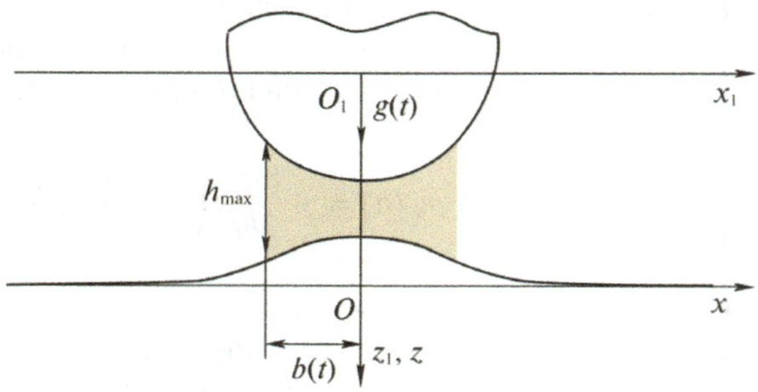

Fig. 12.2 Influence of adhesion on the deformation of the membrane before the onset of mechanical contact

shape of the indenter in coordinate system $O_1x_1z_1$ is described by some function $f(x)$. When $x = 0$, the value of this function is equal to the distance from the centroid of the indenter to its frontal point. The centroid of the indenter O_1 moves normally to the membrane according to the law $g(t)$ with a velocity $\dot{g}(t)$. The dot here and below denotes the derivative of the function with respect to time. At the initial time moment, we take the value of the minimum gap between the bodies equal to the critical value ($h(0, \tau) = h_{\max}$), at which, according to the Maugis model, adhesive attraction $p_a(x, \tau)$ will occur. Let us denote the size of the region within which the adhesive attraction acts as $2b(t)$, where $b(t)$ is the distance from the origin of the fixed coordinate system to the boundary of the adhesive interaction (we will consider the indenter is symmetrical with respect to the central axis passing through the frontal point of the indenter) (Fig. 12.2).

After the onset of the time point at which $h(0, t) = 0$ the next phase of interaction begins. The contact pressures arise in the area of contact between the indenter and the membrane $x \in [-c(t), c(t)]$. The adhesive interaction will affect outside the contact zone of the bodies (Fig. 12.3).

The following dimensionless quantities are used in equations and relations written below:

$$w' = \frac{w}{h_{\max}},\ \tau = \frac{t}{h_{\max}},\ p' = \frac{ph_{\max}}{a^2},\ x' = \frac{x}{h_{\max}},\ \sigma'_{ij} = \frac{\sigma_{ij}}{\lambda + 2\mu},\ h'_{\max} = 1,$$

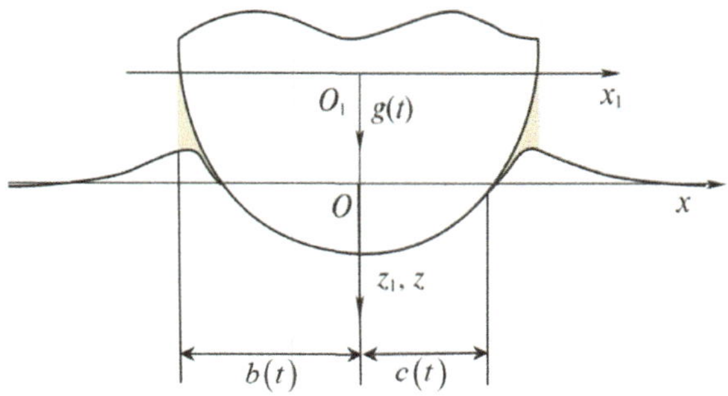

Fig. 12.3 Contact problem of the interaction of an indenter with a membrane, taking into account adhesion

where τ—dimensionless time, a—bending waves propagation velocity in the membrane, λ, μ—Lame elastic parameters. Below, apostrophes for dimensionless quantities are omitted.

The mathematical formulation of the problem includes the following relations:

- The rigid indenter law of motion

$$z = g(\tau). \tag{12.3}$$

- The membrane equation of motion in a plane formulation

$$\ddot{w}(x, \tau) = \frac{\partial^2 w(x, \tau)}{\partial x^2} + p(x, \tau). \tag{12.4}$$

- Expression for pressure considering both phases of interaction

$$p(x, \tau) = -p_0 H(b(\tau) - |x|) + (p_c + p_0) H(\tau - \tau_c) H(c(\tau) - |x|). \tag{12.5}$$

where $H(\cdot)$—Heaviside function, τ_c—contact interaction phase start time, p_c—contact pressure in the area of mechanical interaction.

Expression (12.5) is composed in such a way that the first term describes the phase of interaction between the indenter and the membrane without regard to mechanical contact, while the second term takes into account the contact pressure p_c arising between the considered bodies after the time point τ_c. Due to the absence of disturbances at the initial time moment, the following initial conditions are included in the problem statement:

$$w(x, 0) = \left.\frac{dw(x, \tau)}{d\tau}\right|_{\tau=0} = 0. \tag{12.6}$$

Also, we use a condition describing the behavior of displacements at the point in infinity:

$$\lim_{|x| \to \infty} w(x, \tau) = 0. \tag{12.7}$$

12.3 Solution Method

To determine the membrane deflection, we use the principle of superposition [12]

$$w(x,\tau) = p(x,\tau) ** G(x,\tau) = \int_0^{\tau} \int_{-\infty}^{\infty} G(x-\xi,\tau-t)p(\xi,t)\,d\xi dt. \quad (12.8)$$

$G(x,\tau)$—influence function describing the membrane deflection under the action of concentrated force $p(x,\tau) = \delta(x)\delta(\tau)$. In the plane formulation of the problem, $G(x,\tau)$ is defined as follows [14]:

$$G(x,\tau) = \frac{1}{2}H(\tau - |x|). \quad (12.9)$$

Taking into account (12.5), (12.8), and (12.9), the membrane deflection function can be written as

$$w(x,\tau) = p(x,\tau) ** G(x,\tau) = w_a(x,\tau) + w_c(x,\tau)H(\tau-\tau_c), \quad (12.10)$$

$$w_a(x,\tau) = -\frac{p_0}{2}\int_0^{\tau}\int_{-\infty}^{\infty} \Omega_G(x-\xi,\tau-t)\,\Omega_a(\xi,t)d\xi dt,$$

$$w_c(x,\tau) = \frac{1}{2}\int_0^{\tau}\int_{-\infty}^{\infty} (p_c(\xi,t)+p_0)\,\Omega_c(\xi,t)\,\Omega_G(x-\xi,\tau-t)d\xi dt,$$

$$\Omega_a(x,\tau) = H[b(\tau)-|x|],\ \Omega_c(x,\tau) = H[c(\tau)-|x|],\ \Omega_G(x,\tau) = H(\tau-|x|).$$

Introducing the following notation:

$$\Omega_{Ga}(x,\tau) = \{(\xi,t):\ \Omega_G(x-\xi,\tau-t)\,\Omega_a(\xi,t) = 1\},$$
$$\Omega_{Gc}(x,\tau) = \{(\xi,t):\ \Omega_G(x-\xi,\tau-t)\,\Omega_c(\xi,t) = 1\},$$

we obtain the calculation formula for membrane deflection

$$\begin{gathered} w(x,\tau) = w_a(x,\tau) + w_c(x,\tau)H(\tau-\tau_c). \\ w_a(x,\tau) = -\frac{p_0}{2}\iint_{\Omega_{Ga}} d\xi dt, \\ w_c(x,\tau) = \frac{1}{2}\iint_{\Omega_G} [p_c(\xi,t)+p_0]\,d\xi dt. \end{gathered} \quad (12.11)$$

Note that the integration domains Ω_{Ga}, Ω_{Gc} are not known in advance and must be determined in the course of solving the problem. To find them, a numerical–analytical algorithm has been developed and implemented. In this case, the areas $\Omega_{Ga}(x,\tau)$

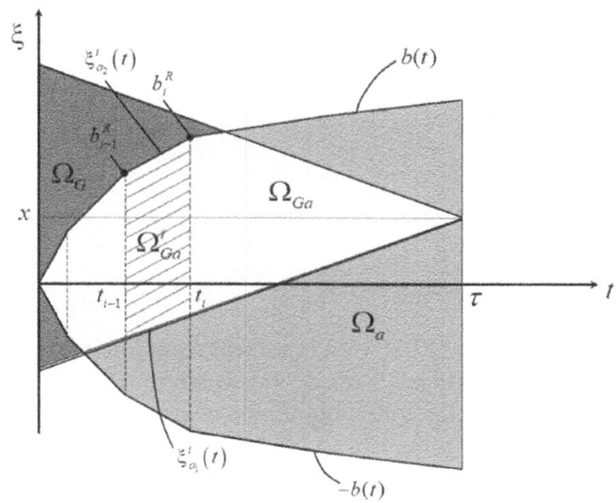

Fig. 12.4 Graph-analytical method for 1st phase of interaction

and $\Omega_{Gc}(x,\tau)$ are presented as a union of subareas:

$$\Omega_{Ga} = \bigcup_{i=1}^{N} \Omega^{i}_{Ga}, \ \Omega_{Gc} = \bigcup_{i=1}^{N} \Omega^{i}_{Gc}, \tag{12.12}$$

where $\Omega^{i}_{Ga}, \Omega^{i}_{Gc}$ are the areas bounded by vertical lines $t = t_i,, (t_i = i\Delta_t, (i = 0..N)$, Δ_t—time step) and oblique straight lines $\xi^{i}_{as}(t) : \xi = k^{i}_{as}(x)t + b^{i}_{as}(x)$, $\xi^{i}_{cs}(t) : \xi = k^{i}_{cs}(x)t + b^{i}_{cs}(x)$ $(s = 1, 2)$, respectively. The index s values "1" and "2" correspond to the upper and lower boundaries of the areas $\Omega^{i}_{Ga}, \Omega^{i}_{Gc}$, (Fig. 12.4).

Taking into account the approximation described above, (12.11) becomes suitable for calculations:

$$\begin{gathered} w(x,\tau) = w_a(x,\tau) + w_c(x,\tau)\,H\left(\tau - \tau_{i_c}\right), \\ w_a(x,\tau) = -\frac{p_0}{2}\sum_{i=1}^{N}\left(\frac{1}{2}\left(k^{i}_{a2}(x) - k^{i}_{a1}(x)\right)t^2 + \left(b^{i}_{a2}(x) - b^{i}_{a1}(x)\right)t\right)\Bigg|_{t_{i-1}}^{t_i}, \\ w_c(x,\tau) = \frac{1}{2}\sum_{i=i_c+1}^{N}\int_{t_{i-1}}^{t_i}\int_{\xi^{i}_{c1}(t)}^{\xi^{i}_{c2}(t)}\left(p_c(\xi,t) + p_0\right)d\xi dt, \end{gathered} \tag{12.13}$$

where i_c corresponds to the time moment $\tau = \tau_c$. Expressions (12.13) make it possible to determine the membrane deflection at any time. Thus, the problem is reduced to determining the adhesive force carrier boundary for the non-contact phase and finding the contact pressure at the contact interaction phase.

Let us consider the solution method of the problem in phases. To determine the boundaries of integration $\xi^{i}_{as}(t)$, $\xi^{i}_{cs}(t)$ in (12.13), we will sequentially consider the moments of time $\tau = \tau_l = \Delta_t l \ \ (l = 1 \ldots N)$. The integration boundaries $\xi^i_{aj}(t)$ are determined from the following condition: the boundary of the adhesive force carrier appears in point where the gap between the surface of the indenter and the membrane is 1:

$$b(\tau_l) = 1. \tag{12.14}$$

Condition (12.14) can be written as follows:

$$f\left[b(\tau_l)\right] - g(\tau_l) - w\left[b(\tau_l), \tau_l\right] = 0 \tag{12.15}$$

Expression (12.15) is a nonlinear equation with respect to $b(\tau_l)$, as $w\left[b(\tau_l), \tau_l\right]$ contains functions $\xi^l_{as}(t)$, $\xi^l_{cs}(t)$ which in turn depend on $b(\tau_l)$. We solve equation (12.15) using the following iterative scheme (r is the iteration number):

$$\begin{gathered} f\left(b^r_l\right) - g(\tau_l) - w\left(b^{r-1}_l, \tau_l\right) = 0, \ (r = 0 \ldots R), \\ b^0_l = b^R_{l-1}, b^0_1 = 0, \end{gathered} \tag{12.16}$$

Here, we assume b^R_l as the value of the adhesive force boundary at time τ_l , obtained with satisfactory accuracy at the iteration with number R . At the zero iteration, we assume $b^0_1 = 0$.

The boundaries of integration $\xi^l_{as}(t)$, $\xi^l_{cs}(t)$ are determined using the graph-analytical method [17]. This method is illustrated by the example of the non-contact phase of interaction ($\tau_l \le \tau_c$) (Fig. 12.4).

Figure 12.4 shows carriers of the adhesive force Ω_a, influence functions Ω_G, and their intersection at a fixed value of the coordinate and time. Knowing the values of the boundaries of the carrier b^R_i, one can determine the curves $\xi^i_{as}(t)$ at a fixed point x and time τ from the above graph-analytical method.

For the contact interaction phase, the integration boundaries must be determined together with the procedure for constructing the contact pressure distribution $p_c(x, \tau)$. To do this, we will use the following algorithm:

1. The contact condition at the time $\tau_l \ge \tau_c$ is used:

$$w(x, \tau_l) = 1 + f(x) - g(\tau_l) \tag{12.17}$$

2. The zero approximation of the area of contact interaction $c_0(\tau_l) = c^0_l$ is determined. At the time moment $\tau_l = \tau_{i_c+1}$, the area of contact interaction is determined assuming the absence of contact pressure $p_c(\xi, t)$.
3. The integral equation (12.17) is solved for the contact pressure $p_c(x, \tau_l) = p^1_{cl}$ at the first iteration.
4. Membrane deflections are specified according to formula (12.13) using the obtained contact pressure.
5. Based on the refined contact pressure, the boundaries of the adhesive interaction carrier b^1_l are refined according to formula (12.16). The boundaries of the contact interaction carrier are refined from condition (12.17). Further, based on the refined

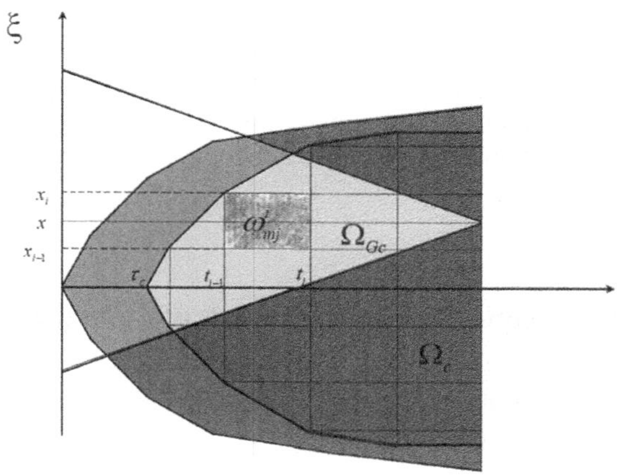

Fig. 12.5 Graph-analytical method for second phase of interaction

boundaries of the carriers, the contact pressure p^2_{cl} is determined from the solution of Eq. (12.17).

6. The iterative process is repeated until a satisfactory accuracy of the results p^R_{cl}, c^R_l, b^R_l is obtained.

Equation (12.17) is integral with respect to $p_c(\xi, \tau_l)$. The solution of the integral equation is carried out numerically assuming the contact pressure p^i_{cj} is constant on segments ω^i_j limited in time $t \in [t_{i-1}, t_i]$ and coordinate $x \in [x_{j-1}, x_j]$, $j = 1..M, x_0 = -c(t_i), x_M = c(t_i)$ with step Δx . These assumptions allow us to reduce (17) to a system of linear algebraic equations for the contact pressure $p^l_{cj} = p_c(\tilde{x}_j, \tau_l), \tilde{x}_j = x_j + \Delta x/2$.

The graphic-analytical method for the contact interaction phase is shown in Fig. 12.5. In Fig. 12.5, the integration area Ω_{Gc} is defined at $x = \tilde{x}_m$, $(m = 1..M)$. One can see in the figure: the contact pressure p^i_{cj} contributes to the displacements at the point $x = \tilde{x}_m$ not on the entire segment ω^i_j, but on the areas defined taking into account Ω_G. In light of the above, we use the following notation here and below: $\omega^i_{mj} = \omega^i_j(x_m)$.

Taking into account (12.13), we write down the system of linear algebraic equations based on (12.17) in matrix form:

$$\begin{gathered}\mathbf{Ap} = \mathbf{h} \\ \mathbf{A} = \{a_{mj}\}_{M\times M}, \mathbf{p} = \{p^l_{cj}\}_{M\times 1}, \mathbf{h} = \{h_m\}_{M\times 1},\end{gathered} \tag{12.18}$$

$$h_m = 1 - g(\tau_l) + f(x_m) - \frac{1}{2}\sum_{i=i_c}^{l-1}\sum_{j=1}^{M} p^i_{cj} \iint\limits_{\omega^i_{mj}} dt d\xi + \frac{p_0}{2} \iint\limits_{\Omega_{Ga}/\Omega_{Gc}} dt d\xi,$$

$$a_{mj} = \frac{1}{2} \iint\limits_{\omega^l_{mj}} dt d\xi.$$

Thus, using the above algorithm and formula (12.18), it is possible to determine the distribution of the contact pressure under the indenter at time moment $\tau = \tau_l$ with satisfactory accuracy. Note that the accuracy of the obtained solution depends on the approximation of the contact pressure. This approximation is determined by the partition number M, which can be selected in any convenient way at the considered time point τ_l.

12.4 Rigid Indenter Motion at Constant Velocity

Let us consider the motion of a rigid indenter of the form $f(x) = L + kx^2$. Indenter moves according to the law $g(\tau) = V_0\tau,\ V_0 = 10^{-2}$ towards an infinite membrane. The initial gap between the bodies is $h_{\max} = 1$. The adhesive force is assumed $p_0 = 10^{-3}$.

Figure 12.6 shows the dependence of the membrane deflections $w(x, \tau)$ on the coordinate at various points in time. The indenter shape is flat ($k = 0$). The solid line shows the deflection of the membrane at $\tau = 0.032$, dashed—$\tau = 0.06$, dotted—$\tau = 0.12$, respectively.

Figure 12.7 shows the deflections of the membrane from the action of the adhesive force, the indenter has a parabolic shape ($k = 0.1$). The solid line corresponds to the deflection of the membrane at $\tau = 0.02$, dashed—$\tau = 0.04$, and dotted—$\tau = 0.05$.

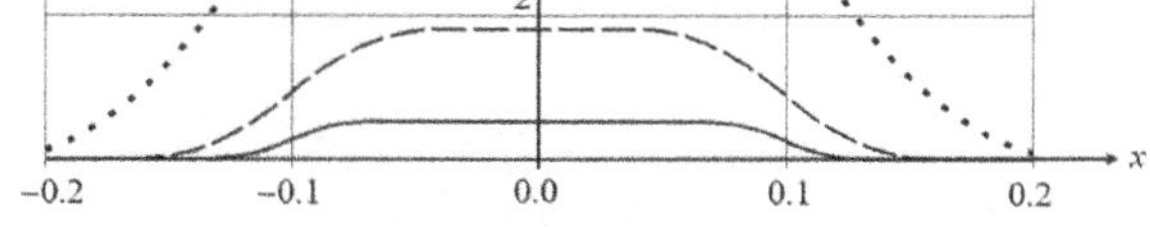

Fig. 12.6 Membrane deflection caused by flat indenter

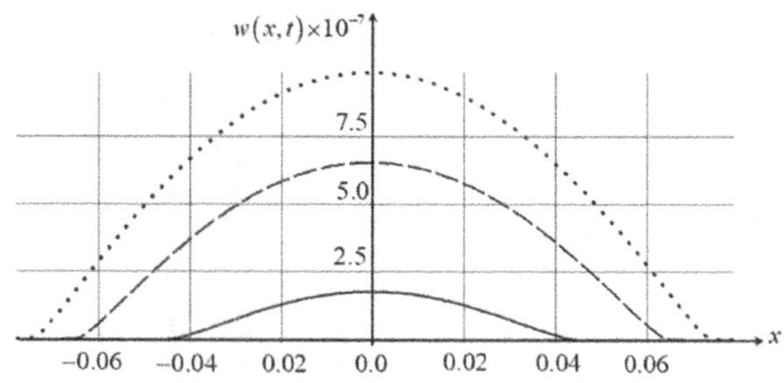

Fig. 12.7 Membrane deflection comparison for various indenter shapes

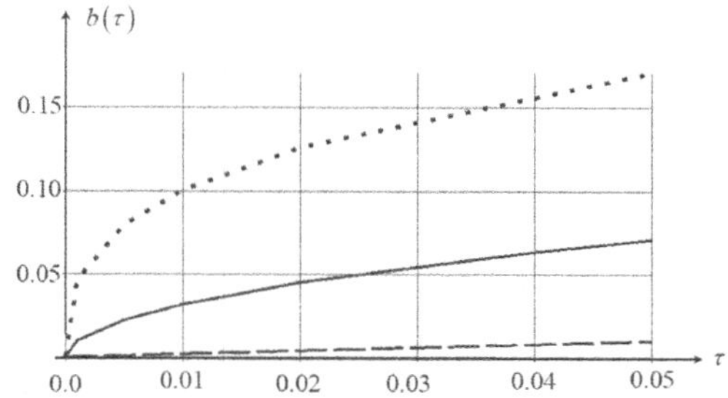

Fig. 12.8 Adhesive force carrier expansion

Figure 12.8 shows the expansion of the adhesive force carrier for various indenter shapes. The solid line referred to a parabolic indenter, the dashed line—conical indenter, the dotted line indicates the carrier corresponding to the hyperbolic indenter.

12.5 Conclusion

The paper proposes an original mathematical problem formulation of the transient interaction of a rigid indenter with an infinite membrane, taking into account adhesive forces. This formulation takes into account both non-contact and contact phases of interaction. A solution approach for both phases of interaction has been developed. A numerical–analytical method has been implemented for the phase of non-contact interaction. A solution algorithm for the phase of contact interaction has been developed. The developed solution approach to the problem makes it possible to study the interaction process for an arbitrary-shaped indenter moving according to an arbitrary law. Graphical results for the non-contact stage of interaction are presented.

Acknowledgements The research is completed under the financial support of the Russian Fund for Fundamental Research (project 20-08-01042).

References

1. Izmailov VV (2014) Discrete contact including adhesion. Vestnik Tver′ State Univ 25:57–60
2. Lyashenko IA, Popov VL (2019) Transition between modes of adhesion and sliding friction in contacts of axially symmetric bodies. J Frict Wear 40(1):39–45
3. Makhovskaya YY (2003) Diskretnyj kontakt uprugih tel pri nalichii adgezii. Izvestiya akademii nauk. mekhanika tverdogo tela 49–61 [In Russian]
4. Maugis D (1991) Adhesion of spheres: the JKR-DMT transition using a Dugdale model. J Colloid Interface Sci 5(150):243–269
5. Goryacheva IG, Makhovskaya YY, Elastic contact between nominally plane surfaces in the presence of roughness and adhesion. Mech Solids. ISSN: 0025-6544, eISSN: 1934-7936
6. Mahovskaya YY (2021) Modelirovanie diskretnogo adgezionnogo kontakta uprugih tel pri nalichii poverhnostnyh plenok zhidkosti: Uprugost' i neuprugost', materialy Mezh-

dunarodnogo nauchnogo simpoziuma po problemam mekhaniki deformiruemyh tel, posvyashchennogo 110-letiyu so dnya rozhdeniya A. A. Il'yushina. Moscow, 499–506 [In Russian]

7. Mahovskaya YY (2020) Adgezionnoe vzaimodejstvie uprugih tel c regulyarnym poverhnostnym rel'efom. Prikladnaya matematika i mekhanika 84(2):242–255
8. Tkachuk M (2018) A numerical method for axisymmetric adhesive contact based on Kalker's variational principle. East-Eur J Enterp Technol 3(7)(93):34–41
9. Goryacheva IG, Makhovskaya YY, Sliding of a wavy indentor on a viscoelastic layer surface in the case of adhesion. Mech Solids. ISSN: 0025-6544, eISSN: 1934-7936
10. Meshcheryakova AR (2021) Kontaktnoe vzaimodejstvie i nakoplenie ustalostnyh povrezhdenij pri kachenii deformiruemyh tel: dissertaciya kandidata fiziko-matematicheskih nauk: 01.02.04. Moskva, 101 [In Russian]
11. Aleksandrov VM, Goryacheva IG, Torskaya EV, Sliding contact of a smooth indenter and a viscoelastic half-space (3d problem). Doklady Phys. Issn: 1028-3358, eissn: 1562-6903
12. Gorshkov AG, Medvedsky AL, Rabinsky LN, Tarlakovsky DV (2004) Volny v sploshnykh sredakh. Moscow, Fizmatlit, Russia [In Russian]
13. Fedotenkov GV, Gritskov AV, Levitskiy DY, Vahterova YA, Sun Y (2021) Timoshenko beam and plate non-stationary vibrations. INCAS Bull 13(Special Issue):41–56
14. Fedotenkov GV, Makarevskii DI, Vahterova YA, Thang TQ (2021) The inverse non-stationary problem of identification of defects in an elastic rod. INCAS Bull 13(Special Issue):57–66
15. Serdyuk AO, Serdyuk DO, Fedotenkov GV (2021) Napryazhyonno-deformirovannoe sostoyanie kompozitnoj plastiny pod vozdejstviem nestacionarnoj podvizhnoj nagruzki. Mekhanika kompozitnyh materialov 57(4):705–720 [In Russian]
16. Lokteva NA, Serdyuk DO, Skopintsev PD (2020) Non-stationary influence function for an unbounded anisotropic Kirchoff-Love shell. J Appl Eng Sci 18(4):737–744
17. Okonechnikov AS, Tarlakovsky DV, Fedotenkov GV (2019) Transient interaction of rigid indenter with elastic half-plane with adhesive force. Lobachevskii J Math 40(4):489–498
18. Okonechnikov AS, Fedotenkov GV, Vahtrerova YA (2021) Unsteady motion of a rigid stamp along the boundary of an elastic half-space. J Balkan Tribol Assoc 27(3):343–358

Chapter 13
Transforming Deformation Model of Flat Beams with Finite Length Fastening Areas Located on One of the Front-Face Surfaces

Vitaly N. Paimushin, Victor M. Shishkin, Vyacheslav A. Firsov, and Ruslan K. Gazizullin

Abstract On the example of a plane problem of the mechanics of a beam with finite length fastening areas located on one of the front-face surfaces, it is shown that in the study of deformation processes with additional account of the fixed area compliance requires the introduction of the concept of the stress–strain state type transformation at the transition across the boundary from an unfastened to a fastened section. Appropriate mathematical models are also required to describe such a phenomenon. Within the framework of the classical Kirchhoff–Love model it is impossible to take into account the presence of such fixed sections. At the same time, within the framework of the simplest S.P. Timoshenko shear model, this transformation is possible if the section is fixed only on one of the front-face surfaces. The kinematic and force conditions for coupling of fastened and unfastened beam sections have been formulated. Based on the derived relations, an exact analytical solution to the problem of static bending of a cantilevered beam under the action of a constant surface load has been found. This solution is in good agreement with the results obtained by modeling of a beam using rectangular finite elements in a plane stress state, as well as using the ANSYS software package based on the equations of a plane problem of elasticity theory. An exact analytical solution has been obtained for the problem of transverse bending vibrations of a flat beam with two cantilevered parts and a finite length

V. N. Paimushin (✉) · V. A. Firsov · R. K. Gazizullin
Tupolev Kazan National Research Technical University, Kazan, Russia
e-mail: vpajmushin@mail.ru

V. A. Firsov
e-mail: vafirsov_49@mail.ru

R. K. Gazizullin
e-mail: gazizullin.rk@yandex.ru

V. N. Paimushin · R. K. Gazizullin
N.I. Lobachevsky Institute of Mathematics and Mechanics, Kazan Federal University, Kazan, Russia

V. M. Shishkin
Vyatka State University, Kirov, Russia
e-mail: tism1@rambler.ru

H. Altenbach et al. (eds.), *Deformation and Destruction of Materials and Structures Under Quasi-static and Impulse Loading*, Advanced Structured Materials 186,
https://doi.org/10.1007/978-3-031-22093-7_13

section between them under vibration loading by a transverse force acting on one of the unfastened section.

Keywords Flat beam · Finite length fastening area · Stress–strain state transformation · Transverse bending · Vibration

13.1 Introduction

To study the deformation mechanics of thin-walled structures, there is an approach when the problem is solved in the spatial formulation with the construction of their numerical solutions based on modern commercial application software packages. Such an approach is desirable, but ineffective for small relative thicknesses of the plate. In this regard, considerable attention in the scientific literature has been paid to the problems of reducing the three-dimensional equations of deformation mechanics to two-dimensional equations of beam, plate, and shell theories taking into account transverse shear and transverse compression. The scientific literature devoted to these problems is very extensive and multifaceted. In particular, it is possible to distinguish the works devoted to the construction of higher-order geometrically linear and nonlinear deformation models of thin-walled structural elements taking into account transverse shear and transverse compression, as well as the development on their basis of analytical and numerical methods for solving the corresponding static and dynamic problems [1–4, etc.]. A critical analysis of the static equations of composite plate theories obtained in the geometrically linear approximation was carried out, in particular, in [5, 6]. In the first of them, a review of methods for reducing the three-dimensional equations of thin-walled structures' mechanics to two-dimensional relations is given. A detailed critical analysis of static equations has been carried out in the formulation of two-dimensional problems for composite plates obtained using two approaches, namely, the method of weighted residuals and the method on the basis of the application of variational principles of anisotropic elasticity theory.

It should be noted that in all works devoted to the development of refined versions of the high-order theory, practically no attention is paid to the problems of formulating the boundary conditions for different variants of connection with other structural elements or their fixation to rigid support elements. For example, when formulating the problems of beam mechanics, the real conditions for fixing their end sections of a finite length, as a rule, are replaced by the conditions of either hinged support or by pinching. However, this approach certainly introduces errors into their solutions.

Due to the insufficient degree of accuracy of the currently existing theoretical foundations for the study of the physical phenomena described above (practically even their absence), the issues discussed below related to the development of refined mathematical models for the deformation of thin beams and plates with finite length fastening areas, as well as the statement and the solution of the corresponding problems on their basis should be considered as urgent and priority areas of scientific research.

13.2 General Equations

The simplest deformation model of a flat beam with fixation section of length l on the $z = -\frac{t}{2}$ surface (Fig. 13.1) can be built based on the Timoshenko model by representing the displacements of an arbitrary point of the cross section in the following form

$$U = u + z\gamma,\ W = w,\ 0 \le x \le a,\ U_0 = u_0 + z\gamma_0,\ W_0 = w_0 = 0,\ -l \le x \le 0 \tag{13.1}$$

and by imposing restrictions on functions u_0, γ_0

$$\gamma_0 = \frac{2u_0}{t},\ -l \le x \le 0 \tag{13.2}$$

where u_0, w_0, γ_0 and u, w, γ are axial displacement, deflection and angle of rotation of the cross section for the fastened and unfastened parts of the beam, respectively. This restriction follows from the condition $U_0\left(z = -\frac{t}{2}\right) = 0$ formulated on the fixing section of length l. In this case, from the condition of displacements U and U_0 equality in the section $x = 0$, formulated as $U|_{x=0} = U_0|_{x=0}$, we obtain the equalities

$$\gamma = \frac{2u_0}{t},\quad u = u_0 \quad \text{at} \quad x = 0. \tag{13.3}$$

Note that when the model (13.1) is replaced by the classical Kirchhoff–Love model $U = u - zw'$, $W = w$, and any of the $z = \frac{\pm t}{2}$ surfaces is fixed in the $-l \le x \le 0$ area, it is possible to formulate only the pinch condition $w|_{x=0} = 0$, $w'\big|_{x=0} = 0$ in $x = 0$ section. This does not allow taking into account the compliance of the beam's fastened section. Here and below the upper symbol ...′ means the differentiation of the corresponding quantity with respect to the x coordinate.

In accordance with (13.1) and (13.2), in the case of small displacements the following geometric dependences take place

$$\varepsilon_x = u' + z\gamma',\ \gamma_{xz} = w' + \gamma,\ 0 \le x \le a, \tag{13.4}$$

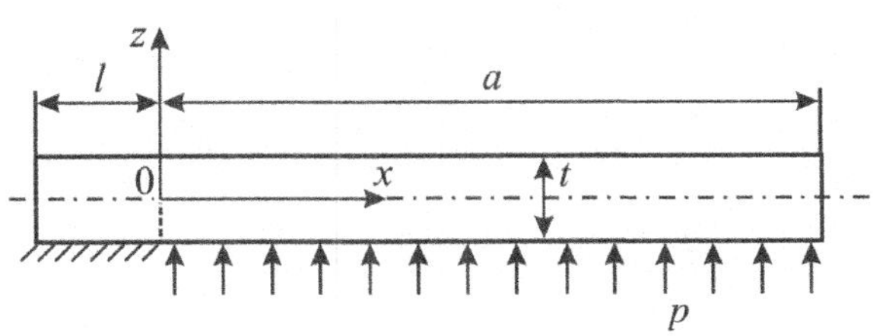

Fig. 13.1 Beam with finite length l fastening section

$$\varepsilon_x^0 = U_{0,x} = \left(1 + \frac{2z}{t}\right)u'_0, \gamma_{xz}^0 = U_{0,z} + W_{0,x} = \frac{2u_0}{t}, -l \le x \le 0, \tag{13.5}$$

where $\varepsilon_x, \gamma_{xz}$ and $\varepsilon_x^0, \gamma_{xz}^0$ are deformation and shear angle of the unfastened and fastened parts of the beam, respectively. Here the lower indices after the comma are the partial derivatives with respect to the coordinates x and z. For normal and shear stresses σ_x, σ_{xz}, the following physical dependencies take place

$$\sigma_x = E_1\left(u' + z\gamma'\right), \sigma_{xz} = G_{13}\left(w' + \gamma\right), 0 \le x \le a, \tag{13.6}$$

$$\sigma_x^0 = \left(1 + \frac{2z}{t}\right)E_1 u_0', \sigma_{xz}^0 = \frac{2u_0 G_{13}}{t}, -l \le x \le 0. \tag{13.7}$$

Here E_1 and G_{13} are the elasticity moduli of the beam material under its deformation along the $0x$ axis and under transverse shear, respectively.

Under the action of a transverse load p, using dependences (13.4), (13.5), it is possible to compose a variational equation

$$\begin{aligned}\delta\Pi_0 + \delta\Pi - \delta A = &\int_{-l}^{0}\int_{-\frac{t}{2}}^{\frac{t}{2}}\left[\sigma_x^0\left(1 + \frac{2z}{t}\right)\delta u_0' + \frac{2\sigma_{xz}^0\delta u_0}{t}\right]dzdx + \int_0^a\int_{-\frac{t}{2}}^{\frac{t}{2}}\left[\sigma_x(\delta u' + z\delta\gamma') + \sigma_{xz}(\delta w' + \delta\gamma)\right]dzdx \\ &- \int_0^a p\delta w dx = \int_{-l}^{0}\left(T_{11}^0\delta u_0' + T_{13}^0\delta u_0\right)dx + \int_0^a\left[T_{11}\delta u' + M_{11}\delta\gamma' + T_{13}(\delta w' + \delta\gamma) - p\delta w\right]dx \\ &= 0.\end{aligned} \tag{13.8}$$

where, due to dependencies (13.6), (13.7)

$$T_{11}^0 = \int_{-\frac{t}{2}}^{\frac{t}{2}} \sigma_x^0\left(1 + \frac{2z}{t}\right)dz = \frac{4Bu_0'}{3}, T_{13}^0 = \frac{4G_{13}u_0}{t}, \tag{13.9}$$

$$T_{11} = E_1 t u', M_{11} = D\gamma', T_{13} = B_{13}\left(w' + \gamma\right), B = E_1 t, B_{13} = G_{13}t, D = \frac{E_1 t^3}{12}. \tag{13.10}$$

Equation (13.8), taking into account conditions (13.3) and (13.5), is transformed to the form

$$\begin{aligned}&\left(T_{11}^0 - T_{11} - \frac{2M_{11}}{t}\right)\delta u_0\Big|_{x=0} - T_{11}^0\delta u_0\big|_{x=-l} + (T_{11}\delta u + M_{11}\delta\gamma)\big|_{x=a} \\ &+ T_{13}\delta w\big|_{x=0}^{x=a} - \int_{-l}^{0}(T_{11,x}^0 - T_{13}^0)\delta u_0 dx \\ &- \int_0^a\left[T_{11,x}\delta u + \left(M_{11,x} - T_{13}\right)\delta\gamma + \left(T_{13,x} + p\right)\delta w\right] = 0.\end{aligned}$$

From these equations follow the equilibrium equations

$$T^0_{11,x} - T^0_{13} = 0, -l \le x \le 0, \tag{13.11}$$

$$T_{11,x} = 0, M_{11,x} - T_{13} = 0, T_{13,x} + p = 0, 0 \le x \le a, \tag{13.12}$$

boundary conditions

$$T^0_{11}\big|_{x=-l} = 0 \quad \text{at} \quad \delta u_0 \ne 0, \tag{13.13}$$

$$T_{11}|_{x=a} = 0 \quad \text{at} \quad \delta u \ne 0, \quad M_{11}|_{x=a} = 0 \quad \text{at} \quad \delta\gamma \ne 0, \quad T_{13}|_{x=a} = 0 \quad \text{at} \quad \delta w \ne 0, \tag{13.14}$$

And in addition to conditions (13.3), the static conjugation condition in the $x = 0$ section is formulated by virtue of $\delta u_0 \ne 0$

$$\left(T^0_{11} - T_{11} - \frac{2M_{11}}{t}\right)\bigg|_{x=0} = 0. \tag{13.15}$$

13.3 Equations Corresponding to the Description of the Unfastened Section of the Beam by the Classical Kirchhoff–Love Model

The simplest equations for describing the deformation process of a beam with a fixed section $-l \le x \le 0$ on the front surface $z = -\frac{t}{2}$ can be obtained based on the representations

$$U_0 = \left(1 + \frac{2z}{t}\right)u_0, W_0 = 0, -l \le x \le 0, \tag{13.16}$$

$$U = u - zw', W = w, 0 \le x \le a. \tag{13.17}$$

Of these, relations (13.17) correspond to the classical Kirchhoff–Love model. Substituting (13.16), (13.17) in the cross section $x = 0$ to the conjugation conditions for displacements $U_0|_{x=0} = U|_{x=0}$, , we obtain instead of (13.3) the dependencies.

$$u = u_0, \ w' = -\frac{2u_0}{t} \quad \text{at} \quad x = 0. \tag{13.18}$$

In the considered approximation instead of (13.8) we come to the variational equation.

$$\int_{-l}^{0} (T_{11}^0 \delta u'_0 + T_{13}^0 \delta u_0) + dx + \int_{0}^{a} (T_{11}\delta u' - M_{11}\delta w'' - p\delta w)dx = 0 \quad (13.19)$$

in which, in contrast to (13.10),

$$M_{11} = -Dw'', D = \frac{E_1 t^3}{12}. \quad (13.20)$$

Taking into account (13.16), Eq. (13.19) after standard transformations is reduced to the form

$$\left(T_{11}^0 - T_{11} - \frac{2M_{11}}{t}\right)\delta u_0\Big|_{x=0} - T_{11}^0 \delta u_0\big|_{x=-l} + (T_{11}\delta u - M_{11}\delta w')\big|_{x=a} + M_{11,x}\delta w\big|_{x=a}$$
$$- M_{11,x}\delta w\big|_{x=0} - \int_{-l}^{0} (T_{11,x}^0 - T_{13}^0)\delta u_0 dx$$
$$- \int_{0}^{a} [T_{11,x}\delta u + (M_{11,xx} + p)\delta w]dx = 0. \quad (13.21)$$

Equation (13.21) still implies the equilibrium Eq. (13.11) for the fixed part of the beam and the corresponding boundary condition (13.13) as well as the equilibrium equation for its unfixed part

$$T_{11,x} = 0, M_{11,xx} + p = 0, 0 \le x \le a \quad (13.22)$$

with boundary conditions

$$T_{11}|_{x=a} = 0 \text{ at } \delta u \ne 0, \; M_{11}|_{x=a} = 0 \text{ at } \delta w' \ne 0,$$

$$M_{11,x}\big|_{x=a} = 0 \text{ at } \delta w \ne 0, \quad (13.23)$$

In the section $x = 0$, , in addition to the kinematic conjugation conditions (13.18), by virtue of the condition $\delta u_0 \ne 0$, the static conjugation condition for Sects. (13.15) is formulated, in which M_{11} is determined by formula (13.20).

13.4 Construction of Analytical Solutions for the Simplest Problems

Based on the derived Eqs. (13.11) and (13.12), let us construct solutions to two simple problems.

13.4.1 *Bending of Cantilevered Fixed Flat Beam Under Transverse Load* p = const

Let us consider the bending problem shown on Fig. 13.1. For the fastening section, using (13.9) and (13.11) is reduced to the form

$$u_0'' - k^2 u_0 = 0, -l \le x \le 0, \tag{13.24}$$

where

$$k^2 = \frac{3G_{13}}{\left(E_1 t^2\right)}. \tag{13.25}$$

The general solution of Eq. (13.24) has the following form.

$$u_0 = c_1 e^{kx} + c_2 e^{-kx},$$

where c_1 and c_2 are integration constant. Subjecting this solution to condition (13.13) and taking into account (13.9), we obtain

$$c_1 = c_2 e^{2kl}, u_0 = c_2\left(e^{k(2l+x)} + e^{-kx}\right), u_0|_{x=0} = c_2\left(1 + e^{2kl}\right). \tag{13.26}$$

For an unfastened section of the beam, using (13.10), we obtain the integrals of the equilibrium Eqs. (13.12)

$$u = d_1, \gamma = d_2 - \left(a^2 x - a x^2 + \frac{x^3}{3}\right)\frac{p}{2D}, \tag{13.27}$$

$$w = \left(ax - \frac{x^2}{2}\right)\frac{p}{G_{13}t} - d_2 x + \left(\frac{a^2x^2}{2} - \frac{ax^3}{3} + \frac{x^4}{12}\right)\frac{p}{2D}, \tag{13.28}$$

where it is taken into account that $w|_{x=0} = 0$ and in addition to c_1, c_2 integration constant d_1, d_2 are introduced into consideration. Note that in solutions (13.27), (13.28) the presence of summand with constants d_1 and d_2 is entirely caused by taking into account the deformability of the fixed section of the beam. At $G_{13} \to \infty$ the solution found corresponds to the use of relations and Eqs. (13.16)–(13.23).

Using further relations (13.9), (13.10), (13.26)–(13.28) and conjugation conditions (13.3), (13.15), we obtain the equalities

$$d_1 = (1 + e^{2kl})c_2; \; d_2 = \tfrac{2}{t}(1 + e^{2kl})c_2; \; -2E_1tk(1 - e^{2kl})c_2 + \frac{3a^2}{2t}p = 0$$

from which it follows

$$c_2 = \frac{3a^2}{4E_1 t^2 k\left(1 - e^{2kl}\right)} p,\ d_1 = \frac{3a^2\left(1 + e^{2kl}\right)}{4E_1 t^2 k\left(1 - e^{2kl}\right)} p,\ d_2 = \frac{3a^2\left(1 + e^{2kl}\right)}{2E_1 t^3 k\left(1 - e^{2kl}\right)} p. \tag{13.29}$$

Thus, in accordance with (13.7), (13.26) and (13.29), to determine the σ_x^0 and σ_{xz}^0 stress components in the fastened section we obtain the following formulas

$$\sigma_x^0 = \left(1 + \frac{2z}{t}\right) E_1 c_2 k\left(e^{k(x+2l)} - e^{-kx}\right) = \left(1 + \frac{2z}{t}\right) \frac{3a^2}{4t^2} \frac{e^{k(x+2l)} - e^{-kx}}{1 - e^{2kl}} p,$$

$$\sigma_{xz}^0 = \frac{2G_{13}}{t} c_2\left(e^{k(x+2l)} + e^{-kx}\right) = \frac{a^2}{2t^2}\sqrt{\frac{3G_{13}}{E_1}} \frac{e^{k(x+2l)} + e^{-kx}}{1 - e^{2kl}} p.$$

13.4.2 Forced Bending Vibrations of a Flat Beam with Two Cantilever Parts and a Fixation Section Between Them

To determine the elastic and damping properties of structural materials, specially made cantilevered test specimens in the form of single- and multi-layer elongated plates (flat beams) are used [7, 8]. During the dynamic tests of such test specimens, the phenomenon of vibration transmission from the working part of the test specimen to its non-working part was found. Such transmission of vibrations can be explained by the compliance of the fixing section of the finite length. This phenomenon requires an appropriate theoretical substantiation. In this connection, we consider the problem of forced bending vibrations of a flat beam with two cantilever parts fixed in a $-l \le x \le 0$ section along the $z = -\frac{t}{2}$ surface (Fig. 13.2) under the action of vibration load on one of these parts in the form of a concentrated force $P = \tilde{P}e^{i\omega\tau}$ with the amplitude $\tilde{P}$ and circular frequency ω (Fig. 13.2).

To describe the dynamic behavior of unfastened sections of the beam, one can use the equations written in the approximation

$$T_{11,x}^{(k)} - \rho t \ddot{u}_{(k)} = 0,\ M_{11,x}^{(k)} - T_{13}^{(k)} = 0,\ T_{13,x}^{(k)} - \rho t \ddot{w}_{(k)} = 0,\ k = 1, 2, \tag{13.30}$$

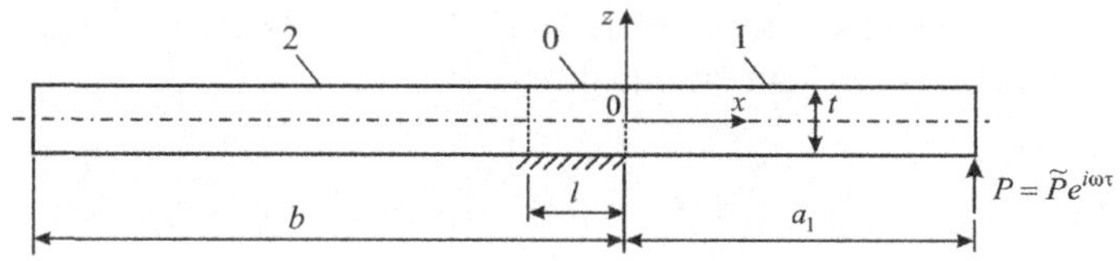

Fig. 13.2 Scheme of fastening and loading of the beam: 1—loaded cantilever part, 2—free cantilever part, 0—fastened section

which are obtained from Eqs. (13.12) at $p = 0$ by adding inertial forces $\rho t \ddot{u}_{(k)}$ and $\rho t \ddot{w}_{(k)}$ (ρ is the density of the beam material). For them, in the cross sections $x = a_1$ and $x = -b$ the force boundary conditions are formulated

$$T_{11}^{(1)}\Big|_{x=a_1} = 0,\ M_{11}^{(1)}\Big|_{x=a_1} = 0,\ T_{13}^{(1)}\Big|_{x=a_1} = P,$$

$$T_{11}^{(2)}\Big|_{x=-b} = 0,\ M_{11}^{(2)}\Big|_{x=-b} = 0,\ T_{13}^{(2)}\Big|_{x=-b} = 0, \tag{13.31}$$

and in sections $x = 0$ and $x = -l$ the following conjugation conditions must be fulfilled

$$\left(\gamma_{(1)} - \tfrac{2u_0}{t}\right)\Big|_{x=0} = 0,\ \left(u_{(1)} - u_0\right)\Big|_{x=0} = 0,\ \left(T_{11}^0 - T_{11}^{(1)} - \tfrac{2M_{11}^{(1)}}{t}\right)\Big|_{x=0} = 0,$$

$$\left(\gamma_{(2)} - \frac{2u_0}{t}\right)\Big|_{x=-l} = 0,\ \left(u_{(2)} - u_0\right)\Big|_{x=-l} = 0,\ \left(T_{11}^0 - T_{11}^{(2)} - \frac{2M_{11}^{(2)}}{t}\right)\Big|_{x=-l} = 0. \tag{13.32}$$

Taking into account (13.10), Eqs. (13.30) are transformed to the following form

$$Bu_{(k)}'' - \rho t \ddot{u}_{(k)} = 0, \tag{13.33}$$

$$D\gamma_{(k)}'' - B_{13}\left(w_{(k)}' + \gamma_{(k)}\right) = 0,\ B_{13}\left(w_{(k)}'' + \gamma_{(k)}'\right) - \rho t \ddot{w}_{(k)} = 0. \tag{13.34}$$

By introducing displacement functions $\phi_{(k)}$ in accordance with the dependencies

$$w_{(k)} = \left(-1 + \frac{D}{B_{13}}\frac{d^2}{dx^2}\right)\phi_{(k)},\ \gamma_{(k)} = \phi_{(k)}' \tag{13.35}$$

the first equation in (13.34) is satisfied identically, and the second takes the form

$$D\phi_{(k)}^{IV} - \frac{\rho t D}{B_{13}}\ddot{\phi}_{(k)}'' + \rho t \ddot{\phi}_{(k)} = 0. \tag{13.36}$$

Solutions to Eqs. (13.33), (13.36) are sought in the form $u_{(k)} = \tilde{u}_{(k)}e^{i\omega\tau}$, $\phi_{(k)} = \tilde{\phi}_{(k)}e^{i\omega\tau}$. It gives the equations

$$\tilde{u}_{(k)}'' + \Omega_u^2 \tilde{u}_{(k)} = 0,\ \tilde{\phi}_{(k)}^{IV} + \Omega_w^2 r \tilde{\phi}_{(k)}'' - \Omega_w^2 \tilde{\phi}_{(k)} = 0, \tag{13.37}$$

where

$$\Omega_u^2 = \frac{\rho\omega^2}{E_1},\ \Omega_w^2 = \frac{\rho t\omega^2}{D},\ r = \frac{D}{B_{13}}.$$

General solutions of Eqs. (13.37) have the following form

$$\tilde{\phi}_{(k)} = g_1^{(k)} e^{r_1 x} + g_2^{(k)} e^{r_2 x} + g_3^{(k)} e^{r_3 x} + g_4^{(k)} e^{r_4 x}, \tag{13.38}$$

$$\tilde{u}_{(k)} = g_5^{(k)} e^{i\Omega_u x} + g_6^{(k)} e^{-i\Omega_u x}, \tag{13.39}$$

where

$$r_1 = \left(-\frac{\Omega_w^2 r}{2} + \sqrt{\frac{\Omega_w^4 r^2}{4} + \Omega_w^2}\right)^{\frac{1}{2}}, r_2 = -\left(-\frac{\Omega_w^2 r}{2} + \sqrt{\frac{\Omega_w^4 r^2}{4} + \Omega_w^2}\right)^{\frac{1}{2}},$$

$$r_3 = \left(-\frac{\Omega_w^2 r}{2} - \sqrt{\frac{\Omega_w^4 r^2}{4} + \Omega_w^2}\right)^{\frac{1}{2}}, r_4 = -\left(-\frac{\Omega_w^2 r}{2} - \sqrt{\frac{\Omega_w^4 r^2}{4} + \Omega_w^2}\right)^{\frac{1}{2}}, \tag{13.40}$$

and $g_j^{(k)} (j = \overline{1,6}; k = 1, 2)$ are integration constants. Using the solution (13.38) according to (13.35), we obtain the following expressions

$$\tilde{\gamma}_{(k)} = \sum\nolimits_{j=1}^{4} g_j^{(k)} r_j e^{r_j x}, \tilde{w}_{(k)} = \sum\nolimits_{j=1}^{4} g_j^{(k)} (-1 + r r_j^2) e^{r_j x}. \tag{13.41}$$

For Eqs. (13.30), boundary conditions (13.31) at $x = a_1$ and $x = -b$ are formulated as

$$\tilde{u}_{(1)}'\big|_{x=a_1} = 0, \tilde{\gamma}_{(1)}'\Big|_{x=a_1} = 0, \left(\tilde{w}_{(1)}' + \tilde{\gamma}_{(1)}\right)\Big|_{x=a_1} = \frac{\tilde{P}}{B_{13}},$$

$$\tilde{u}_{(2)}'\big|_{x=-b} = 0, \tilde{\gamma}_{(2)}'\Big|_{x=-b} = 0, \left(\tilde{w}_{(2)}' + \tilde{\gamma}_{(2)}\right)\Big|_{x=-b} = 0, \tag{13.42}$$

Six algebraic equations are obtained by substituting expressions (13.39) and (13.41) into the boundary conditions.

$$g_5^{(1)} e^{i\Omega_u a_1} - g_6^{(1)} e^{-i\Omega_u a_1} = 0, \sum\nolimits_{j=1}^{4} g_j^{(1)} r_j^2 e^{r_j a_1} = 0, \sum\nolimits_{j=1}^{4} g_j^{(1)} r r_j^3 e^{r_j a_1} = \frac{\tilde{P}}{B_{13}},$$

$$g_5^{(2)} e^{-i\Omega_u b} - g_6^{(2)} e^{i\Omega_u b} = 0, \sum\nolimits_{j=1}^{4} g_j^{(2)} r_j^2 e^{-r_j b} = 0, \sum\nolimits_{j=1}^{4} g_j^{(2)} r_j^3 e^{-r_j b} = 0. \tag{13.43}$$

To describe the dynamic behavior of the fixed section of the beam, instead of (13.11), one should use the equation

$$T_{11,x}^0 - T_{13}^0 - \frac{4}{3}\rho t \ddot{u}_0 = 0, -l \le x \le 0. \tag{13.44}$$

Taking into account (13.9) and representing the axial displacement u_0 of the beam in the form $u_0 = \tilde{u}_0 e^{i\omega\tau}$, from (13.44) we pass to the equation

$$\tilde{u}_0'' - k_*^2 \tilde{u}_0 = 0,\ k_*^2 = k^2 - \frac{\rho\omega^2}{E_1},\ k^2 = \frac{3G_{13}}{E_1 t^2}. \tag{13.45}$$

Calculations have shown that the first summand in k_*^2 for real structural materials and values of ω^2 and t is approximately four orders of magnitude higher than the second one. Therefore, one can take the value $k^2 > 0$ instead k_*^2. In this case, the general solution of Eq. (13.45) has the form

$$\tilde{u}_0 = c_1 e^{kx} + c_2 e^{-kx}. \tag{13.46}$$

In accordance with (13.9), (13.10), (13.39) and (13.46) the amplitudes of the longitudinal forces in the fastened and unfastened parts of the beam will be calculated by the formulas

$$\tilde{T}_{11}^0 = \frac{4E_1 tk}{3}\left(c_1 e^{kx} - c_2 e^{-kx}\right),\ \tilde{T}_{11}^{(k)} = iE_1 t\Omega_u\left(g_5^{(k)} e^{i\Omega_u x} - g_6^{(k)} e^{-i\Omega_u x}\right). \tag{13.47}$$

Using conjugation conditions (13.32) in addition to (13.43), we obtain six more equations.

$$\begin{gathered}
\sum\nolimits_{j=1}^4 g_j^{(1)} r_j - \tfrac{2}{t}(c_1 + c_2) = 0,\ \sum\nolimits_{j=1}^4 g_j^{(2)} r_j e^{-r_j l} - \tfrac{2}{t}\left(c_1 e^{-kl} + c_2 e^{kl}\right) = 0.\\
g_5^{(1)} + g_6^{(1)} - c_1 - c_2 = 0,\ g_5^{(2)} e^{-i\Omega_u l} + g_6^{(2)} e^{i\Omega_u l} - c_1 e^{-kl} - c_2 e^{kl} = 0,\\
\tfrac{4}{3}k(c_1 - c_2) - i\Omega_u\left(g_5^{(1)} - g_6^{(1)}\right) - \tfrac{t}{6}\sum\nolimits_{j=1}^4 g_j^{(1)} r_j^2 = 0,\\
\frac{4}{3}k\left(c_1 e^{-kl} - c_2 e^{kl}\right) - i\Omega_u\left(g_5^{(2)} e^{-i\Omega_u l} - g_6^{(2)} e^{i\Omega_u l}\right) - \frac{t}{6}\sum\nolimits_{j=1}^4 g_j^{(1)} r_j^2 e^{-r_j l} = 0.
\end{gathered} \tag{13.48}$$

The resulting systems (13.43) and (13.48) contain twelve equations and fourteen constants of integration $c_1, c_2, g_j^{(k)} (j = \overline{1,6}; k = 1, 2)$. The missing two equations follow from the conditions $\tilde{w}_{(1)}\big|_{x=0} = 0,\ \tilde{w}_{(2)}\big|_{x=-l} = 0$. These conditions, taking into account the second expression in (13.41), take the following form

$$\sum\nolimits_{j=1}^4 g_j^{(1)}\left(-1 + rr_j^2\right) = 0,\ \sum\nolimits_{j=1}^4 g_j^{(2)}\left(-1 + rr_j^2\right)e^{-r_j l} = 0. \tag{13.49}$$

13.5 Calculation Results and Their Analysis

The stress–strain state of a cantilevered flat beam under the action of constant surface load p (Fig. 13.1) was determined using the relations obtained in Sect. 13.4.1 at the analytical solution of the problem. The material of the beam is a unidirectional fiber reinforced composite, based on the ELUR-P carbon fiber and KhT-118 epoxy with

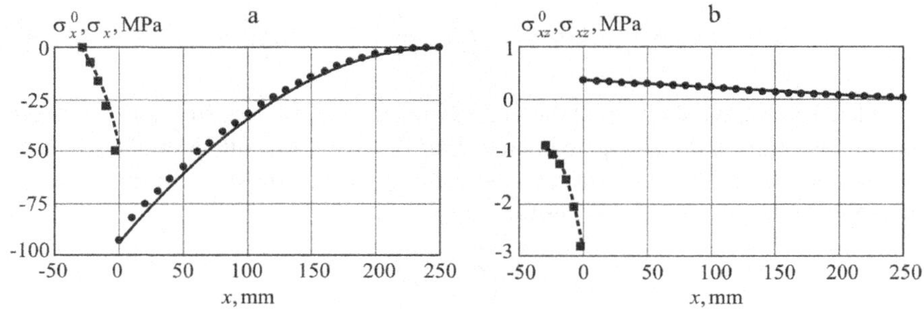

Fig. 13.3 Normal stresses σ_x^0, σ_x on the $z = \frac{t}{2}$ surface (**a**) and shear stress σ_{xz}^0, σ_{xz} (**b**): dashed lines are σ_x^0, σ_{xz}^0 (S.P. Timoshenko model); solid lines are σ_x, σ_{xz} (Timoshenko model); squares are σ_x^0, σ_{xz}^0 (FEM); dots are σ_x, σ_{xz} (FEM)

elastic moduli $E_1 = 100$ GPa, $G_{13} = 1$ GPa. Load and geometrical parameters of the beam are the following: $p = 4500$ N/m2; $l = 30$ mm; $a = 250$ mm; $t = 3$ mm.

Figure 13.3 shows the normal stresses σ_x^0, σ_x on the $z = \frac{t}{2}$ surface and shear stress σ_{xz}^0, σ_{xz} in the fastened and unfastened parts of the considered beam (superscript 0 instresses corresponds to the fastened part of the beam). For the comparison, there are analogous stresses, which were received based on the finite-element method (FEM). The beam was modeled by rectangular finite-elements in the conditions of the plane stressed state. The beam was divided into 25 elements through the thickness, the fastened and unfastened parts were divided along the 0x axis into 250 and 2000 elements, respectively. The shear stresses σ_{xz}^0 and σ_{xz} obtained on the basis of the FEM correspond to their averaged values over the beam thickness.

From Fig. 13.3 one can see that normal and shear stresses corresponding to the S.P. Timoshenko shear model are close enough to the results obtained on the basis of FEM. This fact proves applicability of the given model for the stress–stain state analysis of beams having the finite length fastened sections on one of the face surfaces. A substantial transformation of the stress–strain state parameters is observed at transition over the boundary $x = 0$ from an unfastened to a fastened section of finite length l. The normal stresses σ_x are reduced by a factor of two in absolute value, and the shear stresses σ_{xz} change sign and exceed in absolute value where the values were obtained in section x = 0 of the unfastened part of the beam by almost 7.7 times. This implies the conclusion that taking into account the flexibility of fastening sections of finite length instead of setting kinematic boundary conditions for pinching in the end sections can lead to a fundamental refinement of the results in the calculation of plates and bar systems. This is especially important in strength analysis of plates and beams made of fiber reinforced composite materials, since their failure mainly occurs due to transverse shear stresses σ_{xz}.

In addition to this, the distributions of shear stresses $\sigma_{xz}^0(z)$ over the thickness of the beam in three x = const sections are shown in Fig. 13.4a. The ratios s of maximum shear stresses σ_{xz}^0, σ_{xz} modulus to their averaged values over the thickness of the beam as a function of x are also shown in Fig. 13.4b. Both results were obtained

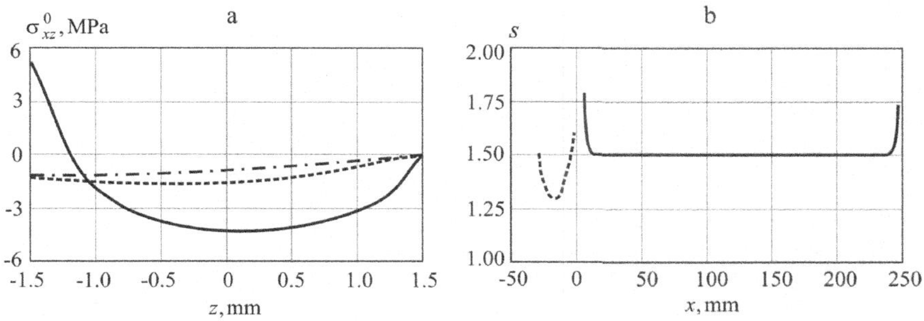

Fig. 13.4 Distribution of shear stresses σ_{xz}^0 over the thickness of the beam **a** solid line corresponds to the section $x = -1.5$ mm, dashed lines corresponds to the section $x = -15$ mm, dot-dash line corresponds to the section $x = -28.5$ mm; the ratio s of the maximum modulus values of stresses σ_{xz}^0, σ_{xz} to the moduli of their averaged values over the thickness of the beam, depending on x **b** dotted line for σ_{xz}^0 stresses; solid line for σ_{xz} stresses

using the ANSYS software package. Calculations have shown that in the sections $x = -l$, $x = 0$ and $x = a$ (see Fig. 13.1) $s \to \infty$. Therefore, the regions x located in the small vicinity of these sections are excluded from consideration.

The Table 13.1 shows the shear stresses σ_{xz}^0 in three x = const sections, found on the basis of the Timoshenko model, which are in good agreement with the results calculated by averaging the distributions obtained using the ANSYS software package.

The dynamic response of a flat beam (Fig. 13.2) under the action of a harmonic force $P = \tilde{P}e^{i\omega\tau}$ with an amplitude of $\tilde{P} = 240$ N/m (per beam's width unit) and circular frequency $\omega = 2\pi f$ at $f = 60$ Hz (this frequency is close to the lowest natural frequency of $f_1 = 61,046$ Hz) have been performed using the relations obtained in Sect. 13.4.2 at the analytical solution of the problem. Beam material is a unidirectional fiber reinforced composite on the basis of ELUR-P carbon fiber and KhT-118 epoxy. This material has the following characteristics: $E_1 = 100$ GPa; $G_{13} = 1$ GPa; $\rho = 1500$ kg/m^3. Beam geometrical parameters are the following: $l = 50$ mm; $b = 300$ mm; $a_1 = 200$ mm; $t = 3$ mm.

Figure 13.5 shows the amplitudes of deflections w of the beam at $f = 60$ Hz frequency. Thus, the deflections are observed not only in the loaded part of the beam, but also in the unloaded part due to the deformability of the fastened section of

Table 13.1 Shear stresses σ_{xz}^0 in three x = const sections

x (mm)	σ_{xz}^0 in case of using the Timoshenko model (MPa)	Averaged $\sigma_{xz}^0(z)$ in case of using ANSYS (MPa)
−28.5	−0.992	−0.923
−15.0	−1.383	−1.381
−1.5	−2.657	−2.758

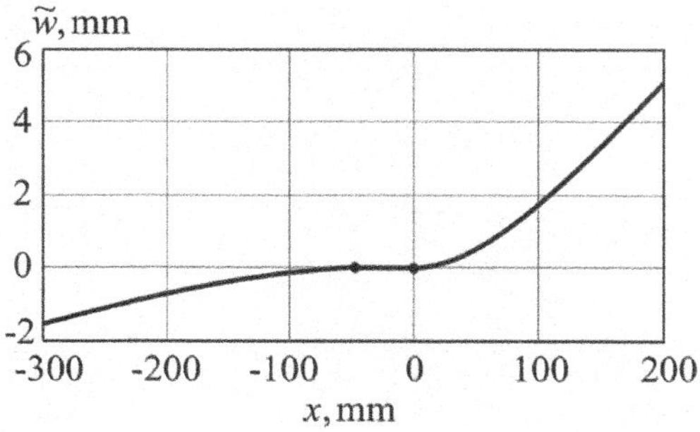

Fig. 13.5 Deflection amplitudes $\tilde{w}$ (points correspond to the boundaries of the fastened part of the beam)

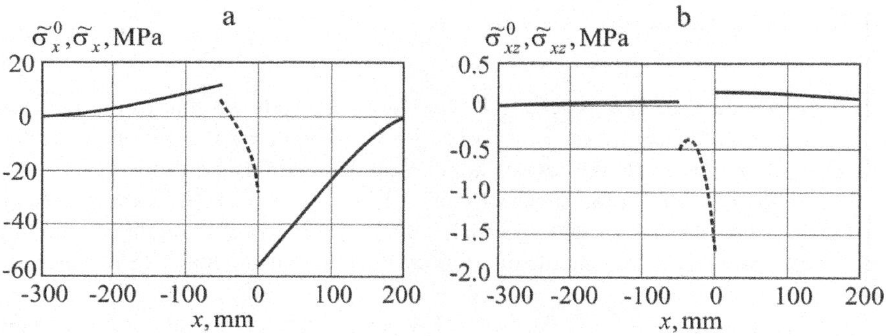

Fig. 13.6 Amplitudes of normal stresses $\overset{\sim 0}{\sigma}_x$, $\tilde{\sigma}_x$ on the $z = \frac{t}{2}$ surface (**a**) and amplitudes of shear stress $\overset{\sim 0}{\sigma}_{xz}$, $\tilde{\sigma}_{xz}$ (**b**): dashed lines are $\sigma_x^0, \sigma_{xz}^0$; solid lines are $\tilde{\sigma}_x$, σ_{xz}

finite length l. Figure 13.6 shows amplitudes of normal stresses $\overset{\sim 0}{\sigma}_x$, $\tilde{\sigma}_x$ on the $z = \frac{t}{2}$ surface and amplitudes of shear stresses $\overset{\sim 0}{\sigma}_{xz}$, $\tilde{\sigma}_{xz}$ in the cross sections of the beam. The results show the presence of a significant stress–strain state type transformation at the transition across the boundary from an unfastened beam's parts to the fastened one.

13.6 Conclusion

The basic relations of mechanics of flat beam with finite length fastening areas located on one of the front-face surfaces are derived based on the S.P. Timoshenko shear model. Equilibrium equations for the unfastened and fastened parts of the beam, as well as boundary conditions for them, are obtained based on the variational Lagrange equation. The kinematic and force conditions for coupling of fastened and unfastened beam sections are formulated. Based on the derived relations, an exact analytical solution to the problem of static bending of a cantilevered beam under the action of a constant surface load has been found. This solution is in good agreement with the results obtained by modeling of a beam using rectangular finite elements in a plane stress state, as well as using the ANSYS software package based on the equations of a plane problem of elasticity theory. An exact analytical solution has been obtained for the problem of transverse bending vibrations of a flat beam with two cantilevered parts and a fixed length section between them under vibration loading by a transverse force acting on one of the unfastened section. The results obtained on the basis of this solution show the transmission of vibrations from the loaded part of the beam to its unloaded part due to the deformability of the fixing section of finite length l.

The results obtained in the article are related to the study of the processes of deformation of flat beams with fixed sections of finite length under static and dynamic loading. They should be considered as a methodology for setting the corresponding problems of mechanics with additional account of the compliance of the beam's fixed section. Within the framework of the classical Kirchhoff–Love model it is possible to formulate only generally accepted boundary conditions in the transition section from an unfastened section to a fastened one. At the same time, within the framework of well-known refined Timoshenko model it is possible to take into account the fastening of a beam with an absolutely rigid supporting element only on one of the front surfaces.

The main result of this article should be considered as the proposed methodology for studying the process of deformation of thin-walled structural elements that have fixed sections of finite sizes on the boundary surfaces. It requires the use of refined deformation models with a high order of accuracy, which are transformed in the transition from an unfastened to a fastened section.

Acknowledgements This paper has been supported by the Russian Science Foundation (Project No 23-19-00021) (Sections 13.1–13.3) and Kazan Federal University Strategic Academic Leadership Program ("PRIORITY-2030") (Sections 13.4–13.6).

References

1. Reddy J (1984) A simple higher-order theory for laminated composite plates. J Appl Mech 51:745–752
2. Librescu L (1987) Refined geometrically non-linear theories of anisotropic laminated shells. Quart Appl Math 45:1–22
3. Schmidt R, Reddy J (1988) A refined small strain and moderate rotation theory of elastic anisotropic shells. J Appl Mech 55:611–617
4. Librescu L, Schmidt R (1988) Refined theories of elastic anisotropic shells accounting for small strains and moderate rotations. Int J Nonlinear Mech 23:217–229
5. Yankovskii A (2020) Critical analysis of the equations of statics in the bending theories of composite plates obtained on the basis of variational principles of elasticity theory 1. General Theories of high order. Mech Compos Mater 56(3):271–290

6. Yankovskii A (2020) Critical analysis of the equations of statics in the bending theories of composite plates obtained on the basis of variational principles of elasticity theory 2. Particular low-order theories. Mech Compos Mater 56(4):437–454
7. Paimushin VN, Firsov VA, Gynal I, Shishkin VM (2016) Theoretical-experimental method for evaluating the elastic and damping characteristics of soft materials based on studying the resonance flexsural vibrations of test specimens. Mech Compos Mater 52(5):571–582
8. Paimushin VN, Firsov VA, Shishkin VM (2019) Identification of the dynamic elasticity characteristics and damping properties of the OT-4 titanium alloy based on study of damping flexural vibrations of the test specimens. J Mach Manuf Reliab 48(2):119–129

Chapter 14
Active Damping of Transverse Vibrations of Console Beam by Piezoelectric Layer with Different Electrode Shapes

Egor V. Petrakov and Dmitry V. Balandin

Abstract In the problems of damping vibration, the question often arises on the practical implementation of damping actuators. The damping efficiency is considered for a console beam described by a linear viscosity Bernoulli–Euler model. The article presents the methods of damping transverse vibrations implemented by a dynamic damper from a piezoelectric layer distributed symmetrically along the axis of symmetry of the beam. Piezoelectric layers with a triangular and rectangular shape of electrode plates are considered, which affect the nature of mechanical stresses upon application of electrical voltage. The electrode plates are thin layers made of nickel or silver several microns thick and located normal to the polarization axis, that is, along the length of the piezoceramic plate. The control of the piezoelectric layers is realized by changing the potential difference between the electrode plates, while the piezoelectric material uncoated by the electrode plate on both sides is useless to use as an active material. In turn, mathematical models of the effect of piezoelectric elements on the cantilever beam are derived from the Hamilton principle. The Pareto-efficiency of quenching by piezoelectric plates with different electrode shapes is evaluated relative to two criteria: the level of control voltage and the maximum deflection of the beam. Also, for a more general analysis, the quenching efficiency is also given for a beam with a piezoelectric plate applied along the entire length and an electrode layer. In addition to Pareto sets, efficiency is also considered in a more applied and particular example—time history. It is worth noting that the synthesis of Pareto-optimal controls is based on the Germeier convolution, and the search for optimal feedback is based on the application of the theory of linear matrix inequalities and effective algorithms for solving them.

E. V. Petrakov (✉)
National Research Lobachevsky State University of Nizhny Novgorod, 23 Gagarin Avenue, Nizhny Novgorod 603022, Russian Federation
e-mail: pet.e@mail.ru

D. V. Balandin
Research and Education Mathematical Center, National Research Lobachevsky State University of Nizhny Novgorod, 23 Gagarin Avenue, Nizhny Novgorod 603022, Russian Federation
e-mail: dbalandin@yandex.ru

H. Altenbach et al. (eds.), *Deformation and Destruction of Materials and Structures Under Quasi-static and Impulse Loading*, Advanced Structured Materials 186,
https://doi.org/10.1007/978-3-031-22093-7_14

Keywords Vibration damping · Distributed system · Piezoelectrics · Bernoulli–Euler model · Generalized H_2-norm · Pareto set · Linear matrix inequalities

14.1 Introduction

The problem of reducing the level of vibrations affects many areas of engineering, instrumentation and construction. Often the solution is to increase the strength and stability of protected objects, but apart from the widely known methods of vibration (vibration, balancing, balancing machines, changing the materials of the protected object) [1] there is a less known method of applying a mass damper (MD). The use of MD takes a special place because their use can be implemented not only at the design and construction stage, but also in the case when unsatisfactory dynamic characteristics of the structure are identified already during operation. In addition, the advantage of dampers is that at a relatively low cost of additional material, they make it relatively easy to get the effect of reducing the level of vibrations.

To create forces that dampen harmful vibrations of the protected object, it is important to choose the optimal actuator that will meet various requirements to the desired extent, for example, such as compactness or scale of force generation. Known components of actuators (electromagnetic devices, pneumatic drives, rotors and linear motors [2], etc.) can be used effectively in vibration damping tasks. In contrast to the mentioned devices, modern technical materials, which are often referred to as smart materials, have a number of advantages, since they can be lightweight and, more importantly, can be seamlessly integrated structurally into an already designed system without changing the inertial characteristics of the system. Smart materials are materials that have one or more properties that can be changed significantly in a controlled way by external perturbations such as voltage, temperature, humidity, pH, electric or magnetic fields. There are many types of smart materials, some of which are already common. Due to the possibility of generating forces, piezoelectric materials are preferable for solving this problem. Among the many types of piezoelectric devices, bending piezoactuators are of particular interest, the use of which is considered in this article for damping vibrations.

A striking example of such integration is the composite aeroelastic wing, equipped with thin piezoelectric plates in the wing structure, which allow to suppress unwanted vibrations, without adding significant mass and without changing the shape of the wing. A similar quenching method is used in the wing of an FA-18 deck aircraft equipped with piezoelectric plates for vertical stabilization [3] (Fig. 14.1).

Fig. 14.1 Wing carrier-based aircraft FA-18

14.2 Statement of the Problem

To dampen vibrations of the structure (Fig. 14.2a), which is under the influence of external perturbations, we consider piezoelectric layers that partially cover the surface of the beam and have different forms of electrode plates (Fig. 14.2b, c). In other words, we consider a piezoelectric layer consisting of a piezoceramic plate, which is covered with electrode plates along the length. A mathematical model of vibration damping of an elastic beam with piezoelectric layer actuators is derived. A comparative analysis of the action of piezoelectric layers with identical piezoceramic plates, but with electrode plates of different shapes, is performed. The location of the piezoelectric layer is selected from the conditions for the most effective control of vibrations. The effectiveness of quenching is considered relative to two criteria: the control voltage applied to the piezoelectric layer and the deflection of the beam at the end. In other words, it is necessary to find a Pareto set of solutions with respect to two criteria, while the maximum value of the control voltage does not exceed the maximum allowable value of the voltage for piezoelectric layers.

The main object of research is a cantilever metal beam (Fig. 14.2a), the size and weight of which are so small in comparison with the size and weight of the base that the influence of the beam on the base can be ignored. Piezoelectric layers are represented with triangular and rectangular electrode forms of plates, the length of the layers is 5 times less than the length of the beam, and the thickness is less by an order of magnitude. The behavior of the beam is considered in the framework of the Bernoulli–Euler hypothesis, the influence of the thickness of the piezoelectric layer at the moment of inertia of the beam is ignored, since it does not significantly

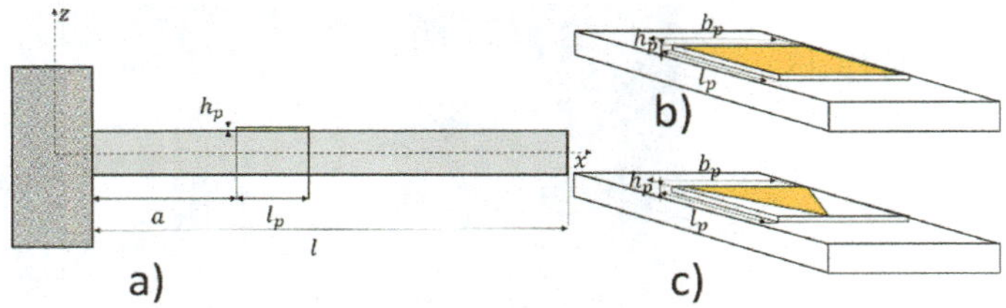

Fig. 14.2 Protected object with piezoelectric layers: **a** general view, with **b** rectangular and **c** triangular plates attached to the beam

change the stiffness and inertia characteristics of the system, the hysteresis of the piezoelectric layer is not taken into account.

14.3 Mathematical Model of Active Damping of Transverse Beam Vibrations

A linear viscoelastic model is used to describe the transverse vibrations [4] of a beam in the framework of the Bernoulli–Euler hypotheses in dimensionless variables:

$$\frac{\partial^2\omega}{\partial t^2} + a^2\left(\frac{\partial^4\omega}{\partial x^4} + \vartheta_0\frac{\partial^5\omega}{\partial x^4\partial t}\right) = -\frac{u_0}{\rho A} + v_0(t) \tag{14.1}$$

where $\omega = \omega(x, t)$ is dimensionless deflection of the beam, $K = \frac{\partial^4}{\partial x^4}$ is the differential operator of the fourth degree, $\beta = \frac{a}{l^2}\vartheta$ is dimensionless damping coefficient of the system ϑ is the coefficient of internal viscosity, $a^2 = EJ/\rho A$, E is elastic modulus, J is moment of inertia of the section, ρ is density, A is cross-sectional area of the beam, l is beam length, $v(t)$ is acceleration acting on the base, t is time, $f(u, x_1, .., x_n)$ is control function applied at certain points of the beam $x_1, .., x_n$ or at intervals between points, u is the control, which will be mainly discussed later.

Equation (14.1) in partial derivatives is reduced to the equation in principal coordinates (relative to the time function). For this purpose, the methods of separated variables and normal forms are used, described in more detail in [5].

$$\omega(x, t) = \sum_{i=1}^{\infty} X_i(x)T_i(t), \int_0^1 X_iX_jdx = 1, \int_0^1 KX_iX_jdx = \begin{cases} \lambda_i^4, i = j \\ 0, i \neq j \end{cases} \tag{14.2}$$

λ_i is eigenvalues of a beam, analogous to the eigenfrequency of a beam in dimensionless variables. X_i are parameters that are functions of the beam shape. The values of eigenvalues and form functions are derived from the boundary conditions of the cantilever beam described by the Krylov function [6]. The transformation (14.1) to the main coordinates is represented in the equation:

$$\ddot{T}_i + \beta\lambda_i^4\dot{T}_i + \lambda_i^4 T_i = -\int_0^1 X_i f(u, x_1, .., x_n)dx + \int_0^1 X_i dx v(t) \tag{14.3}$$

Equation (14.3) is a mathematical model for controlling transverse vibrations of a beam under external perturbation. However, when considering a practical problem, the control is implemented using actuators, which makes very impressive changes to the mathematical model for controlling transverse vibrations, which is described in more detail later.

14.4 Active Damping of the Cantilever Beam with a Piezoelectric Layer

The piezoelectric layer is a piezoceramic plate with electrode plates attached to it on both sides (Fig. 14.3), through which control is performed by changing the potential difference between them. Piezoceramic material that is not covered with an electrode lining on both sides is useless to use as an active material.

Linear equations of electrostatics and deformation for piezoceramics can be written as:

$$\begin{aligned} [S] &= [s]\{T\} + [d]\{E\} \\ [D] &= [d]^T\{T\} + [\varepsilon]\{E\} \end{aligned} \tag{14.4}$$

where $\{S\}$, $\{T\}$ is strain and stress vectors, $[s]$ is the matrix of elastic compliance, $[d]$ is tensor of piezoelectric constants, $[\varepsilon]$ is the tensor of the dielectric constants, $\{E\}$ is vector of electric field strength, $[D]$ is vector of electrical induction. In Eq. (14.4), the first equation describes the reverse effect, and the second one describes the direct effect. Since the variables are interconnected in pairs, any pair of mechanical and electrical variables can be selected as independent [7].

The electrical energy of a piezoelectric element is found from the expression:

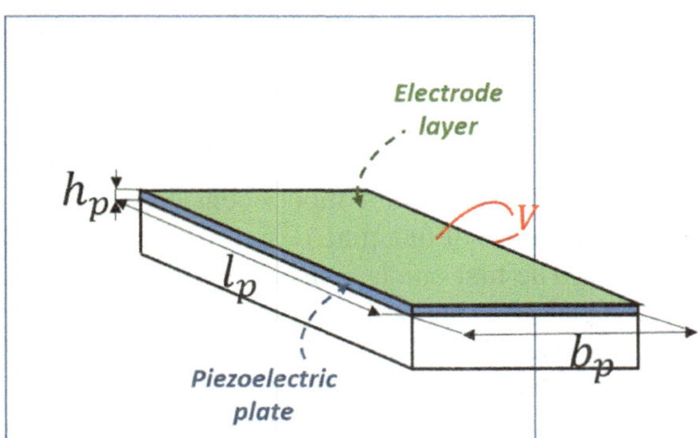

Fig. 14.3 Cantilever beam with a piezoelectric layer along its entire length

$$W_e = \frac{1}{2}\int\limits_{\Omega} (\{E\}^T[\varepsilon]\{E\} + 2\{S\}[e]\{E\} - \{S\}^2[c]\{S\})d\Omega$$

For the case of a piezoelectric plate polarized along the z axis operating on a pure bend $(S_1 = -z\omega'')$ applied to the beam at the interval $[x_1, x_2]$ written as:

$$W_e^* = \frac{1}{2}\int_0^l \int_A (\varepsilon_{33}E_3^2 - 2\omega'' z H(x_1^0, x_2^0)e_{31}E_3 - c_{11}(\omega'')^2)dAdx \tag{14.5}$$

где c_{11} is the modulus of elasticity of the piezoelectric ceramics, $e_{31} = d_{31}c_{11}$, $H(x_1^0, x_2^0)$ is Heaviside function describing the location of the piezoelectric layer. Applying the Hamilton least action principle for the considered beam the following equation is derived:

$$\int_{t_0}^{t_1}\int_0^l \frac{1}{2}(\rho A\delta\dot{\omega}^2 - 2e_{31}z_m b_p(x)H(x_1^0, x_2^0)V\delta\omega'' + EI\delta\omega'^2)dxdt = 0 \tag{14.6}$$

where $V = E_3 h_p$ is the voltage applied to the piezoelectric layer, h_p is the piezolayer height, $b_p(x)$ is the width of the electrode plate, in the future we will consider cases of changing the width, so it is accepted $b_p^0 = \max(b_p(x))$, also $b_p(x) = b_p^0\overline{b_p(x)}$, $\overline{b_p(x)}$ is a dimensionless function of changing the width of the electrode lining along the length. Thus, a dimensionless model of transverse vibrations of a beam with consideration for friction and an applied piezoelectric layer with an arbitrary shape electrode plate at the interval will take the following form:

$$\ddot{\omega} + \beta K\dot{\omega} + K\omega = \gamma(\overline{b_p(x)}H(x_1, x_2))''U \tag{14.7}$$

where $\gamma = lb_p^0 z_m e_{31}V_0/EI$ is a dimensionless coefficient of influence of the piezoelectric layer, $(x_1, x_2) = \frac{(x_1^0, x_2^0)}{l}$ is the dimensionless interval of the application, U is the dimensionless voltage applied to the piezoelectric layer that is the control.

$$U = \sum_{i=1}^{m} \theta_i^{(1)}T_i + \theta_i^{(2)}\dot{T}_i \tag{14.8}$$

where $\theta_i^{(1)}, \theta_i^{(2)}$ are the feedback coefficients for movement and speed, respectively, m is the number of the first modes considered.

Quantification of damping design is usually based on the results of solving two problems. The first one is connected with consideration of free oscillations, where the dissipation of the system is manifested in the damping of the oscillations and the decay rate quantifies the dissipative properties of the system. The second task is focused on dealing with forced steady-state oscillation [8]. In addition, the damping of the system's vibrations is manifested in the restriction of resonant amplitudes.

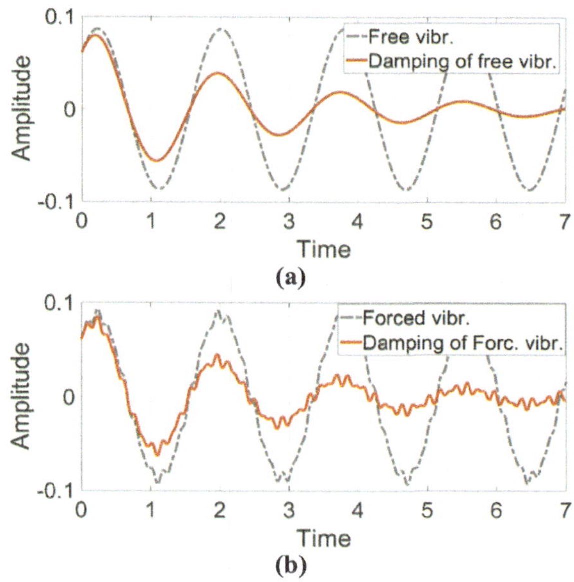

Fig. 14.4 Amplitude history: **a** free and **b** forced oscillations of the system

Figure 14.4 shows amplitude history for a piezoelectric layer with an electrode lining distributed along the entire length of the beam (Fig. 14.3).

From the above amplitude histories of natural and forced vibrations of the beam, with a piezoelectric layer distributed along the entire length, the effect of control is obvious. However, choosing such a feedback with minimal control costs that has the maximum possible amplitude reduction is not a trivial task, and it is obvious that it is more difficult than the task of single-criteria optimization.

14.5 Statement of the Two-Criteria Problem of Control Theory

The need to transfer the problem in point 1 to the problem of control theory is due to the fact that for this problem, the search for an optimal set of solutions with respect to the quenching efficiency criteria (moving the end of the beam and controlled stress), at least for three modes, is extremely cumbersome. Moreover, it is difficult to implement, and the solution by iteration cannot always give optimal solutions. Hence, the system of Eqs. (14.3) reduces to the canonical form of a controlled linear system with two outputs:

$$\begin{aligned}\dot{T} &= AT + B_u V + B_v v \\ z_1 &= CT \\ z_2 &= Du\end{aligned} \tag{14.9}$$

where $T = \left(T_1 \ldots T_m, \dot{T}_1 \ldots \dot{T}_m\right)^T$ is the state vector, A is the control matrix, B_u, B_v are the control and perturbation vectors, v is the external impact on the system, z_1, z_2 are the output values of the system and indicate the movement of the free end of the beam and the voltage generated by the feedback (quenching efficiency criteria). The theory of linear matrix inequalities and effective algorithms for solving them, implemented in the MATLAB package, is used to find optimal feedback about the two criteria.

Despite the fact that over the past two decades, progress has been made in solving optimal control problems with criteria such as H_∞- and H_2-norms that have correct physical interpretations in the form of levels of damping of deterministic or stochastic perturbations from various classes, the consideration of multi-criteria problems with these criteria causes significant difficulties. These difficulties are mainly related to the complexity of characterization of the Pareto set and finding the corresponding scalar multi-purpose function that would define this set. In addition, the task is complicated by the fact that each of the criteria is characterized by its quadratic Lyapunov function with a matrix which is the solution of Riccati equations or linear matrix inequalities and scalar optimization multi-purpose function in the form of a particular convolution leads in general to a bilinear system of inequalities matrices of these functions that are Lyapunov matrices and controller feedback. To solve such a system, an additional condition was often introduced about the equality of all Lyapunov functions, which introduced conservatism to the problem under consideration [7, 9–13]. At the same time, the main question remained unanswered, as to what extent the obtained control laws differ from the Pareto optimal ones.

In recent works [14–18] on multi-criteria optimization with criteria in the form of H_∞-and γ_0-norms in deterministic and stochastic settings, we have found Pareto suboptimal control laws whose relative losses in comparison with Pareto optimal ones do not exceed $1 - \frac{\sqrt{N}}{N}$, where N is the number of criteria. Therefore, we will use the generalized H_2 norm. [19, 20] Using two competing outputs is associated with bringing the task to practical implementation, since the limited control resource imposes certain restrictions on the feedback values.

For solving two-criterion problem of minimization, it is necessary to use functionality that combines both criteria, generally used for this convolution. In this problem, the Germeyer convolution is implemented, since the solution within the Germeyer convolution for the generalized H_2 -norm gives optimal solutions:

$$J_\alpha(\theta) = \max\left\{\frac{J_1(\theta)}{\alpha}; \frac{J_2(\theta)}{1-\alpha}\right\}, \alpha \in (0, 1) \tag{14.10}$$

The search for an optimal set of solutions with respect to two criteria is implemented by linear matrix inequalities describing the finding of a generalized H_2-norm:

$$\begin{pmatrix} AY + YA^T + B_u Z + Z^T B_u^T & B_v \\ B_v^T & -I \end{pmatrix} < 0 \begin{pmatrix} Y & C_\alpha^T \\ C_\alpha & \alpha^2 \gamma^2 \end{pmatrix} \geq 0 \begin{pmatrix} Y & Z^T D_\alpha^T \\ D_\alpha Z & (1-\alpha)^2 \gamma^2 \end{pmatrix} \geq 0$$

where $Z = \theta Y$, $Y = X^{-1}$ is the inverse Lyapunov matrix. The result of the solution will be a Pareto set of optimal solutions with respect to the two outputs of the deflection of the end of the beam and the control value, which are found from the expressions:

$$\begin{aligned} J_1 &= \|H_{z_1 v}\|_{g_2} = \sup_{v \in L_2} \frac{\|z_1(t)\|_{g\infty}}{\|v\|_2} = \sup_{v \in L_2} \frac{sup_{t \geq 0} |z_1(t)|_{g\infty}}{\|v\|_2} = \lambda_{max}^{\frac{1}{2}} \left(C Y C^T \right) \\ J_2 &= \|H_{z_2 v}\|_{g_2} = \sup_{v \in L_2} \frac{\|z_2(t)\|_{g\infty}}{\|v\|_2} = \lambda_{max}^{\frac{1}{2}} \left(D \theta Y \theta^T D^T \right) \end{aligned} \tag{14.11}$$

This describes the mathematical apparatus for searching for optimal quenching values, and it is necessary to find a mathematical model of active quenching of a beam by piezoelectric layers with different forms of electrode plates.

14.6 Mathematical Model of Active Damping of a Beam by Piezolelectric Layers of Various Shapes

Let's consider special cases of application of piezoelectric layers, using the model of active vibration damping of the piezoelectric layer with an arbitrary shape of the electrode lining (14.7) (Figs. 14.5 and 14.6).

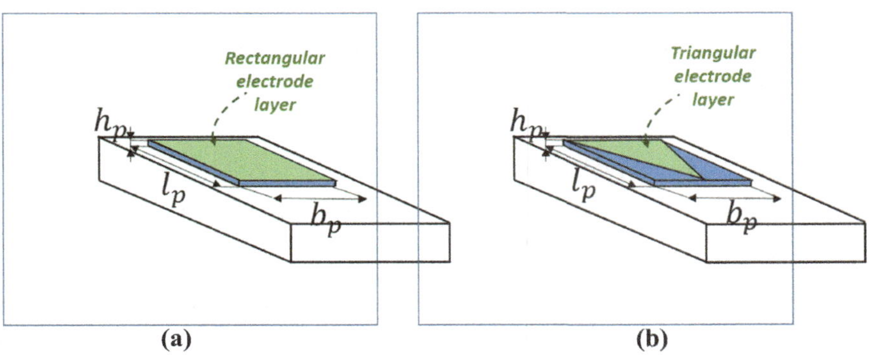

Fig. 14.5 Application of piezoelectric layers with different shapes of electrode plates: **a** rectangular, **b** triangular

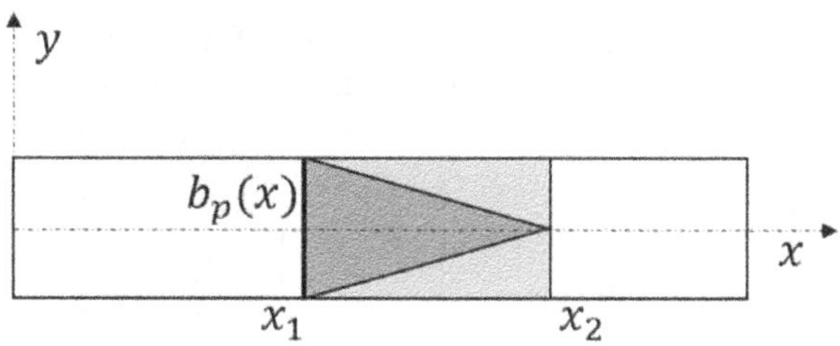

Fig. 14.6 Piezoelectric layer with a triangular electrode plate

For a piezoelectric layer with an electrode lining, based on the results obtained using (14.7), we can derive the equation in the main coordinates:

$$\ddot{\omega} + \beta K\dot{\omega} + K\omega = \gamma(-\delta'(x - x_1) + \delta'(x - x_2))U \tag{14.12}$$

After the discovery of the piezoelectric effect in the study of the electrical properties of solid dielectrics of the crystal structure, Pierre Curie formulated a general principle, which is now called the Curie principle. Its meaning is that the phenomenon has all the attributes of symmetry that the cause that gave them birth has; the asymmetry of the phenomenon is predetermined by the asymmetry of the cause. If we consider that the piezoelectric material is self-balancing, then the question arises about the behavior of the piezoelectric element for the case of transverse asymmetry. For example, what will be the behavior and influence of the piezoelectric layer when changing the shape of the electrode plates and, consequently, the generated mechanical loads. The shape of the surface electrode plates determines the nature of the mechanical load represented by the piezoelectric layer. Therefore, consider a piezoelectric layer covered with a triangular electrode plate on both sides.

The equations of transverse vibrations of a beam with a triangular overlay are written as follows:

$$\ddot{\omega} + \beta K\dot{\omega} + K\omega = \gamma\left(-\delta'(x - x_1) - \frac{l}{l_p}\delta(x - x_1) + \frac{l}{l_p}\delta(x - x_2)\right)U \tag{14.13}$$

Thus, piezoelectric layers, depending on the shape of the plates when applying voltage, can be described as mechanical loads that are controlled by an external energy source with a limited resource:

$$\begin{aligned}
\ddot{T}_i + \beta\lambda_i^4\dot{T}_i + \lambda_i^4 T_i &= -M_p X_i'(1)U + v(t)\\
\ddot{T}_i + \beta\lambda_i^4\dot{T}_i + \lambda_i^4 T_i &= -M_p\big(X_i'(x_2) - X_i'(x_1)\big)U + v(t)\\
\ddot{T}_i + \beta\lambda_i^4\dot{T}_i + \lambda_i^4 T_i &= -(F_p(X_i(x_2) - X_i(x_1)) - M_p X_i'(x_1))U + v(t)
\end{aligned} \tag{14.14}$$

where the first equation describes the behavior of a piezoelectric layer evenly distributed along the entire length of the beam (Fig. 14.7a), a kind of the most efficient and limiting case, the second and third equations describe the behavior of a piezoelectric layer with rectangular and triangular facings, respectively (Fig. 14.7b, c).

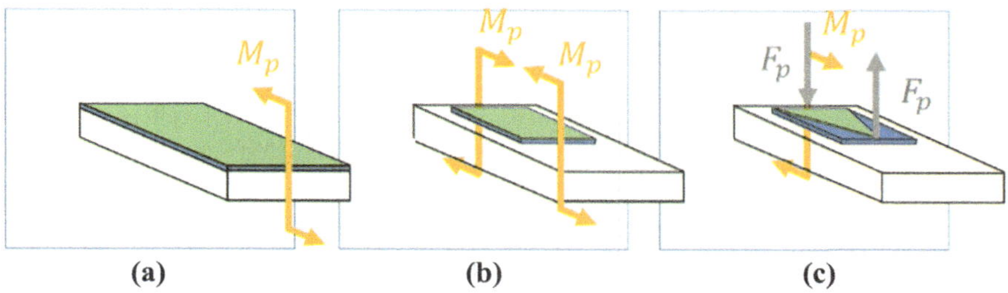

Fig. 14.7 Mechanical loads of piezolayers with different plates: **a** distributed along the entire length; **b** rectangular and **c** triangular forms of electrode plates

14.7 Results

All calculations were given for the coefficient $\gamma = 0.0516$, (see (14.7)). In Fig. 14.8 optimal sets of solutions for the criteria are given.

Judging by the optimal Pareto sets, the efficiency of a piezo layer with a rectangular lining is higher than that of a triangular one in both operating modes. It would seem that it is possible to talk about the inefficient use of a piezo layer with a triangular lining for the tasks of active damping of the cantilever beam, but the following is an amplitude history for both cases of operation with respect to forced vibrations of the system (Fig. 14.9).

From amplitude history obviously, when an operating voltage (Fig. 14.9a) the difference between amplitude history for different plates is minimal, but extreme voltage differences amplitude history (Fig. 14.9b) has more noticeable, but still not much. It is worth noting that the length of the piezoelectric layers was 5 times less than the length of the beam, and the thickness of the layers is less by an order of magnitude. That is, when using thin piezoelectric layers, it is possible to reduce significantly the amplitude of vibrations in the case of external influence on the cantilever metal beam, both for the case of the limit voltage of the piezoelectric layer and for the operating voltage.

Fig. 14.8 Optimal Pareto sets of solutions

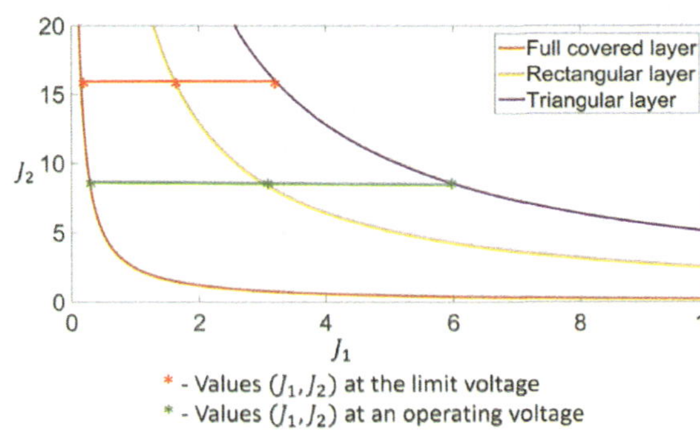

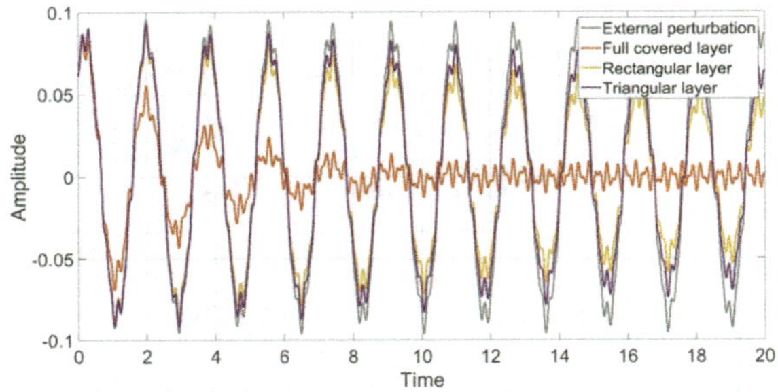

(a)

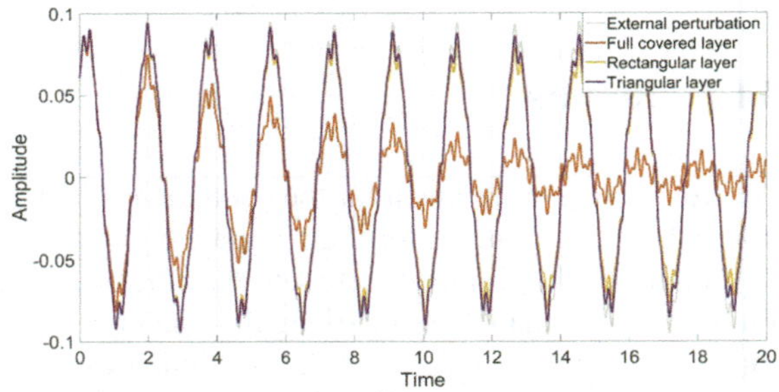

(b)

Fig. 14.9 Amplitude history: **a** at the limit and **b** at the operating voltage

14.8 Conclusions

The use of piezoelectric layers in the system of active vibration damping of a cantilever beam is considered. To solve the problem of the efficiency of using piezoelectric layers as an active damping device, two related criteria are introduced: the level of voltage applied to the piezoelectric layer and the level of deflection of the beam end. A comparison was made with respect to the selected criteria for rectangular and triangular forms of plates, which were compared with a variant of a piezo layer evenly distributed over the entire length.

Based on the Hamilton Principle, the influence of the electrode shape on the generation of various mechanical loads by the piezoelectric layer was derived (14.7). It was found that with respect to two Pareto criteria, the efficiency of a piezoelectric plate of standard sizes with a rectangular shape of the electrode plate shows a better result compared to a triangular plate.

Acknowledgements The work is supported financially by the Ministry of Science and Higher Educa-tion of the Russian Federation, project no. 0729-2020-0055.

References

1. http://www.obayashi.co.jp/english/services/technologies/#anc02
2. Bhaskararao A, Jangid R (2006) Seismic analysis of structures connected with friction dampers. Eng Struct 28(5):690–703
3. Xu YL, He Q, Ko J (1999) Dynamic response of damper-connected adjacent buildings under earthquake excitation. Eng Struct 21(2):135–148
4. Xu YL, Zhan S, Ko J, Zhang W (1999) Experimental investigation of adjacent buildings connected by fluid damper. Earthquake Eng Struct Dynam 28(6):609–631
5. Zhang WS, Xu YL (2000) Vibration analysis of two buildings linked by maxwell model-defined fluid dampers. J Sound Vib 233(5):775–796
6. Bharti S, Dumne S, Shrimali M (2010) Seismic response analysis of adjacent buildings connected with MR dampers. Eng Struct 32(8):2122–2133
7. Christenson RE, Spencer B Jr, Johnson EA (2007) Semiactive connected control method for adjacent multidegree-of-freedom buildings. J Eng Mech 133(3):290–298
8. Xu YL, Ng C (2008) Seismic protection of a building complex using variable friction damper: experimental investigation. J Eng Mech 134(8):637–649
9. Asano M, Yamano Y, Yoshie K, Koike Y, Nakagawa K, Murata T (2003) Development of active-damping bridges and its application to triple high-rise buildings. JSME Int J Ser C 46(3):854–860
10. Bian Y, Gao Z (2017) Nonlinear vibration absorption for a flexible arm via a virtual vibration absorber. J Sound Vib 399:197–215
11. Christenson RE, Spencer B Jr, Hori N, Seto K (2003) Coupled building control using acceleration feedback. Comput Aided Civ Infrastruct Eng 18(1):4–18
12. Kasagi M, Fujita K, Tsuji M, Takewaki I (2016) Automatic generation of smart earthquake-resistant building system: hybrid system of base isolation and building-connection. Heliyon 2:69–86
13. Zhang YF, Iwan WD (2003) Statistical performance analysis of seismic excited structures with active interaction control. Earthquake Eng Struct Dynam 32(7):1039–1054
14. Balandin DV, Kogan MM (2017) Pareto optimal generalized H2-control and vibroprotection problems. Autom Remote Control 78(8):1417–1429
15. Balandin DV, Kogan MM (2007) Sintez zakonov upravlenia na osnove lineinyh matrichnyh neravenstv [Control law synthesis based on linear matrix inequalities]. Fizmatlit Publ, Moscow [In Russian]
16. Hazanov HS (2002) Mechanicheskie kolebaniya sistem s raspredelennymi parametrami [Mechanical vibrations of systems with distributed parametrs]. Samara. SSAU Publ [In Russian]
17. Hlebnikov MV, Sherbakov PS Ogranichennoe lineinoe upravlenie, optimal'noe po kvadratichnomu kriteriu special'nogo vida [Limited linear control optimal by a quadratic criterion of special type]. [ISA RAN work]. 63:85–89 [In Russian]
18. Szmidt T, Pisarskin D, Bajer C, Dyniewicz B (2017) Double-beam cantilever structure with embedded intelligent damping block: dynamics and control. J Sound Vib 401:127–138
19. Balandin DV, Ezhov EN, Petrakov EV, Fedotov IA (2018) Multi-criteria problems for optimal protection of elastic constructions from vibrations. Mater Phys Mech 40:239–245
20. Korenev BG, Reznikov DM (1988) Dinamicheskie gasiteli kolebanii [Mass Damper]. Nauka Publ, Moscow [In Russian]

Chapter 15
Investigation of the Dynamic Response in a One-Dimensional Partially Saturated Poroelastic Medium

Andrey N. Petrov and Leonid A. Igumnov

Abstract In this paper, the dynamic response of a poroelastic material to an impulse load has been analytically investigated. The poroelastic material is represented by a three-phase elastically deformable partially saturated porous medium, one phase is a deformable solid skeleton, and the other two are fluid and gas that fill the pore space. The mathematical formulation is a system of equations including the laws of conservation of mass for fluid phases and a solid frame, an equilibrium equation for a three-phase continuum, a constitutive relation for a three-phase medium, a relation for determining the stress tensor, and Darcy's law of filtration. The solution of the equations of motion of a one-dimensional poroelastic medium is written in terms of the variables of the displacement of a solid skeleton, and the pore pressure of fluid and gas in the Laplace domain. The time domain solution is calculated using the numerical inversion stepped method of the Laplace transform. The effect of material parameters on the dynamic response is analyzed in a series of numerical experiments. The results of this study may help to gain better insight into one-dimensional wave propagation in unsaturated soils.

Keywords Poroelastic medium · Equilibrium equation for a three-phase continuum · Constitutive relation for a three-phase medium · Waves propagation in unsaturated soils

15.1 Introduction

Porous materials are widely distributed in nature and engineering as exemplified by fluid-saturated soils, rocks, biological tissues, foam metals, etc. The mechanics of porous media is of great importance in many branches of science, such as geotech-

A. N. Petrov (✉) · L. A. Igumnov
Research Institute for Mechanics, National Research Lobachevsky State University of Nizhny Novgorod, 23, Gagarin av., bld. 6, 603950 Nizhny Novgorod, Russian Federation
e-mail: andrey.petrov@mech.unn.ru

L. A. Igumnov
e-mail: igumnov@mech.unn.ru

H. Altenbach et al. (eds.), *Deformation and Destruction of Materials and Structures Under Quasi-static and Impulse Loading*, Advanced Structured Materials 186,
https://doi.org/10.1007/978-3-031-22093-7_15

nics, geomechanics, engineering geology, biomechanics, mechanical engineering, and materials science.

In recent years, the dynamic response of saturated soils under non-stationary loads has been widely studied analytically. For example, Philippacopoulos [8] obtained explicit solutions for displacements and stresses to the problem of deformation of a homogeneous poroelastic half-space in the frequency domain using Helmholtz expansions and Fourier transform. Tabatabaie Yazdi et al. [10] analyzed the dynamic response of a porous-elastic layer subject to loading on a strip which simulates a strip foundation. They found an exact solution in the frequency domain and proposed a finite element scheme for solving complex geometry problems. Jin and Liu [3] investigated the motion of a three-dimensional poroelastic half-space created by a horizontal buried load using integral transformation methods.

The use of the model of a fluid-saturated porous material when considering wave processes, even with account for substantial simplifications, significantly complicates the computational scheme of the boundary value problem compared to elastic or viscoelastic formulations. The contribution of the dynamic behavior of the filler fundamentally changes the form of wave patterns, which can be predicted by using advanced computational methods. Nevertheless, analytical solutions are of great importance, since they make it possible to reveal the mechanisms of interaction between solid and fluid phases, to carry out a posteriori evaluation of numerical results, and to study transient processes in a porous medium subjected to dynamic loading. An overview of available analytical solutions to the problems of wave propagation in saturated porous media can be found in the work by Schanz [9]. Li and Schanz [4] presented an analytical solution for a one-dimensional partially saturated poroelastic rod.

15.2 Analytical Solution of One-Dimensional Wave Propagation

In this study, the partially saturated poroelastic medium consists of three constituents, that is, solid skeleton, pore fluid, and pore gas. All the three phases are assumed to be compressible. Temperature variations are neglected. Such a poroelastic material can be considered as partially saturated. In the following, the mathematical model suggested by Zhang et al. [11] will be used to describe the dynamic behavior of a partially saturated poroelastic medium.

The relative proportions of constituent volumes are characterized by its porosity ϕ and saturation degree S_f of each phase $f \in \{w, a\}$ as

$$\phi = \frac{V_v}{V}\,, S_f = \frac{V_f}{V_v}\,, \tag{15.1}$$

where V_v is the volume of interconnected pores in the specimen, V is the total volume of the material, and V_f is the volume of the f-phase. This gives the average density

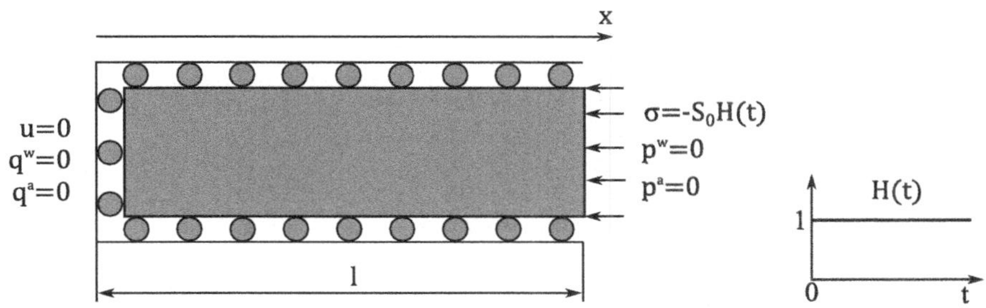

Fig. 15.1 Problem description

of the mixture ρ as

$$\rho = (1 - \phi)\rho_s + \phi S_w \rho_w + \phi S_a \rho_a \ , \tag{15.2}$$

where ρ_s, ρ_w, and ρ_a are partial densities of each phase. Consider a case where the pores are completely filled:

$$S_a + S_w = 1 \ . \tag{15.3}$$

The commonly applied [1] model of capillary pressure p_c function is utilized in this study, i.e.,

$$S_e = \left(1 + (\chi p_c)^d\right)^{-m} \ , \tag{15.4}$$

where S_e is the effective fluid saturation given as $S_e = (S_w - S_{rw})/(S_{ra} - S_{rw})$, S_{ra} and S_{rw} are the residual saturations, χ, m, and d are the material parameters of the van Genuchten model.

Based on the theoretical pore size distribution model of Mualem [7], the relative permeabilities of fluid K_{rw} and gas K_{ra} as a function of fluid saturation have the following form:

$$K_{rw} = \sqrt{S_e}\left(1 - \left(1 - S_e^{\frac{1}{m}}\right)^m\right)^2 \ , \tag{15.5}$$

$$K_{ra} = \sqrt{1 - S_e}\left(1 - S_e^{\frac{1}{m}}\right)^{2m} \ . \tag{15.6}$$

Consider the problem of the load impact in the form of a Heaviside step function $H(t)$ with respect to time on a one-dimensional partially saturated poroelastic rod of length l (Fig. 15.1).

The boundary conditions are given as

$$\hat{p}^w(s, x = l) = 0 \ , \tag{15.7}$$

$$\hat{p}^a(s, x = l) = 0 \ , \tag{15.8}$$

$$\hat{\sigma}(s, x = l) = -S_0 \,, \tag{15.9}$$

$$\hat{q}^w(s, x = 0) = 0 \,, \tag{15.10}$$

$$\hat{q}^a(s, x = 0) = 0 \,, \tag{15.11}$$

$$\hat{u}(s, x = 0) = 0 \,, \tag{15.12}$$

where $\hat{q}^w$ and $\hat{q}^a$ are the fluid and gas fluxes, $\hat{\sigma}$ is the total stress, S_0 denotes the loading amplitude, symbol denotes Laplace transform with complex variable s.

The solutions for the displacement of solid skeleton $\hat{u}$ and pressures of fluid $\hat{p}^w$ and gas $\hat{p}^a$ in the rod can then be derived in the Laplace domain as

$$\hat{u} = \frac{S_0}{Ms^2B_1}\sum_{i=1}^{3}\left(\frac{e^{-\lambda_i s(x+l)} - e^{-\lambda_i s(x-l)}}{1+e^{-2\lambda_i sl}}t_i\right) , \tag{15.13}$$

$$\hat{p}^w = \frac{-S_0}{Ms^2B_1}\sum_{i=1}^{3}\left(\frac{e^{-\lambda_i s(x+l)} + e^{-\lambda_i s(x-l)}}{1+e^{-2\lambda_i sl}}a_i t_i\right) , \tag{15.14}$$

$$\hat{p}^a = \frac{-S_0}{Ms^2B_1}\sum_{i=1}^{3}\left(\frac{e^{-\lambda_i s(x+l)} + e^{-\lambda_i s(x-l)}}{1+e^{-2\lambda_i sl}}b_i t_i\right) , \tag{15.15}$$

where the coefficients M, B_1, λ_i, t_i, a_i, and b_i are defined in appendix. For more details on the derivation process of solution, please refer to Li and Schanz [4]. In this work, the response in the time domain is calculated using the numerical inversion method of the Laplace transform.

15.3 Laplace Transform Inversion

Consider a method based on the fundamental integration theorem—the stepped method of numerical inversion of the Laplace transform. Consider the following integral:

$$y(t) = \int_0^t f(\tau)\mathrm{d}\tau, 0 \le t \le T. \tag{15.16}$$

Integral (15.16) gives rise to the Cauchy problem for an ordinary differential equation:

$$\frac{\mathrm{d}}{\mathrm{d}t}x(t) = sx(t) + C, x(0) = 0. \tag{15.17}$$

Integral (15.16) is substituted for by a quadrature sum, weighting factors of which are determined using Laplace representation and the linear multistep method [2]. The traditional stepped method yields approximations of (15.16) at discrete time points

$$t_n = n\Delta t \ , \ \Delta t = T/N \ , n = 0, \ldots, N \ , \tag{15.18}$$

via the quadrature formula

$$y(0) = 0 \ , \ y(n\Delta t) = \sum_{k=1}^{n} \omega_k(\Delta t) \ , n = 1, \ldots, N \ , \tag{15.19}$$

For a parameter $0 < R < 1$, the quadrature weights are defined by

$$\omega_n(\Delta t) = \frac{R^{-n}}{L} \sum_{l=0}^{L-1} \hat{f}\left(\frac{\gamma(R e^{il\frac{2\pi}{L}})}{\Delta t}\right) e^{-inl\frac{2\pi}{L}} , \tag{15.20}$$

where is γ the quotient of the characteristic polynomials of a linear multistep method according to Lubich [5, 6], the backward differentiation formula of second-order $\gamma(z) = 3/2 - 2z + z^2/2$ is adopted in this paper.

15.4 Numerical Simulations

In order to investigate the effects of fluid saturation, a series of parametric studies are conducted. The length of the rod l is taken equal to 10 m. The material properties are provided in the appendix, and a Heaviside step function with $S_0 = 1\mathrm{N/m}^2$ is assumed for the time history of loading.

Figure 15.2 shows the displacement graphs at the loaded rod end with the saturation changing from 0.4 to 0.9. Firstly, the figure shows that the time dependences of displacement exhibit an oscillating character. Secondly, when saturation is less than 0.9, the displacements differ insignificantly from each other in magnitude and propagation velocity of the longitudinal wave suggests little changes. In this case, a decrease in the displacement amplitude by approximately 1.16 times and a slight increase in the longitudinal wave velocity are observed in comparison with the same characteristics for $0.4 \leq S_w \leq 0.9$.

With increasing saturation from 0.93 to 0.99, the displacement amplitude decreases by several times and the longitudinal wave velocity exhibits a slight increase (Fig. 15.3). In this case, the amplitude at $S_w = 0.99$ is already three times less than that at $S_w = 0.93$.

An increase in saturation has a similar effect on the behavior of pore pressures except that their amplitude also increases (Figs. 15.4, 15.5, 15.6, and 15.7). In this case, the pressure amplitude grows faster with increasing saturation than the

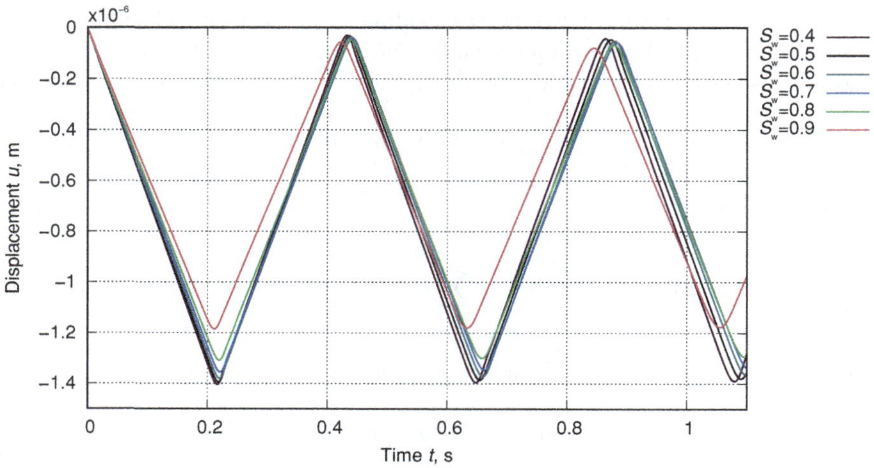

Fig. 15.2 Distributions of the displacement of solid skeleton u with time using different fluid saturation values (S_w from 0.4 to 0.9)

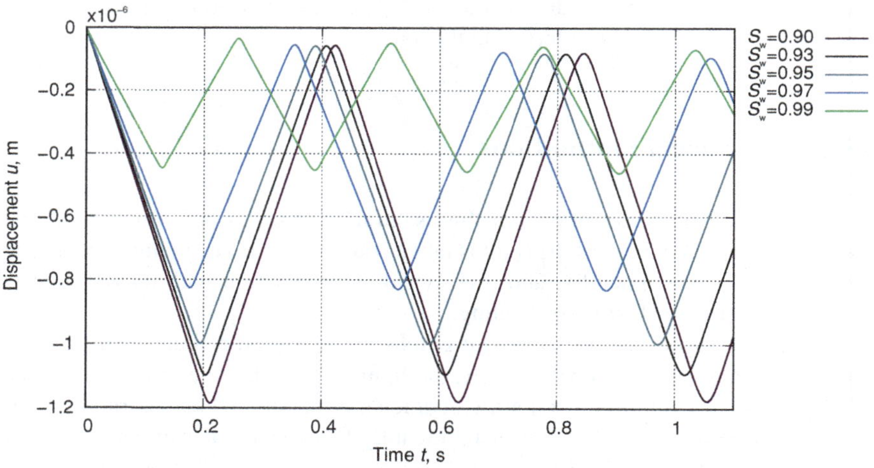

Fig. 15.3 Distributions of the displacement of solid skeleton u with time using different fluid saturation values (S_w 0.9–0.99)

displacement amplitude. Thus, the amplitude of the pore fluid pressure at $S_w = 0.8$ differs twice from the pressure amplitude at $S_w = 0.4$ and almost by the same time from the pressure amplitude at $S_w = 0.4$ (Fig. 15.4).

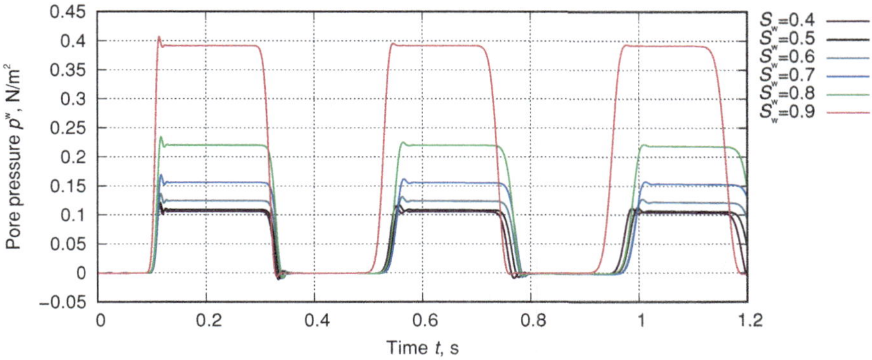

Fig. 15.4 Distributions of fluid pressure p^w with time using different fluid saturation values (S_w 0.4–0.9)

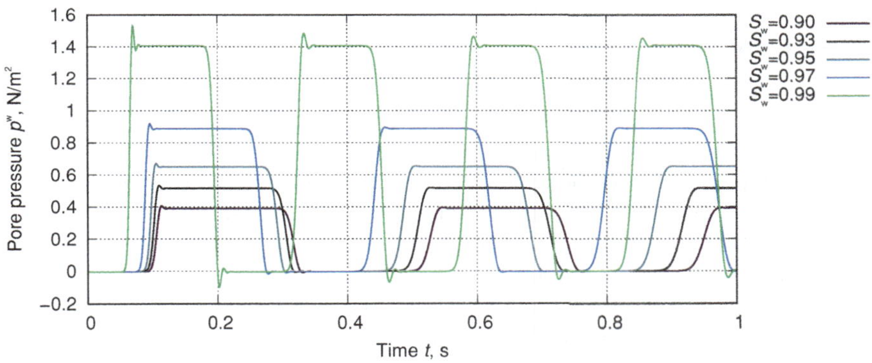

Fig. 15.5 Distributions of the fluid pressure p^w with time using different fluid saturation values (S_w 0.9–0.99)

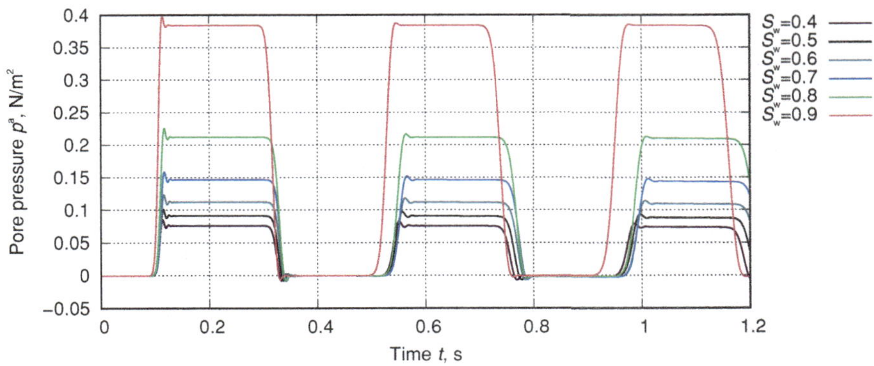

Fig. 15.6 Distributions of the gas pressure p^a with time using different fluid saturation values (S_w 0.4–0.9)

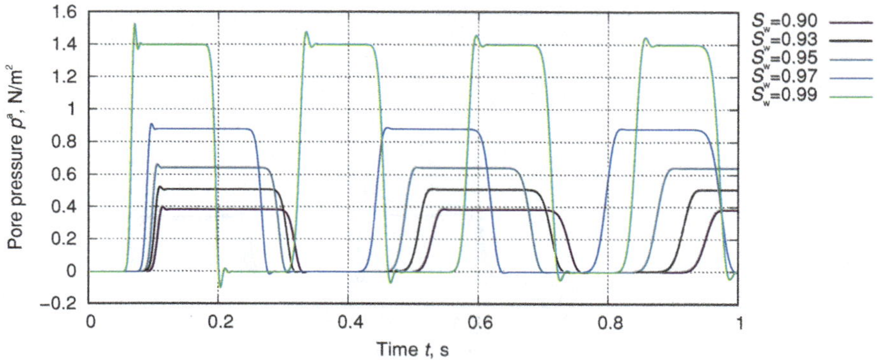

Fig. 15.7 Distributions of the gas pressure p^a with time using different fluid saturation values (S_w 0.9–0.99)

15.5 Conclusion

An analytical solution to the problem of the dynamic load affecting a one-dimensional partially saturated poroelastic rod is presented. The results of calculations obtained using the step method of numerical inversion of the Laplace transform are presented. The effect of material parameters on dynamic responses of displacement and pore pressures is demonstrated. An increase in saturation results in a gradual decrease in the frame displacement amplitude and leads to an increase in the fluid and gas pressure amplitude. An increase in saturation above 0.9 gives rise to a significant acceleration of the longitudinal wave propagation.

Acknowledgements The work was carried out with the financial support of the Ministry of Science and Higher Education of the Russian Federation (task 0729-2020-0054).

Appendix

See Table 15.1.

$$\alpha = 1 - \frac{K}{K_s}, \quad S_w + S_a = 1, \quad S_e = \frac{S_w - S_{rw}}{S_{ra} - S_{rw}}, \quad \varsigma = \frac{\alpha - n}{K_s}, \tag{15.21}$$

$$p_c = \frac{1}{\chi}\left(S_e^{-\frac{1}{m}} - 1\right)^{\frac{1}{d}}, \quad S_u = -\chi m d (S_{ra} - S_{rw}) S_e^{\frac{m+1}{m}} \left(S_e^{-\frac{1}{m}} - 1\right)^{\frac{d-1}{d}}, \tag{15.22}$$

$$S_{ww} = S_w + p_c S_u, \quad S_{aa} = S_a - p_c S_u, \tag{15.23}$$

Table 15.1 General material parameters

Parameter type	Symbol	Value	Unit
Porosity	n	0.45	–
Density of the solid skeleton	ρ_s	2650	$\mathrm{kg/m^3}$
Density of the fluid	ρ_w	1000	$\mathrm{kg/m^3}$
Density of the gas	ρ_a	1.28	$\mathrm{kg/m^3}$
Drained bulk modulus of the mixture	K	8.33×10^6	$\mathrm{N/m^2}$
Shear modulus of the mixture	G	3.85×10^6	$\mathrm{N/m^2}$
Bulk modulus of the solid grains	K_s	3.5×10^{10}	$\mathrm{N/m^2}$
Bulk modulus of the fluid	K_w	2.25×10^9	$\mathrm{N/m^2}$
Bulk modulus of the gas	K_a	1.45×10^5	$\mathrm{N/m^2}$
Intrinsic permeability	k	5.3×10^{-12}	$\mathrm{m^2}$
Viscosity of the fluid	η_w	1.0×10^{-3}	$\mathrm{N\,s/m^2}$
Viscosity of the gas	η_a	1.8×10^{-7}	$\mathrm{N\,s/m^2}$
Fitting parameter of van Genuchten model	χ	1×10^{-4}	$\mathrm{m^2/N}$
Fluid saturation	S_w	0.9	–
Residual fluid saturation	S_{rw}	0	–
Residual gas saturation	S_{ra}	1	–
Fitting parameter of van Genuchten model	m	0.5	–
Fitting parameter of van Genuchten model	d	2	–

$$\kappa_w = \frac{K_{rw}k}{\eta_w}\,, \quad \kappa_a = \frac{K_{ra}k}{\eta_a}\,, \tag{15.24}$$

$$\rho = (1-n)\rho_s + nS_w\rho_w + nS_a\rho_a\,, \tag{15.25}$$

$$\beta = \frac{\kappa_w n\rho_w s}{nS_w + \kappa_w\rho_w s}\,, \quad \gamma = \frac{\kappa_a n\rho_a s}{nS_a + \kappa_a\rho_a s}\,, \tag{15.26}$$

$$B_1 = K + \frac{4}{3}G\ , \quad B_2 = \rho - \beta S_w \rho_w - \gamma S_a \rho_a\ , \quad B_3 = (\alpha - \beta)\, S_w\ , \tag{15.27}$$

$$B_4 = (\alpha - \gamma)\, S_a\ , \quad B_5 = \zeta S_{ww} S_w + \frac{n}{K_w} S_w - S_u n\ , \quad B_6 = \frac{\beta S_w}{\rho_w s}\ , \tag{15.28}$$

$$B_7 = \zeta S_{aa} S_w + S_u n\ , \quad B_8 = \zeta S_{ww} S_a + S_u n\ , \tag{15.29}$$

$$B_9 = \zeta S_{aa} S_a + \frac{n}{K_a} S_w - S_u n\ , \quad B_{10} = \frac{\gamma S_a}{\rho_a s}\ , \tag{15.30}$$

$$C_1 = B_1 B_6 B_{10}\ , \tag{15.31}$$

$$C_2 = -\left(B_1 B_5 B_{10} + B_1 B_6 B_9 + B_2 B_6 B_{10} + B_3^2 B_{10} + B_4^2 B_6\right)\ , \tag{15.32}$$

$$\begin{aligned} C_3 = {} & B_1(B_5 B_9 - B_7 B_8) + B_2(B_5 B_{10} - B_6 B_9) \\ & - B_3 B_4 (B_7 + B_8) + B_3^2 B_9 + B_4^2 B_5\ , \end{aligned} \tag{15.33}$$

$$C_4 = B_2(B_7 B_8 - B_5 B_9)\ , \tag{15.34}$$

$$N_1 = -\frac{C_2}{3C_1}\ , \quad N_2 = \frac{\sqrt[3]{2}}{N_3}\ , \quad N_3 = \sqrt[3]{N_4 + \sqrt{4N_5^3 + N_4^2}}\ , \tag{15.35}$$

$$N_4 = -2C_2^3 + 9C_1 C_2 C_3 - 27 C_1^2 C_4\ , \quad N_5 = -C_2^2 + 3C_1 C_3\ , \tag{15.36}$$

$$\lambda_1 = \sqrt{N_1 + \frac{N_2 C_2^2}{3C_1} - N_2 C_3 + \frac{1}{3N_2 C_1}}\ , \tag{15.37}$$

$$\lambda_2 = \sqrt{N_1 + \frac{N_5 N_2}{3C_1}\left(1 - i\sqrt{3}\right) - \frac{1}{6N_2 C_1}\left(1 + i\sqrt{3}\right)}\ , \tag{15.38}$$

$$\lambda_3 = \sqrt{N_1 + \frac{N_5 N_2}{3C_1}\left(1 + i\sqrt{3}\right) - \frac{1}{6N_2 C_1}\left(1 - i\sqrt{3}\right)}\ , \tag{15.39}$$

$$a_i = \frac{(B_1 B_7 + B_3 B_4)\lambda_i^2 - B_2 B_7}{(B_4 B_6 \lambda_i^2 + B_3 B_7 - B_4 B_5)\lambda_i}\ , \quad b_i = \frac{(B_1 B_8 + B_3 B_4)\lambda_i^2 - B_2 B_8}{(B_3 B_{10} \lambda_i^2 + B_4 B_8 - B_3 B_9)\lambda_i}\ , \tag{15.40}$$

$$t_1 = a_2 b_3 - a_3 b_2\ , \quad t_2 = a_3 b_1 - a_1 b_3\ , \quad t_3 = a_2 b_1 - a_1 b_2\ , \tag{15.41}$$

$$M = t_1 \lambda_1 + t_2 \lambda_2 + t_3 \lambda_3\ . \tag{15.42}$$

References

1. van Genuchten MT (1980) A closed-form equation for predicting the hydraulic conductivity of unsaturated soils. Soil Sci Soc Am J 44(5):892–898
2. Igumnov LA, Petrov AN (2016) Dynamics of partially saturated poroelastic solids by boundary-element method. PNRPU Mech Bull 3:47–61
3. Jin B, Liu H (2001) Dynamic response of a poroelastic half space to horizontal buried loading. Int J Solids Struct 38(44):8053–8064
4. Li P, Schanz M (2011) Wave propagation in a 1-D partially saturated poroelastic column. Geophys J Int 184(3):1341–1353
5. Lubich C (1988) Convolution quadrature and discretized operational calculus. I. Numer Math 52(2):129–145
6. Lubich C (1988) Convolution quadrature and discretized operational calculus. II. Numer Math 52(4):413–425
7. Mualem Y (1976) A new model for predicting the hydraulic conductivity of unsaturated porous media. Water Resour Res 12(3):513–522
8. Philippacopoulos AJ (1988) Lamb's problem for fluid-saturated, porous media. Bull Seismol Soc Am 78(2):908–923
9. Schanz M (2009) Poroelastodynamics: linear models, analytical solutions, and numerical methods. Appl Mech Rev 62(3)
10. Tabatabaie Yazdi J, Valliappan S, Zhao C (1994) Analytical and numerical solutions for wave propagation in water-saturated porous layered half-space. Soil Dyn Earthq Eng 13(4):249–257
11. Zhang M, Wang X, Yang G, Xie L (2014) Solution of dynamic Green's function for unsaturated soil under internal excitation. Soil Dyn Earthq Eng 64:63–84

Chapter 16
Dynamic Problems for Piecewise Homogeneous Viscoelastic Bodies

Sergey G. Pshenichnov, Ekaterina A. Korovaytseva, Emilia Bazhlekova, and Maria D. Datcheva

Abstract A non-stationary dynamic problem of linear viscoelasticity for a piecewise homogeneous body is considered for the case when the disturbed domain is finite. The interrelation between such a problem and the spectral problem of piecewise homogeneous body free oscillations is established. The structure of the eigenvalues set of the spectral problem is investigated. A method of searching for eigenvalues near the limit points of the spectral set is proposed. The integral Laplace transform in time is applied to the non-stationary dynamic problem for a linear viscoelastic piecewise homogeneous structure. For the case when each of the hereditary kernels is a finite sum of exponentials, the solution in the originals is presented as a series of residues at the points of the spectrum. Thus, the constructing of the non-stationary solution is reduced to the search for the elements of the spectral set. As an example, the solution to the plane axisymmetric problem of transient longitudinal waves' propagation in a cross-section of a hollow infinitely long cylinder consisting of two coaxial elastic layers and a viscoelastic layer between them is constructed and discussed. As a result, it becomes possible to investigate and to reveal the influence of the piecewise inhomogeneity on the non-stationary waves' propagation in a cylinder with viscoelastic and elastic layers.

S. G. Pshenichnov (✉) · E. A. Korovaytseva
Institute of Mechanics, Lomonosov Moscow State University, Michurinsky Prospect, 1, Moscow 119192, Russia
e-mail: serp56@yandex.ru

E. A. Korovaytseva
e-mail: katrell@mail.ru

E. Bazhlekova
Institute of Mathematics and Informatics, Bulgarian Academy of Sciences, Acad. G. Bonchev St, Bl. 8, Sofia 1113, Bulgaria
e-mail: e.bazhlekova@math.bas.bg

M. D. Datcheva
Institute of Mechanics, Bulgarian Academy of Sciences, Acad. G. Bonchev St, Bl. 4, Sofia 1113, Bulgaria
e-mail: datcheva@imbn.bas.bg

H. Altenbach et al. (eds.), *Deformation and Destruction of Materials and Structures Under Quasi-static and Impulse Loading*, Advanced Structured Materials 186,
https://doi.org/10.1007/978-3-031-22093-7_16

Keywords Dynamic problems of viscoelasticity · Piecewise homogeneous bodies · Wave processes · Layered cylinder

16.1 Introduction

Piecewise homogeneous bodies with viscoelastic inclusions are widely used in modern technology. One of the directions in the field of modeling the behavior of such structures under dynamic excitations is related to the study of transient wave processes in piecewise homogeneous viscoelastic structures using analytical and semi-analytical methods. Results of such investigations employing analytical methods were reported several decades ago [1–4]. In recent years, considerable attention has been paid to investigations of these problems with analytical methods in combination with numerical ones [5, 6]. Among others, an example of such an approach is the application of the boundary integral equations method [7, 8]. The applications of this method allow to solve a number of dynamic problems for composite viscoelastic media [9, 10]. It has to be also noted the existence of recent publications devoted to the study of harmonic waves in viscoelastic layered media [11, 12], as well as the dynamics of viscoelastic layered thin-walled structures [13]. At the same time, despite the achievements in the field under consideration there are still many issues that require the attention of researchers.

Among various methods of solving linear viscoelastic dynamic problems, the most common method is the integral Laplace transform in time. It is used often in combination with other methods. However, due to the complexity of the inversion, the solution in the originals is normally constructed asymptotically or by accepting significant restrictions on the hereditary properties of the material. In this regard, it is worth noting that in [14] the general properties of the solution to nonstationary dynamic problems for viscoelastic piecewise homogeneous bodies in the Laplace domain were established.

The aim of this work is based on the results of the work [14] to reduce the construction of the solution of non-stationary dynamic problems for piecewise homogeneous viscoelastic body to searching for the eigenvalues of the free oscillations spectral problem under certain conditions imposed on the initial parameters. In addition, the aim is to propose and verify a method for finding the elements of the spectrum near its final limit points and to demonstrate such an approach on a concrete example.

16.2 Non-stationary Dynamic Problem Statement

Consider a non-stationary dynamic problem of linear viscoelasticity with the assumption that the disturbances propagations domain Ω and its boundary Σ are bounded. Herewith note that the body itself can be infinite, but, for example, can be considered as being in a plane strain state. Thus, the dimension of the vector $\boldsymbol{x} = \{x_k\} \in \Omega$

depends on the specific problem and can vary from 1 to 3. The domain Ω consists of N homogeneous isotropic linear-viscoelastic subdomains $\Omega = \Omega_1 \cup \Omega_2 \cup \cdots \cup \Omega_N$. Herewith the subdomains Ω_i and Ω_j do not intersect at internal points for any i and j $(i \neq j)$. On the subdomain contact surfaces, the continuity conditions for the displacement and stress vectors are imposed. The mathematical formulation of the problem includes the dynamic equations ($n = 1, 2, .., N$):

$$\left(\widehat{\lambda}_n + \widehat{\mu}_n\right)\text{grad div}\,\boldsymbol{u}^{(n)}(\boldsymbol{x}, t) + \widehat{\mu}_n \Delta \boldsymbol{u}^{(n)}(\boldsymbol{x}, t) + \boldsymbol{f}^{(n)}(\boldsymbol{x}, t) = \rho_n \ddot{\boldsymbol{u}}^{(n)}(\boldsymbol{x}, t),$$
$$\boldsymbol{x} \in \Omega_n, \tag{16.1}$$

the constitutive relation:

$$\overset{\sim}{\boldsymbol{\sigma}}^{(n)}(\boldsymbol{x}, t) = 2\widehat{\mu}_n \text{def}\,\boldsymbol{u}^{(n)}(\boldsymbol{x}, t) + \widehat{\lambda}_n \text{div}\,\boldsymbol{u}^{(n)}(\boldsymbol{x}, t)\,\overset{\sim}{\boldsymbol{I}}, \quad \boldsymbol{x} \in \Omega_n, \tag{16.2}$$

generalized boundary conditions for the m -th subdomain ($1 \leq m \leq N$), with a boundary Σ:

$$\overset{\sim}{\boldsymbol{\alpha}}^{(m)}(\boldsymbol{x})\overset{\sim}{\boldsymbol{\sigma}}^{(m)}(\boldsymbol{x}, t)\boldsymbol{n} + \overset{\sim}{\boldsymbol{\beta}}^{(m)}(\boldsymbol{x})\boldsymbol{u}^{(m)}(\boldsymbol{x}, t) = \boldsymbol{p}^{(m)}(\boldsymbol{x}, t), \quad \boldsymbol{x} \in \Sigma, \quad t > 0, \tag{16.3}$$

continuity conditions between the adjacent subdomains p and q:

$$\boldsymbol{u}^{(p)}(\boldsymbol{x}, t) = \boldsymbol{u}^{(q)}(\boldsymbol{x}, t), \quad \overset{\sim}{\boldsymbol{\sigma}}^{(p)}(\boldsymbol{x}, t)\boldsymbol{n} = \overset{\sim}{\boldsymbol{\sigma}}^{(q)}(\boldsymbol{x}, t)\boldsymbol{n}, \quad \boldsymbol{x} \in \Sigma_{pq} \tag{16.4}$$

and initial conditions:

$$\boldsymbol{u}^{(n)}(\boldsymbol{x}, 0) = \boldsymbol{b}_1^{(n)}(\boldsymbol{x}), \quad \dot{\boldsymbol{u}}^{(n)}(\boldsymbol{x}, 0) = \boldsymbol{b}_2^{(n)}(\boldsymbol{x}), \quad \boldsymbol{x} \in \Omega_n, \tag{16.5}$$

where $\widehat{\lambda}_n$, $\widehat{\mu}_n$ are the operators

$$\widehat{\lambda}_n = \frac{1}{3}[3K_0^{(n)}(1 - \widehat{T}_v^{(n)}) - 2G_0^{(n)}(1 - \widehat{T}_s^{(n)})], \quad \widehat{\mu}_n = G_0^{(n)}\left(1 - \widehat{T}_s^{(n)}\right),$$

$$\widehat{T}_j^{(n)}\xi(t) = \int_0^t T_j^{(n)}(t - \tau)\xi(\tau)d\tau, \quad j = v, s \tag{16.6}$$

The dot above the variables denotes the derivative with respect to time t; ρ_n is density; $\overset{\sim}{\boldsymbol{\sigma}}^{(n)}$ is the stress tensor; $\boldsymbol{u}^{(n)}$, $\boldsymbol{p}^{(m)}$, $\boldsymbol{f}^{(n)}$, $\boldsymbol{b}_1^{(n)}$, $\boldsymbol{b}_2^{(n)}$ are vectors of displacements, external loading, body forces, initial displacements and velocities; $\overset{\sim}{\boldsymbol{\alpha}}^{(m)}$, $\overset{\sim}{\boldsymbol{\beta}}^{(m)}$ are given second-rank tensors that determine the type of boundary conditions; $G_0^{(n)}$, $K_0^{(n)}$ are the instantaneous values of the shear and bulk moduli; $T_v^{(n)}(t)$, $T_s^{(n)}(t)$ are the volumetric and shear relaxation kernels, characterizing the n-th subdomain; $\overset{\sim}{\boldsymbol{I}}$

is the unit tensor; $\boldsymbol{n}$ is an exterior unit normal to the respective domain boundary; Δ is Laplace operator; def is the operator with which the tensor of small deformations $\overset{\sim(n)}{\boldsymbol{\varepsilon}}$ is expressed by $\boldsymbol{u}^{(n)}$.

After applying the integral Laplace transform in time to Eqs. (16.1)–(16.4), we obtain the problem in images expressed in terms of the following equations

$$(\Lambda_n(s) + \mathrm{M}_n(s))\mathrm{grad\,div}\boldsymbol{U}^{(n)}(\boldsymbol{x}, s) + \mathrm{M}_n(s)\Delta\boldsymbol{U}^{(n)}(\boldsymbol{x}, s) - \rho_n s^2\boldsymbol{U}^{(n)}(\boldsymbol{x}, s) + \rho_n\left[s\boldsymbol{b}_1^{(n)}(\boldsymbol{x}) + \boldsymbol{b}_2^{(n)}(\boldsymbol{x})\right] + \boldsymbol{F}^{(n)}(\boldsymbol{x}, s) = 0, \quad \boldsymbol{x} \in \Omega_n \tag{16.7}$$

constitutive relations:

$$\overset{\sim(n)}{\boldsymbol{S}}(\boldsymbol{x}, s) = 2\mathrm{M}_n(s)\mathrm{def}\,\boldsymbol{U}^{(n)}(\boldsymbol{x}, s) + \Lambda_n(s)\mathrm{div}\,\boldsymbol{U}^{(n)}(\boldsymbol{x}, s)\,\tilde{\boldsymbol{I}}, \quad \boldsymbol{x} \in \Omega_n, \tag{16.8}$$

$$\Lambda_n(s) = \frac{1}{3}\left[3K_0^{(n)}\left(1 - \Theta_v^{(n)}(s)\right) - 2G_0^{(n)}\left(1 - \Theta_s^{(n)}(s)\right)\right],$$
$$\mathrm{M}_n(s) = G_0^{(n)}\left(1 - \Theta_s^{(n)}(s)\right),$$

boundary conditions:

$$\overset{\sim(m)}{\boldsymbol{\alpha}}(\boldsymbol{x})\overset{\sim(m)}{\boldsymbol{S}}(\boldsymbol{x}, s)\boldsymbol{n} + \overset{\sim(m)}{\boldsymbol{\beta}}(\boldsymbol{x})\boldsymbol{U}^{(m)}(\boldsymbol{x}, s) = \boldsymbol{P}^{(m)}(\boldsymbol{x}, s), \quad \boldsymbol{x} \in \Sigma, \quad 1 \le m \le N, \tag{16.9}$$

The conditions on the contact between the adjacent subdomains:

$$\boldsymbol{U}^{(p)}(\boldsymbol{x}, s) = \boldsymbol{U}^{(q)}(\boldsymbol{x}, s), \quad \overset{\sim(p)}{\boldsymbol{S}}(\boldsymbol{x}, s)\boldsymbol{n} = \overset{\sim(q)}{\boldsymbol{S}}(\boldsymbol{x}, s)\boldsymbol{n}, \quad \boldsymbol{x} \in \Sigma_{pq}, \tag{16.10}$$

Where

$$\boldsymbol{U}^{(n)}(\boldsymbol{x}, s), \overset{\sim(n)}{\boldsymbol{S}}(\boldsymbol{x}, s), \boldsymbol{F}^{(n)}(\boldsymbol{x}, s), \boldsymbol{P}^{(m)}(\boldsymbol{x}, s), \Theta_v^{(n)}(s), \Theta_s^{(n)}(s), \quad s \in C$$

are correspondingly the Laplace transforms of

$$\boldsymbol{u}^{(n)}(\boldsymbol{x}, t), \overset{\sim(n)}{\boldsymbol{\sigma}}(\boldsymbol{x}, t), \boldsymbol{f}^{(n)}(\boldsymbol{x}, t), \boldsymbol{p}^{(m)}(\boldsymbol{x}, t), T_v^{(n)}(t), T_s^{(n)}(t)$$

Let us assume that the solution $\boldsymbol{U}^{(n)}(\boldsymbol{x}, s)$ to the problem defined with Eqs. (16.7)–16.10) in images is constructed and focus on the issues related to the construction of its original in the time domain.

16.3 Connection Between the Non-stationary Dynamic Problem and the Problem of Free Oscillations

Consider the problem of free oscillations of the discussed viscoelastic piecewise-homogeneous body. We will assume that in the absence of body forces and external loading ($\boldsymbol{f}^{(n)} \equiv 0$, $\boldsymbol{p}^{(m)} \equiv 0$), the body has been freely oscillating for a sufficiently long period of time from an initial moment that the character of the oscillations does not depend on the way of their excitation. The lower limit of integration in the relations (16.6) is taken to be minus infinity. Representing the nontrivial solution of such a problem in the form

$$\boldsymbol{u}^{(n)}(\boldsymbol{x}, t) = \boldsymbol{V}^{(n)}(\boldsymbol{x}, s)e^{st}, \quad s \in C,$$

we obtain the spectral problem that includes the equations listed below [14]:

$$\begin{aligned}&(\Lambda_n(s) + \mathrm{M}_n(s))\operatorname{grad}\operatorname{div}\boldsymbol{V}^{(n)}(\boldsymbol{x}, s) + \mathrm{M}_n(s)\Delta\boldsymbol{V}^{(n)}(\boldsymbol{x}, s)\\&- \rho_n s^2\boldsymbol{V}^{(n)}(\boldsymbol{x}, s) = 0, \quad \boldsymbol{x} \in \Omega_n,\end{aligned} \tag{16.11}$$

boundary conditions:

$$\overset{\sim(m)}{\boldsymbol{\alpha}}(\boldsymbol{x})\overset{\sim(m)}{\boldsymbol{Q}}(\boldsymbol{x}, s)\boldsymbol{n} + \overset{\sim(m)}{\boldsymbol{\beta}}(\boldsymbol{x})\boldsymbol{V}^{(m)}(\boldsymbol{x}, s) = 0, \quad \boldsymbol{x} \in \Sigma \tag{16.12}$$

and the contact conditions:

$$\boldsymbol{V}^{(p)}(\boldsymbol{x}, s) = \boldsymbol{V}^{(q)}(\boldsymbol{x}, s), \quad \overset{\sim(p)}{\boldsymbol{Q}}(\boldsymbol{x}, s)\boldsymbol{n} = \overset{\sim(q)}{\boldsymbol{Q}}(\boldsymbol{x}, s)\boldsymbol{n}, \quad \boldsymbol{x} \in \Sigma_{pq}. \tag{16.13}$$

where

$$\overset{\sim(n)}{\boldsymbol{Q}}(\boldsymbol{x}, s) = 2\mathrm{M}_n(s)\operatorname{def}\boldsymbol{V}^{(n)}(\boldsymbol{x}, s) + \Lambda_n(s)\operatorname{div}\boldsymbol{V}^{(n)}(\boldsymbol{x}, s)\,\tilde{\boldsymbol{I}}, \quad \boldsymbol{x} \in \Omega_n.$$

The eigenvalues $s \in C$ of the spectral problem (16.11)–(16.13) determine the frequencies and damping coefficients of the body free vibrations. Denote the set of all the eigenvalues of this problem as E_s. The relationship between E_s and the branch points, as well as the poles of the solution $\boldsymbol{U}^{(n)}$ to the problem (16.7)–(16.10) are discussed in [14]. In addition to these results, we formulate here the following statements.

Statement 1. Any $s \in \mathrm{E}_s$ is a singular point for $\boldsymbol{U}^{(n)}$—the solution to the problem (16.7)–(16.10).

Based on this fact and the results reported in [14], the following statement can be formulated.

Statement 2. Let the conditions be met:

(1) the set E_s is at most countable;

(2) the images of hereditary kernels $\Theta_v^{(n)}(s)$, $\Theta_s^{(n)}(s)$, as well as components of the vectors of the body and the external forces $\boldsymbol{F}^{(n)}(\boldsymbol{x}, s)$, $\boldsymbol{P}^{(n)}(\boldsymbol{x}, s)$ have no branch points in the complex plane;
(3) for the solution $\boldsymbol{U}^{(n)}(\boldsymbol{x}, s)$ in the Laplace domain the asymptotic conditions known from the contour integration theory are fulfilled in the vicinity of an infinitely distant point and in the small neighborhoods of the finite limit points of the poles set.

Then the solution of the non-stationary dynamic problem (16.1)–(16.6) in the originals $\boldsymbol{u}^{(n)}(\boldsymbol{x}, t)$ can be represented as a sum of residues in the poles of $\boldsymbol{U}^{(n)}(\boldsymbol{x}, s)e^{st}$. Herewith the set of such poles, along with the poles determined by body and external forces, includes E_s.

16.4 Eigenvalue Search Method

Thus, searching for the eigenvalues of the spectral problem (16.11)–(16.13) is necessary not only in the investigation of free oscillations, but is important for constructing solution to a non-stationary dynamic problem for a viscoelastic piecewise homogeneous body. Therefore, in the following we focus our attention on the method of searching for the elements of the set E_s. The domain Ω is assumed to be bounded, so the countability of the set E_s appears naturally. Assume further that the relaxation kernels $T_v^{(n)}(t)$, $T_s^{(n)}(t)$ belong to the class of functions:

$$\sum\nolimits_{k=1}^{K} a_k e^{-b_k t}, \quad 0 \le \sum\nolimits_{k=1}^{K} \frac{a_k}{b_k} < 1, \quad a_k,\ b_k - \text{const}, \quad b_k > 0, \quad k = 1, 2, \ldots K. \tag{16.14}$$

Herewith, for the material of each homogeneous subdomain of Ω, each of the kernels $T_v^{(n)}(t)$, $T_s^{(n)}(t)$ has its own values for the constants a_k, b_k, as well as the number K of terms in the expression (16.14). Functions $\Theta_v^{(n)}(s)$, $\Theta_s^{(n)}(s)$ have no branch points.

Let us express the displacement through the scalar and vector potentials $\boldsymbol{u}^{(n)}(\boldsymbol{x}, t) = \operatorname{grad} \varphi^{(n)}(\boldsymbol{x}, t) + \operatorname{rot} \boldsymbol{\psi}^{(n)}(\boldsymbol{x}, t)$. Taking into account the representation $\varphi^{(n)}(\boldsymbol{x}, t) = \Phi^{(n)}(\boldsymbol{x}, s)e^{st}$, $\boldsymbol{\psi}^{(n)}(\boldsymbol{x}, t) = \boldsymbol{\Psi}^{(n)}(\boldsymbol{x}, s)e^{st}$, we receive the spectral problem (16.11)–(16.13) in potentials. It includes the equations

$$[1 - \Theta^{(n)}(s)]\Delta\Phi^{(n)}(\boldsymbol{x}, s) = \frac{s^2}{[c_1^{(n)}]^2}\Phi^{(n)}(\boldsymbol{x}, s),$$

$$\left[1 - \Theta_s^{(n)}(s)\right]\Delta\boldsymbol{\Psi}^{(n)}(\boldsymbol{x}, s) = \frac{s^2}{[c_2^{(n)}]^2}\boldsymbol{\Psi}^{(n)}(\boldsymbol{x}, s), \quad \boldsymbol{x} \in \Omega_n \tag{16.15}$$

with boundary conditions (16.12) and contact conditions (16.13), where $\boldsymbol{V}^{(n)}(\boldsymbol{x}, s) = \operatorname{grad} \Phi^{(n)}(\boldsymbol{x}, s) + \operatorname{rot}\boldsymbol{\Psi}^{(n)}(\boldsymbol{x}, s)$;

$\Theta^{(n)}(s)$ is the Laplace transform of $T^{(n)}(t) = \Big\{(1+\nu_0^{(n)})T_\upsilon^{(n)}(t) + 2(1-2\nu_0^{(n)})$ $T_s^{(n)}(t)\Big\}\Big\{3(1-\nu_0^{(n)})\Big\}^{-1}$, $\nu_0^{(n)}$ is the instantaneous value of the Poisson's ratio, $c_1^{(n)}$, $c_2^{(n)}$ are the longitudinal and the transverse elastic wave velocities in the n -th subdomain of Ω.

Note that each root s^* of each of the equations

$$1 - \Theta^{(n)}(s) = 0, \quad 1 - \Theta_s^{(n)}(s) = 0, \quad n = 1, 2, ..., N \tag{16.16}$$

converts the left-hand side of the corresponding Eq. (16.15) to zero. Thus s^* may be a finite limit point of the set E_s of the spectral problem. Hereditary kernels $T_\upsilon^{(n)}(t)$, $T_s^{(n)}(t)$ belong to the class (16.14), so each of the Eqs. (16.16) reduces to a polynomial one. Let us introduce the notation:

$$\beta^{(n)} = \frac{s}{\sqrt{1-\Theta^{(n)}(s)}}, \quad \beta_1^{(n)} = \frac{s}{\sqrt{1-\Theta_s^{(n)}(s)}}, \quad n = 1, 2, ..., N.$$

We assume that all zeros of the functions $1-\Theta^{(n)}(s)$, $1-\Theta_s^{(n)}(s)$ are real and simple. Let s^* be one of the roots of one of the Eqs. (16.16), for example, $1 - \Theta^{(j)}(s^*) = 0$. Then $1-\Theta^{(j)}(s) \approx B^{(j)}(s-s^*)$ at $s \to s^*$, where $B^{(j)}$ is a non-zero real constant. Note that when s tends to s^* on the real axis so that the condition $B^{(j)}(s-s^*) < 0$ is met, the value $\beta^{(j)}$ tends to infinity, remaining purely imaginary. In many problems, this situation turns the functions included in the characteristic equation for the elements of the set E_s into trigonometric ones in a small neighborhood of s^*. Let us assume as a hypothesis that the elements of the set E_s in a small neighborhood of its limit point s^* can be searched for in accordance with the condition $\mathrm{Re}\{\beta^{(j)}\} = 0$. The same procedure is applicable to $1-\Theta_s^{(j)}(s^*) = 0$. Thus, if all the roots of Eqs. (16.16) are real and simple, we propose to search for the elements of E_s in a small neighborhood of each of these roots on the real axis. The complex elements of E_s can be searched for using iteration methods, taking elements of the spectrum for an elastic body as an initial approximation.

16.5 Example. Dynamic Problem for a Three-Layer Cylinder

Hereafter, an example is given for application of the method proposed above for finding the elements of the spectral set E_s, followed by constructing a solution to a non-stationary dynamic problem for a piecewise homogeneous body.

Consider a problem of an unsteady longitudinal wave propagation in a cross section of an infinite hollow layered cylinder. The cylinder consists of two coaxial layers of the same elastic material, which are separated by a softer viscoelastic layer.

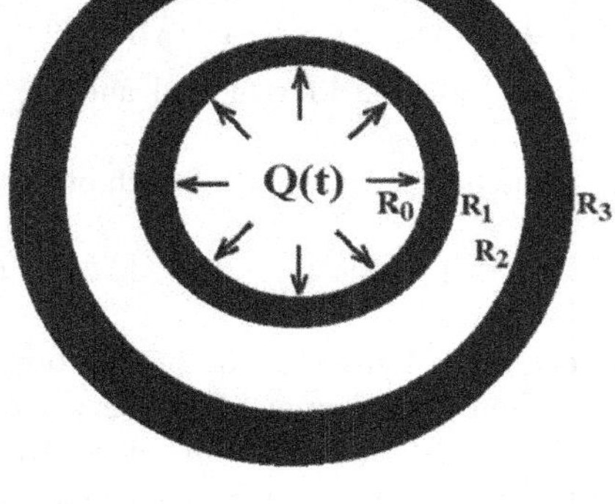

Fig. 16.1 Cross section of the cylinder

The continuity of displacement and radial stress is imposed at the layer' interfaces. We denote R_0, R_3—the inner and the outer radii of the cylinder, R_1, R_2—the radii of the viscoelastic layer ($R_0 < R_1 < R_2 < R_3$). The cylinder is initially in a non-perturbed state. The outer surface of the cylinder is free, while the inner surface is exposed to an evenly distributed radial load $Q(t)$ since the moment $t = 0$ (Fig. 16.1).

We will consider this problem as a special case of the general one of a cylinder with an arbitrary number N viscoelastic layers, whose solution in the Laplace images is presented in [15]. In the present work, for the case when the hereditary kernels belong to the class (16.14), the solution in the originals for the three-layer cylinder is presented as a series of residues.

In the polar coordinate system (R, θ) with the origin at the center of the cross section of the cylinder we present a formulation of the problem with an arbitrary number N layers with boundaries $R = R_m$, $(m = 1, 2, \cdots, N-1)$, $R_0 < R_1 < R_2 < \ldots < R_{N-1} < R_N$. Let us introduce dimensionless values:

$$\tau = \frac{t}{t_0}, \quad r = \frac{R}{R_N}, \quad r_0 = \frac{R_0}{R_N}, \quad r_m = \frac{R_m}{R_N}, \quad Q_0 f(\tau) = \frac{Q(t)}{2G_0^{(1)}},$$

$$\alpha_n = \frac{c_1^{(N)}}{c_1^{(n)}}, \quad w_n = \frac{1-\nu_0^{(n)}}{1-2\nu_0^{(n)}}, \quad u^{(n)}(r,\tau) = \frac{v^{(n)}(R,t)}{R_N},$$

$$\sigma_r^{(n)}(r,\tau) = \frac{P_R^{(n)}(R,t)}{2G_0^{(n)}}, \quad \sigma_\theta^{(n)}(r,\tau) = \frac{P_\theta^{(n)}(R,t)}{2G_0^{(n)}},$$

$$\gamma_s^{(n)}(\tau) = t_0 T_s^{(n)}(t), \quad \gamma_v^{(n)}(\tau) = t_0 T_v^{(n)}(t), \quad n = 1, 2, \ldots, N,$$

where $t_0 = R_N / c_1^{(N)}$, Q_0 is a dimensionless constant; $v^{(n)}(R,t)$ is the radial displacement, $P_R^{(n)}(R,t)$, $P_\theta^{(n)}(R,t)$ are radial and circumferential stresses in the n-th layer. Values $T_v^{(n)}(t)$, $T_s^{(n)}(t)$, $G_0^{(n)}$, $\nu_0^{(n)}$, $c_1^{(n)}$ for the n -th component (layer) of the body were introduced in the previous sections of the article. We will suppose that the creep is bounded for all the layers' material.

The dimensionless formulation of the problem is given by the equations:

$$\left(1-\hat{d}_1^{(n)}\right)\frac{\partial}{\partial r}\left[\frac{\partial u^{(n)}(r,\tau)}{\partial r}+\frac{u^{(n)}(r,\tau)}{r}\right]-a_n^2\frac{\partial^2 u^{(n)}(r,\tau)}{\partial \tau^2}=0,\quad r_{n-1}\le r\le r_n, \tag{16.17}$$

herewith

$$\sigma_r^{(n)}(r,\tau)=w_n\left(1-\hat{d}_1^{(n)}\right)\frac{\partial u^{(n)}(r,\tau)}{\partial r}+(w_n-1)\left(1-\hat{d}_2^{(n)}\right)\frac{u^{(n)}(r,\tau)}{r}, \tag{16.18}$$

$$\sigma_\theta^{(n)}(r,\tau)=w_n\left(1-\hat{d}_1^{(n)}\right)\frac{u^{(n)}(r,\tau)}{r}+(w_n-1)\left(1-\hat{d}_2^{(n)}\right)\frac{\partial u^{(n)}(r,\tau)}{\partial r},$$

$$\hat{d}_j^{(n)}\xi(r,\tau)=\int_0^\tau d_j^{(n)}(\tau-\chi)\xi(r,\chi)d\chi,\quad j=1,2,$$

$$d_1^{(n)}(\tau)=\frac{1}{3\left(1-\nu_0^{(n)}\right)}\left[\left(1+\nu_0^{(n)}\right)\gamma_v^{(n)}(\tau)+2\left(1-2\nu_0^{(n)}\right)\gamma_s^{(n)}(\tau)\right],$$

$$d_2^{(n)}(\tau)=\frac{1}{3\nu_0^{(n)}}\left[\left(1+\nu_0^{(n)}\right)\gamma_v^{(n)}(\tau)-\left(1-2\nu_0^{(n)}\right)\gamma_s^{(n)}(\tau)\right],\quad n=1,2,...,N.$$

The equations are supplemented by initial conditions:

$$u^{(n)}(r,0)=0,\quad \frac{\partial u^{(n)}}{\partial \tau}(r,0)=0, \tag{16.19}$$

the continuity conditions on the contact between the layers ($m = 1, 2, ..., N-1$):

$$u^{(m)}(r_m,\tau)=u^{(m+1)}(r_m,\tau),\quad G_0^{(m)}\sigma_r^{(m)}(r_m,\tau)=G_0^{(m+1)}\sigma_r^{(m+1)}(r_m,\tau) \tag{16.20}$$

and the boundary conditions:

$$\sigma_r^{(1)}(r_0,\tau)=-Q_0 f(\tau),\quad \sigma_r^{(N)}(1,\tau)=0,\ \tau>0. \tag{16.21}$$

For the considered three-layer cylinder (Fig. 16.1) $N = 3$, and $\gamma_s^{(1)}(\tau)\equiv\gamma_v^{(1)}(\tau)\equiv\gamma_s^{(3)}(\tau)\equiv\gamma_v^{(3)}(\tau)\equiv 0$, therefore $d_1^{(1)}\equiv d_2^{(1)}\equiv d_1^{(3)}\equiv d_2^{(3)}\equiv 0$.

In addition, the elastic material of the first and third layers is the same, so $\alpha_1 = 1$, $w_1 = w_3$, $G_0^{(3)}/G_0^{(1)} = 1$.

Let us apply integral Laplace transform in time to Eqs. (16.17), (16.18), (16.20) (16.21) taking into account the initial conditions (16.19), herewith $s \in C$, and

$$U^{(n)}(r,s),\ S_r^{(n)}(r,s),\ S_\theta^{(n)}(r,s),\ F(s),\ D_1^{(2)}(s),\ D_2^{(2)}(s),\ \Gamma_v^{(2)}(s),\ \Gamma_s^{(2)}(s)$$

are respectively the transforms of functions

$$u^{(n)}(r,\tau),\ \sigma_r^{(n)}(r,\tau),\ \sigma_\theta^{(n)}(r,\tau),\ f(\tau),\ d_1^{(2)}(\tau),\ d_2^{(2)}(\tau),\ \gamma_v^{(2)}(s),\ \gamma_s^{(2)}(s).$$

The solution for displacement in transforms takes the form ($n = 1, 2, 3$):

$$U^{(n)}(r,s) = Q_0 F(s)\frac{q_n(s)}{Z(s)}[I_1(y_n)g_1^{(n)}(s) + K_1(y_n)g_2^{(n)}(s)], \quad r_{n-1} \le r \le r_n, \tag{16.22}$$

herewith

$$y_1(r,s) = rs, \quad y_2(r,s) = r\alpha_2\beta^{(2)}(s), \quad y_3(r,s) = rs,$$
$$\beta^{(2)}(s) = \frac{s}{\sqrt{1 - D_1^{(2)}(s)}},$$
$$Z(s) = \left[\frac{I_1(r_0 s)}{r_0 s} - w_1 I_0(r_0 s)\right]g_1^{(1)} + \left[\frac{K_1(r_0 s)}{r_0 s} + w_1 K_0(r_0 s)\right]g_2^{(1)}.$$

Functions $g_1^{(n)}(s)$, $g_2^{(n)}(s)$ are determined from recurrence relations:

$$\begin{pmatrix} g_1^{(m)} \\ g_2^{(m)} \end{pmatrix} = \begin{pmatrix} z_{11}^{(m)} & z_{12}^{(m)} \\ z_{21}^{(m)} & z_{22}^{(m)} \end{pmatrix}\begin{pmatrix} g_1^{(m+1)} \\ g_2^{(m+1)} \end{pmatrix}, \quad m = 1,\ 2,$$

where

$$g_1^{(3)} = w_3 K_0(s) + \frac{K_1(s)}{s}, \quad g_2^{(3)} = w_3 I_0(s) - \frac{I_1(s)}{s},$$

$$z_{11}^{(m)} = I_1\left(b_2^{(m)}\right)\psi_{12}^{(m)} - K_1\left(b_1^{(m)}\right)\psi_{21}^{(m)}, \quad z_{12}^{(m)} = K_1\left(b_2^{(m)}\right)\psi_{12}^{(m)} - K_1\left(b_1^{(m)}\right)\psi_{22}^{(m)},$$

$$z_{21}^{(m)} = I_1\left(b_1^{(m)}\right)\psi_{21}^{(m)} - I_1\left(b_2^{(m)}\right)\psi_{11}^{(m)}, \quad z_{22}^{(m)} = I_1\left(b_1^{(m)}\right)\psi_{22}^{(m)} - K_1\left(b_2^{(m)}\right)\psi_{11}^{(m)},$$

$m = 1, 2,$

$$b_1^{(1)} = r_1 s, \quad b_2^{(1)} = r_1\alpha_2\beta^{(2)}, \quad b_1^{(2)} = r_2\alpha_2\beta^{(2)}, \quad b_2^{(2)} = r_2 s,$$

and also

$$\psi_{11}^{(1)} = \delta\left[w_1 I_0\left(b_1^{(1)}\right) - \frac{I_1\left(b_1^{(1)}\right)}{b_1^{(1)}}\right], \quad \psi_{12}^{(1)} = -\delta\left[w_1 K_0\left(b_1^{(1)}\right) + \frac{K_1\left(b_1^{(1)}\right)}{b_1^{(1)}}\right],$$

$$\psi_{21}^{(1)} = w_2\left(1 - D_1^{(2)}\right)I_0\left(b_2^{(1)}\right) - \frac{(1 - \Gamma_s^{(2)})I_1\left(b_2^{(1)}\right)}{b_2^{(1)}},$$

$$\psi_{22}^{(1)} = -\left[w_2\left(1 - D_1^{(2)}\right)K_0\left(b_2^{(1)}\right) + \frac{(1 - \Gamma_s^{(2)})K_1\left(b_2^{(1)}\right)}{b_2^{(1)}}\right],$$

$$\psi_{11}^{(2)} = \delta^{-1}\left[w_2\left(1 - D_1^{(2)}\right)I_0\left(b_1^{(2)}\right) - \frac{(1 - \Gamma_s^{(2)})I_1\left(b_1^{(2)}\right)}{b_1^{(2)}}\right],$$

$$\psi_{12}^{(2)} = -\delta^{-1}\left[w_2\left(1 - D_1^{(2)}\right)K_0\left(b_1^{(2)}\right) + \frac{(1 - \Gamma_s^{(2)})K_1\left(b_1^{(2)}\right)}{b_1^{(2)}}\right],$$

$$\psi_{21}^{(2)} = w_1 I_0\left(b_2^{(2)}\right) - \frac{I_1\left(b_2^{(2)}\right)}{b_2^{(2)}}, \qquad \psi_{22}^{(2)} = -\left[w_1 K_0\left(b_2^{(2)}\right) + \frac{K_1\left(b_2^{(2)}\right)}{b_2^{(2)}}\right],$$

$I_0,\ I_1, K_0,\ K_1$ are the modified Bessel functions of the first and second kind;

$$\delta = \frac{G_0^{(1)}s}{G_0^{(2)}\alpha_2\beta^{(2)}}, \qquad q_1 \equiv \frac{1}{s}, \qquad q_2 = -\frac{G_0^{(1)}w_1}{G_0^{(2)}r_1\alpha_2\beta^{(2)}s}, \qquad q_3 = \frac{w_1 w_2\left(1 - D_1^{(2)}\right)}{r_1 r_2\alpha_2\beta^{(2)}s^2}.$$

Note that in the expressions above the values r_0, r_1, r_2, w_1, w_2, $G_0^{(1)}$, $G_0^{(2)}$, α_2 are constants. Other values are functions of complex argument s. Besides, unlike the previous section, in this section the value $\beta^{(2)}(s)$ is dimensionless. Transforms of the stresses are obtained using (16.22) and the relations ($n = 1$ or $n = 3$):

$$S_r^{(n)} = w_n\frac{dU^{(n)}}{dr} + (w_n - 1)\frac{U^{(n)}}{r}, \qquad S_\theta^{(n)} = w_n\frac{U^{(n)}}{r} + (w_n - 1)\frac{dU^{(n)}}{dr},$$

$$S_r^{(2)} = w_2\left[1 - D_1^{(2)}\right]\frac{dU^{(2)}}{dr} + (w_2 - 1)\left[1 - D_2^{(2)}\right]\frac{U^{(2)}}{r}, \tag{16.23}$$

$$S_\theta^{(2)} = w_2\left[1 - D_1^{(2)}\right]\frac{U^{(2)}}{r} + (w_2 - 1)\left[1 - D_2^{(2)}\right]\frac{dU^{(2)}}{dr}.$$

Let the external load be defined as the Heaviside function $f(\tau) = h(\tau)$ and the relaxation kernels $\gamma_v^{(2)}(s)$, $\gamma_s^{(2)}(s)$ belong to the class (16.14) with dimensionless parameters and so functions $F(s)$, $\Gamma_v^{(2)}(s)$, $\Gamma_s^{(2)}(s)$, $D_1^{(2)}(s)$, $D_2^{(2)}(s)$ do not have branch points. Then, functions $U^{(n)}$, $S_r^{(n)}$, $S_\theta^{(n)}$ determined by formulae (16.22), (16.23), do not have branch points in the complex plain, despite the fact that they are expressed through the functions which have such points. This becomes clear after cumbersome transformations. At the same time, it is not necessary to carry out these transformations as this fact follows from a general theorem proved in [14].

We assume that all the roots of the equation $1 - D_1^{(2)}(s) = 0$ are real and simple. In this case, after asymptotic investigation of the functions (16.22), (16.23) in the vicinity of the limit points of the poles set, the originals $u^{(n)}$, $\sigma_r^{(n)}$, $\sigma_\theta^{(n)}$ represent in the form of series of residues. We represent only the displacement:

$$u^{(n)}(r,\tau) = u_0^{(n)}(r) + 2\sum\nolimits_{\{z_j\}} \mathrm{Re}\{\underset{z_j}{\mathrm{Res}}[U^{(n)}(r,s)e^{s\tau}]\} + \sum\nolimits_{\{x_l\}} \underset{x_l}{\mathrm{Res}}\big[U^{(n)}(r,s)e^{s\tau}\big], \tag{16.24}$$

$$u_0^{(n)}(r) = \underset{s=0}{\mathrm{Res}}\big[U^{(n)}(r,s)e^{s\tau}\big], \quad n = 1, 2, 3.$$

Here $\{z_j\} \cup \{x_l\} = E_s$, herewith z_j are complex, x_l are real $(j, l = 1, 2, 3, \ldots)$. With simple zeros of the function $Z(s)$ each residue in the formula (16.24) in a point $s_k \in E_s$ is determined as:

$$\underset{s_k}{\mathrm{Res}}[U^{(n)}e^{s\tau}] = \frac{Q_0 q_n(s_k)}{s_k Z'(s_k)}[I_1(y_n(r,s_k))g_1^{(n)}(s_k) + K_1(y_n(r,s_k))g_2^{(n)}(s_k)]e^{s_k\tau},$$

a prime in Z' denotes a derivative. It can be shown that $u_0(r)$ is a solution to the static elastic problem for a layered cylinder with long-term modules that correspond to the viscoelastic kernels in the dynamic problem. Its expression is presented in [15]. The expressions for $\sigma_r^{(n)}(r,\tau)$, $\sigma_\theta^{(n)}(r,\tau)$ are constructed similarly using $S_r^{(n)}(r,s)$, $S_\theta^{(n)}(r,s)$. For arbitrary load function $f(\tau)$ the solution will be based on the formula (16.24) with a convolution.

Thus, the dimensionless initial input values are r_0, r_1, r_2, $\nu_0^{(1)} = \nu_0^{(3)}$, $\nu_0^{(2)}$, $G_0^{(3)}/G_0^{(2)} = G_0^{(1)}/G_0^{(2)}$, $\alpha_2 = c_1^{(3)}/c_1^{(2)}$, $\gamma_v^{(2)}(\tau)$, $\gamma_s^{(2)}(\tau)$, $f(\tau)$. For constructing the original $u^{(n)}(r,\tau)$, the main problem is to find zeros of the function $Z(s)$.

16.6 Numerical Results

In this section, we present the results of calculation of the elements of the spectral set E_s, as well as the results of calculation of the parameters of a non-stationary wave propagation in the cross section of the cylinder with specific initial characteristics:

$$r_0 = 0.4, \quad r_1 = 0.5, \quad r_2 = 0.9, \quad \nu_0^{(1)} = \nu_0^{(3)} = 0.3, \quad \nu_0^{(2)} = 0.2,$$

$$\alpha_2 = 0.25\sqrt{210}, \ \frac{G_0^{(3)}}{G_0^{(2)}} = \frac{G_0^{(1)}}{G_0^{(2)}} = 10, \quad \gamma_v^{(2)}(\tau) \equiv 0, \quad \gamma_s^{(2)}(\tau) = 0.4e^{-\tau} + 0.1e^{-0.5\tau}.$$

The value for the parameter α_2 was chosen as indicated because the densities of all layers were considered the same. The limit points of the spectrum E_s determined from the equation $1 - D_1^{(2)}(s) = 0$ will be: $s_1^* = -0.8265564$, $s_2^* = -0.4234436$ (here and further, up to seven significant digits are indicated). In the process of calculations, we limited ourselves to 8 real elements of the spectrum:

$$x_1 = -0.4960449, \quad x_2 = -0.4591964, \quad x_3 = -0.4269673, \quad x_4 = -0.4244406.$$

$$x_5 = -0.9913209, \quad x_6 = -0.8972754, \quad x_7 = -0.8405825, \quad x_8 = -0.8311154.$$

It is established that in the study of the transient wave process, these 8 real elements are completely sufficient. The complex elements of the spectrum z_j were calculated by means of the Newton iteration method, while the purely imaginary values for the elastic cylinder were chosen as the initial approximation. It is established that in the study of the transient wave process, few dozen complex values z_j are sufficient.

Below are the results for the case when the external load function is $f(\tau) = 1 - e^{-100\tau}, \quad \tau > 0$ (smoothed Heaviside function). Figures 16.2, 16.3, 16.4 and 16.5 show the time dependence of relative stresses:

$$\kappa_r^{(n)} = G_0^{(n)} \sigma_r^{(n)}(r, \tau)/(G_0^{(1)} Q_0) = P_R^{(n)}(R, t)/(2G_0^{(1)} Q_0),$$

$$\kappa_\theta^{(n)} = G_0^{(n)} \sigma_\theta^{(n)}(r, \tau)/(G_0^{(1)} Q_0) = P_\theta^{(n)}(R, t)/(2G_0^{(1)} Q_0)$$

at fixed points r. Negative stresses are compressional (it is believed that $Q_0 > 0$).

In Figs. 16.2, 16.3, 16.4 and 16.5 thick curves 1 show the time dependence of $\kappa_r^{(2)}$ at point $r = 0.9$, $\kappa_\theta^{(2)}$ at point $r = 0.9$, $\kappa_\theta^{(3)}$ at point $r = 0.9$, $\kappa_\theta^{(3)}$ at point $r = r_0 = 0.4$ respectively. Dash-dotted curves 2 in Figs. 16.2, 16.3, 16.4 and 16.5 show similar

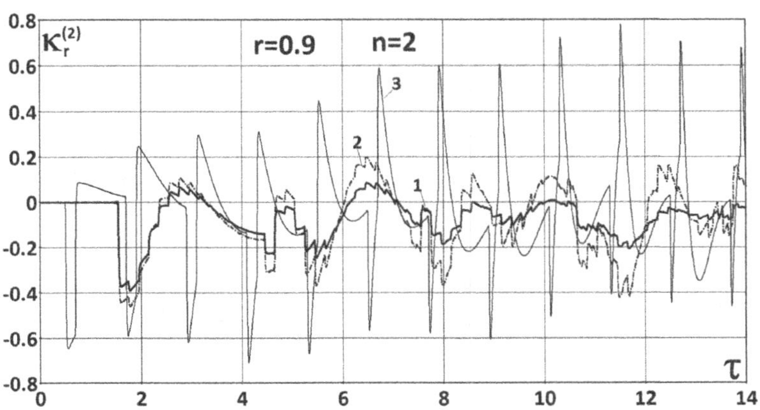

Fig. 16.2 Time dependence of $\kappa_r^{(2)}$ at $r = 0.9$. Curve 1 for three-layer cylinder with viscoelastic layer, 2—for cylinder with three elastic layers, 3—for elastic homogeneous cylinder

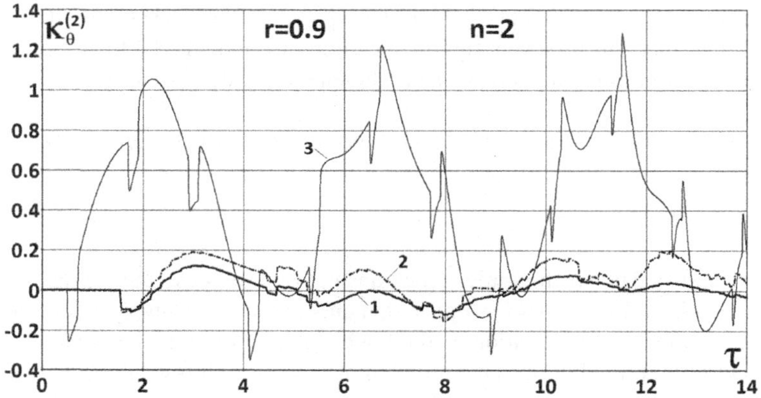

Fig. 16.3 Time dependence of $\kappa_\theta^{(2)}$ at point $r = 0.9$. Curve 1 for three-layer cylinder with viscoelastic layer, 2—for cylinder with elastic layers, 3—for elastic homogeneous cylinder

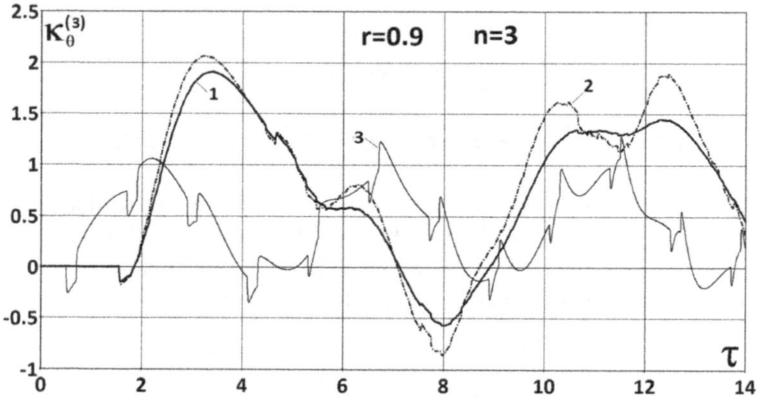

Fig. 16.4 Time dependence of $\kappa_\theta^{(3)}$ at $r = 0.9$. Curve 1 for three-layer cylinder with viscoelastic layer, 2—for cylinder with three elastic layers, 3—for elastic homogeneous cylinder

results for a cylinder consisting of three elastic layers with initial characteristics values:

$$r_0 = 0.4, \quad r_1 = 0.5, \quad r_2 = 0.9, \quad \nu_0^{(1)} = \nu_0^{(3)} = 0.3, \quad \nu_0^{(2)} = 0.2,$$
$$\alpha_2 = 0.25\sqrt{210}, G_0^{(3)}/G_0^{(2)} = G_0^{(1)}/G_0^{(2)} = 10, \quad \gamma_v^{(2)}(\tau) \equiv \gamma_s^{(2)}(\tau) \equiv 0.$$

Thin curves 3 in Figs. 16.2, 16.3, 16.4 and 16.5 illustrate results for the elastic homogeneous cylinder:

$$N = 1, \quad r_0 = 0.4, \quad \nu_0^{(1)} = 0.3, \quad \gamma_v^{(1)}(\tau) \equiv \gamma_s^{(1)}(\tau) \equiv 0.$$

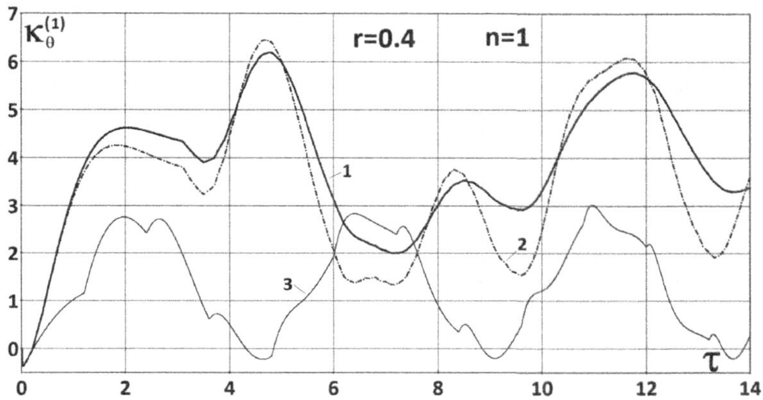

Fig. 16.5 Time dependence of $\kappa_\theta^{(1)}$ at $r = 0.4$. Curve 1 for three-layer cylinder with viscoelastic layer, 2—for cylinder with three elastic layers, 3—for elastic homogeneous cylinder

For the viscoelastic cylinder, the results are obtained with 100 complex and the 8 above-mentioned real elements of the spectrum.

It can be seen from the figures that the presence of a softer viscoelastic inclusion significantly reduces the amplitude of the relative radial stress at the boundary between the layers. Herewith, the amplitude of the relative circumferential stress at the same boundary increases significantly in a harder layer and decreases in a softer one compared to the relative circumferential stress at the corresponding point of a homogeneous cylinder. The validity of the proposed approach to spectrum search and the reliability of the calculation results is confirmed by the accuracy of meeting the condition of the absence of disturbances at the selected point until the arrival of the first front from the loaded boundary.

16.7 Conclusion

It is established that the connection between the solution of the non-stationary dynamic problem of linear viscoelasticity for a piecewise homogeneous body and the solution of the spectral problem of free oscillations of such a body in the case of a finite perturbation propagation domain. Using the integral Laplace transform in time with subsequent reversal the process of constructing a non-stationary solution is reduced to the search for the eigenvalues of the spectral problem.

The properties of the set of eigenvalues of the spectral problem with relaxation kernels in the form of a finite sum of exponentials are discussed. We formulate sufficient conditions, which relaxation kernels must satisfy so that all finite limit points of the spectrum are real. When these conditions are met, it is proposed to search for the elements of the spectrum near its final limit points on the real axis. It is proposed to search for the complex elements of the spectrum by an iterative numerical method. The validity of this approach is confirmed by constructing the solution to the problem of transient longitudinal waves' propagation in a cross-section of a hollow infinitely long three-layered cylinder. The derived solution is valid for the entire time interval and is convenient for numerical implementation. This solution allowed us to investigate the non-stationary waves' propagation in such a cylinder with specific initial parameters.

Acknowledgements The reported study was performed within the bilateral project funded by the Russian Foundation for Basic Research (project number 20-58-18002) and by the Bulgarian National Science Fund (project number KP-06-Russia/5 from 11.12.2020).

References

1. Kozlov VI, Kucher NK (1980) Dynamic behavior of multilayer cylindrical structures with transient loads. Strength Mater 12(5):639–648. (in Russian)
2. Lokshin AA (1994) The Head wave at the boundary of two hereditary-elastic half-spaces. The case of a linear source. J Appl Math Mech 58(1):171–176
3. Nuriev BR (1985) Impact on a viscoelastic layered composite. Izv Akad Nauk AzSSR Ser Fiz-Tekh Mat Nauk 4:35–41. (in Russian)
4. Sabodash PF (1971) Propagation of longitudinal viscoelastic waves in a three-layer medium. Polymer Mech 7(1):124–128. (in Russian)
5. Hashemi Sh, Hosseini, Khaniki HB (2017) Dynamic behavior of multi-layered viscoelastic nanobeam system embedded in a viscoelastic medium with a moving nanoparticle. J Mech 33(5):559–575. https://doi.org/10.1017/jmech.2016.91
6. Lee HS (2014) Viscowave—a new solution for viscoelastic wave propagation of layered structures subjected to an impact load. Int J Pavement Eng 15(6):542-557. https://doi.org/10.1080/10298436.2013.782401
7. Bazhenov VG, Belov AA, Igumnov LA (2009) Boundary-element modeling of the dynamics of piecewise homogeneous media and structures. N. Novgorod. (in Russian)
8. Bazhenov VG, Igumnov LA (2008) Methods of boundary integral equations and boundary elements for solving three-dimensional problems of the dynamic theory of elasticity with conjugated fields. Fizmatlit, Moscow. (in Russian)
9. Belov AA (2008) Boundary-element analysis of dynamics of compound viscoelastic bodies. Probl Strength Plast. N. Novgorod, NNSU. (70):162–168. (in Russian)
10. Petrov AN, Ermolaev MD (2011) Boundary element analysis of the dynamics of composite viscoelastic bodies. Mechanics of deformed solid. Vestnik Nizhegorodskogo universiteta N.I. Lobachevskii 4(4):1694–1696. (in Russian)
11. Shamaev AS, Shumilova VV (2020) Asymptotics of the spectra of one-dimensional natural vibrations in media consisting of solid and fluid layers. Dokl Phys 65(4):153–156. https://doi.org/10.1134/S1028335820040084
12. Zhu JB, Zhao XB, Wu W, Zhao J (2012) Wave propagation across rock joints filled with viscoelastic medium using modified recursive method. J Appl Geophys 86:82–87. https://doi.org/10.1016/j.jappgeo.2012.07.012
13. Khudoynazarov K, Yalgashev BF, Mavlonov T (2021) Mathematical modelling of torsional vibrations of the three-layer cylindrical viscoelastic shell. IOP Conf Ser Mater Sci Eng. https://doi.org/10.1088/1757-899X/1030/1/012098.
14. Pshenichnov SG (2016) Dynamic linear viscoelasticity problems for piecewise homogeneous bodies. Mech Solids 51(1):65–74. https://doi.org/10.3103/S0025654416010076

15. Korovaytseva EA, Pshenichnov SG (2021) Non-stationary dynamic problem for layered viscoelastic cylinder. Adv Struct Materials Multiscale Solid Mech. Springer Nature Switzerland AG 141:261–274. https://doi.org/10.1007/978-3-030-54928-2_20

Chapter 17
Nanoindentation Derived Mechanical Properties of TiN Thin Film Deposited Using Magnetron Sputtering Method

Evgeniy V. Sadyrin, Andrey L. Nikolaev, Regina A. Bardakova, Anzhelika A. Kotova, Ivan O. Kharchevnikov, Igor Yu. Zabiyaka, and Sergei M. Aizikovich

Abstract A study of the mechanical properties of a thin titanium nitride (TiN) film deposited using the magnetron sputtering on a silicon surface has been carried out. The values of indentation hardness and reduced Young's modulus were evaluated using nanoindentation in a course of a series of experiments with increasing indentation load. The microgeometrical characteristics of the film surface (average roughness, maximum roughness height) were determined from the results of atomic force microscopy. The chemical composition and thickness of the film were estimated using scanning electron microscopy.

Keywords Thin film · Nanoindentation · Magnetron sputtering · Mechanical properties · Atomic force microscopy · Scanning electron microscopy

---

E. V. Sadyrin (✉) · A. L. Nikolaev · R. A. Bardakova · A. A. Kotova · I. O. Kharchevnikov · I. Yu. Zabiyaka · S. M. Aizikovich
Don State Technical University, 1 Gagarin Square, 344000 Rostov-on-Don, Russia
e-mail: e.sadyrin@sci.donstu.ru

A. L. Nikolaev
e-mail: andreynicolaev@eurosites.ru

R. A. Bardakova
e-mail: reginabardakova99@gmail.com

A. A. Kotova
e-mail: anzhelickakotowa@yandex.ru

I. O. Kharchevnikov
e-mail: babaduksac@yandex.ru

I. Yu. Zabiyaka
e-mail: zabiyakaigor@gmail.com

S. M. Aizikovich
e-mail: saizikovich@gmail.com

H. Altenbach et al. (eds.), *Deformation and Destruction of Materials and Structures Under Quasi-static and Impulse Loading*, Advanced Structured Materials 186,
https://doi.org/10.1007/978-3-031-22093-7_17

17.1 Introduction

Nanoindentation is a set of techniques that use local precision penetration of a specialized tool—an indenter—into the material with depth detection at nanometer resolution [1, 2]. When working with nanoindentation units—devices that perform nanoindentation—various forms of indenters are used. Thus, spherical [3, 4], conical [5, 6], flat [7, 8] and pyramidal indenters [9–11] are often used in studies of biological materials. Among the latter, Berkovich indenters, which are a trihedral pyramid, are especially popular, especially in the study of thin coatings and films [12–16]. A tetrahedral indenter in the form of a cube corner is used for especially delicate samples, since it can be relatively easily damaged during the experiment [17, 18]. Figure 17.1 shows the general scheme of the experiment.

To select the correct nanoindentation parameters, in practice, one often resorts to assessing the microgeometrical characteristics of coatings using atomic force microscopy (AFM) [19–21] and scanning electron microscopy (SEM) [22–24]. The latter is also used to study the chemical composition of the coating using energy-dispersive X-ray spectroscopy to assess the quality of deposition [25–28].

In the present work, nanoindentation is used to evaluate the properties of a thin TiN film, which is widely used in the manufacture of wear-resistant dental implants [29–31], metal-cutting tools [32], as well as components for microelectronics [33, 34], and in other applications. Films made from TiN demonstrate high hardness, electrical conductivity, chemical inertness and adhesion to substrates. This kind of coatings can significantly improve the mechanical properties of product surfaces [35–37], which leads to a significant increase in the service life of the product, the possibility of their use in experimental facilities and under extreme conditions. The evaluation of the mechanical properties of the coating was accompanied by a study of its microgeometrical characteristics and chemical composition using AFM and SEM.

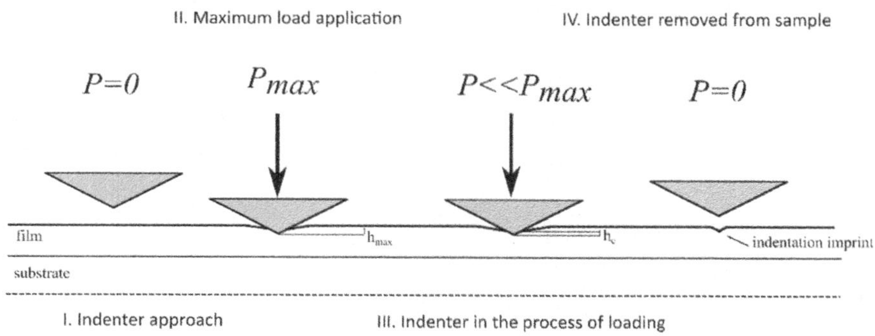

Fig. 17.1 Indentation process of a coating, where *P*—indentation load at the current time *t*; P_{max}—maximum indentation load; h_{max}—maximum indenter depth; h_c—indenter depth without the surface displacement

17.2 Materials and Methods

To create the coating under consideration, we used the magnetron sputtering of Ti targets onto a silicon substrate in nitrogen atmosphere using a VSM100 unit (Aktan Vacuum, Russia). The deposition took place in the mode with a stable power of 600 W, deposition type DC. In the additionally deposition time was 40 min and temperature was 200 °C. The nitrogen pressure of the magnetron sputtering process was set to $7 \cdot 10^{-2}$ mbar.

To study the mechanical characteristics of the resulting coating, a Nanotest 600 Platform 3 nanoindentation unit (Micro Materials, UK) with a Berkovich diamond indenter was used. The maximum indentation force, increasing from experiment to experiment, was set: from 1.5 to 29 mN. The indentation time into the TiN thin film was set to 30 s, the holding period of indenter at maximum load was 30 s, and then the indenter was unloaded for 30 s. The areas for indentation were chosen in such a way as to avoid contact of the indenter with dust particles using an optical microscope device, positional synchronized with the indenter of the measuring-force cell (maximum magnification 400×). It is important that the measurement results are not affected by the presence of sags and depressions in the contact area of the sample boundaries caused by previous indentations in the series. For this purpose, the indentation sites were positioned in such a way as to be at least three times their diameter from the sample boundaries, and the minimum distance between the indentations was at least five times the largest indentation diameter [38]. Before carrying out the experiments, all the necessary calibrations of the nanoindentation unit (load, indenter area function, machine compliance) were performed on a standard sample of fused quartz, followed by testing on a standard sample of sapphire. Before each test, the pendulum was balanced additionally and the depth signal was tested.

The microgeometrical characteristics of the film surface were studied using an AFM Nanoeducator (NT-MDT, Russia) equipped with a tungsten probe in the semi-contact mode. The probe movement speed was 7 μm/s, the estimated scanning time was 40 min. The film thickness and its chemical composition (utilizing energy-dispersive X-ray spectroscopy) were determined using a Crossbeam 340 SEM (Carl Zeiss, Germany). For the present study, an Everhart–Thornley secondary electron emission detector was used at an accelerating voltage of 2 kV.

17.3 Results and Discussion

Figure 17.2 shows the surface topography of a thin TiN film obtained using AFM.

Due to the irregularity of the surface structure, the microgeometrical parameters were determined using 12 profiles (four each in the horizontal, vertical, and diagonal directions). Thus, the average values of the average roughness Ra and the maximum roughness height Rt were 17 ± 2 and 127 ± 35 nm, respectively. Deposition artifacts, presumably dust microparticles, were found on the surface of the sample.

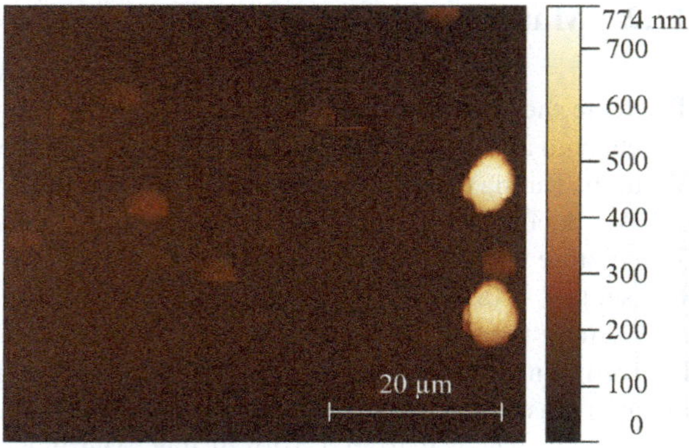

Fig. 17.2 TiN film surface topography obtained by AFM

The indentation occurred under the action of loads in the range from 1.5 to 50 mN. Thus, the maximum indentation depth varied from 51.31 ± 3.93 to 460.29 ± 22.4 nm. In this regard, it is necessary to accurately know the roughness values of the film surface in order to adjust the mathematical apparatus for processing the results of indentation.

By studying the cross section obtained using a focused ion beam on SEM, the thickness of the studied TiN coating was 734.7 nm (Fig. 17.3). Table 17.1 presents the stoichiometric ratios of a thin film of TiN for the three studied spectra. The results testify to the high quality of the deposition. In all spectra, iron is represented to a small extent (0.91–0.95), which presumably got into the composition of the film due to the non-ideal purity of the chemical composition of the Ti target.

Nanoindentation made it possible to obtain the sets of diagrams "indentation load–indentation depth". Figure 17.4 shows these diagrams for the maximum loads of 17 and 26 mN.

Starting from the indentation depth of about 120 nm, a characteristic "step" appears on the loading branches of the diagrams, indicating plastic processes in the coating. This phenomenon is referred as "pop-in" in literature [39]. With increasing strength, the number of such steps increases. Figure 17.5 for the "coating-substrate" package shows the graphs of the effective dependences of the reduced Young's modulus E_r and indentation hardness H on the maximum indentation depth h_{max}. The mechanical characteristics were calculated using the Oliver–Pharr method [40, 41].

Table 17.1 Chemical composition of TiN film

Specter number	N	Ti	Fe	Sum (at%)
1	50.84	48.26	0.91	100.00
2	50.72	48.37	0.91	100.00
3	50.69	48.36	0.95	100.00

Fig. 17.3 Evaluation of TiN thin film thickness after ion etching on SEM

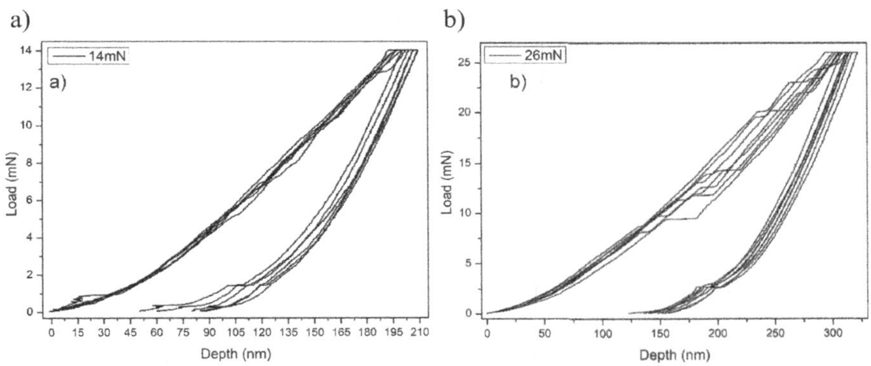

Fig. 17.4 Diagrams "indentation load–indentation depth" for the maximum loads P_{max}: **a** 14 mN; **b** 26 mN

The plots of mechanical properties reveal a rather complex behavior. At small indentation depths (up to 180 nm), underestimated values of mechanical properties are observed due to the influence of surface microgeometry on the process of indenter penetration (e.g. indenter tends to flatten out elements of the micro-roughness instead of penetrating the continuous medium of the coating). Peak values of mechanical properties (256.38 $\pm$ 12.79 GPa reduced Young's modulus, 21.3 $\pm$ 1.64 GPa hardness) were found at depths of h_{max} = 209.26 $\pm$ 6.49 nm, presumably these values are close to the real values of the TiN coating. Taking Poisson's ration of TiN for υ

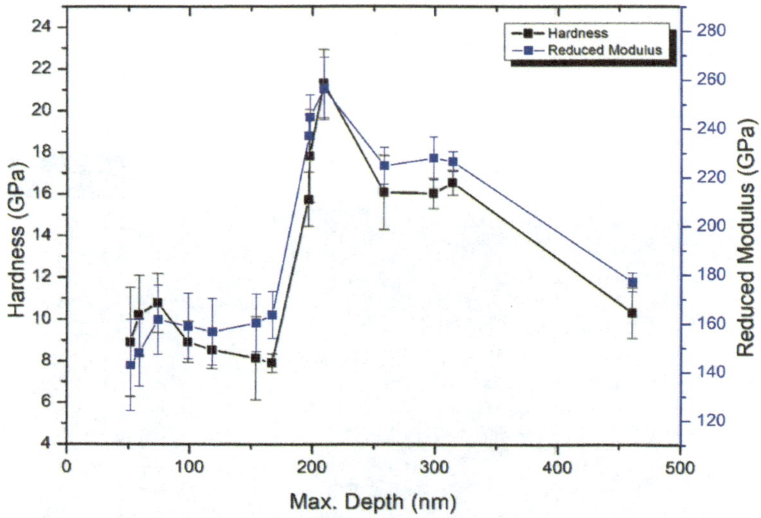

Fig. 17.5 Plots of effective values of reduced Young's modulus (blue) and indentation hardness (black) versus maximum indentation depth with standard deviations

= 0.23 [42] the presumable value of the Young's modulus of the film can be restored $E = 312.75 \pm 12.25$ GPa, which is close to some results reported in the literature previously [42] however, lower than in some other studies [43, 44]. A further drop in the mechanical property values is associated with an increase in plastic processes in the coating, stabilizing at depths of $250 < h_{max} < 330$ nm. At greater depths, a further decrease in the studied characteristics is observed, associated with an increase in the contribution of the substrate to the results. Taking into account the small thickness of the coating and the relatively high value of its surface roughness, for this sample it is not possible to unambiguously determine the value of the real Young's modulus of the film according to the international standard [45] by experimental means, obtaining only sets of effective values for the "film-substrate" package. In the future, for such cases, it is proposed to use mathematical modeling of the indentation process, which allows either analytical [46–51] or finite element tools [52–55] to calculate the value of the desired quantity from experimental data. Such methods have shown high efficiency, in particular, in the study of ZrN coatings [56, 57] on various substrates, two-layered NiO–Ni coatings [58].

17.4 Conclusion

In the present work, using nanoindentation we studied the behavior of the effective reduced Young's modulus and the indentation hardness of TiN thin film deposited by magnetron sputtering of particles on a silicon substrate. AFM, SEM, and energy-dispersive X-ray spectroscopy were used to select the experimental parameters and evaluate the deposition quality. The complex nature of the behavior of mechanical properties with increasing indentation depth, as well as a number of features of the "indentation force–indentation depth" diagrams for the "coating-substrate" package, were found. The hypothetical values of the real reduced Young's modulus and the indentation hardness of the coating were determined, however, to refine these values, it is proposed to use a mathematical model.

Acknowledgements This work was supported by the Russian Science Foundation grant 22-49-08014. The research was carried out at the Nanocenter of the Research and Education Center "Materials" of Don State Technical University (https://nano.donstu.ru/).

References

1. Golovin YI (2008) Nanoindentation and mechanical properties of solids in submicrovolumes, thin near-surface layers, and films: a review. Phys Solid State 50(12):2205–2236
2. Weppelmann E, Swain MV (1996) Investigation of the stresses and stress intensity factors responsible for fracture of thin protective films during ultra-micro indentation tests with spherical indenters. Thin Solid Films 286(1–2):111–121
3. Kontomaris SV (2018) The Hertz model in AFM nanoindentation experiments: applications in biological samples and biomaterials. Micro Nanosyst 10(1):11–22
4. Qian L, Zhao H (2018) Nanoindentation of soft biological materials. Micromachines 9(12):654. https://doi.org/10.3390/mi912065
5. Argatov I, Mishuris G (2018) Indentation testing of biological materials, In: Advanced structured materials, vol 91. Springer, Cham
6. Schäfer E, Vache M, Kliesch TT, Janshoff A (2015) Mechanical response of adherent giant liposomes to indentation with a conical AFM-tip. Soft Matter 11(22):4487–4495
7. Lai Y, Hu Y (2017) Unified solution for poroelastic oscillation indentation on gels for spherical, conical and cylindrical indenters. Soft Matter 13(4):852–861
8. Delaine-Smith RM, Burney S, Balkwill FR, Knight MM (2016) Experimental validation of a flat punch indentation methodology calibrated against unconfined compression tests for determination of soft tissue biomechanics. J Mech Behav Biomed Mater 60:401–415
9. Sadyrin EV (2022) Correlating the mechanical properties to the mineral density of brown spot lesion in dentine using nanoindentation and X-ray micro-tomography. In: Altenbach H, Eremeyev VA, Galybin A, Vasiliev A (eds) Advanced materials modelling for mechanical, medical and biological applications. Advanced structured materials, vol 155. Springer, Cham, pp 389–398
10. Sadyrin E, Swain M, Mitrin B, Rzhepakovsky I, Nikolaev A, Irkha V, Yogina D, Lyanguzov N, Maksyukov S, Aizikovich S (2020) Characterization of enamel and dentine about a white spot lesion: mechanical properties, mineral density, microstructure and molecular composition. Nanomaterials 10(9):1889
11. Kislyakov EA, Karotkiyan RV, Sadyrin EV, Mitrin BI, Yogina DV, Kheygetyan AV, Maksyukov SY (2020) Nanoindentation derived mechanical properties of human enamel and dentine subjected to etching with different concentrations of citric acid. In: Aizikovich SM, Altenbach H, Eremeyev V, Swain MV, Galybin A (eds) Modeling, synthesis and fracture of advanced materials for industrial and medical applications. Springer, Cham, pp 75–83
12. Nikolaev AL, Mitrin BI, Sadyrin EV, Zelentsov VB, Aguiar AR, Aizikovich SM (2020) Mechanical properties of MICROPOSIT S1813 thin layers. In: Modeling, synthesis and fracture

of advanced materials for industrial and medical applications. Advanced structured materials, vol 136. Springer, Cham, pp 137–146
13. Fu K, Chang L, Zheng B, Tang Y, Yin Y (2015) Analysis on cracking in hard thin films on a soft substrate under Berkovich indentation. Vacuum 112:29–32
14. Sadyrin EV, Nikolaev AL, Vasiliev AS, Golushko IY (2021, March) Nanoindentation derived mechanical properties of Au thin film deposited by pulsed laser sputtering on Si substrate. In: International conference on physics and mechanics of new materials and their applications. Springer, Cham, pp 215–222
15. Gilewicz A, Kuznetsova T, Aizikovich S, Lapitskaya V, Khabarava A, Nikolaev A, Warcholinski B (2021) Comparative investigations of AlCrN coatings formed by cathodic arc evaporation under different nitrogen pressure or arc current. Materials 14(2):304. https://doi.org/10.3390/ma14020304
16. Kuprin AS, Gilewicz A, Kuznetsova TA, Lapitskaya VA, Tolmachova GN, Warcholinski B, Aizikovich SM, Sadyrin EV (2021) Structure and properties of ZrON coatings synthesized by cathodic arc evaporation. Materials 14(6):1483. https://doi.org/10.3390/ma14061483
17. Ghosh A, Jin S, Arreguin-Zavala J, Brochu M (2017) Characterization and investigation of size effect in nano-impact indentations performed using cube-corner indenter tip. J Mater Res 32(12):2241–2248
18. Chudoba T, Schwaller P, Rabe R, Breguet JM, Michler J (2006) Comparison of nanoindentation results obtained with Berkovich and cube-corner indenters. Phil Mag 86(33–35):5265–5283
19. Nazemian M, Chamani M, Baghani M (2019) A combined experimental and numerical study of the effect of surface roughness on nanoindentation. Int J Appl Mech 11(07):1950070
20. Costa DO, Allo BA, Klassen R, Hutter JL, Dixon SJ, Rizkalla AS (2012) Control of surface topography in biomimetic calcium phosphate coatings. Langmuir 28(8):3871–3880
21. Melnikova G, Kuznetsova T, Lapitskaya V, Petrovskaya A, Chizhik S, Zykova A, Safonov V, Aizikovich S, Sadyrin E, Sun W, Yakovin S (2021) Nanomechanical and nanotribological properties of nanostructured coatings of tantalum and its compounds on steel substrates. Nanomaterials 11(9):2407. https://doi.org/10.3390/nano11092407
22. Wang C, Farhat Z, Jarjoura G, Hassan MK, Abdullah AM (2017) Indentation and erosion behavior of electroless Ni–P coating on pipeline steel. Wear 376:1630–1639
23. Murray JW, Ahmed N, Yuzawa T, Nakagawa T, Sarugaku S, Saito D, Clare AT (2020) Dry-sliding wear and hardness of thick electrical discharge coatings and laser clads. Tribol Int 150:106392
24. Burlakova VE, Tyurin AI, Drogan EG, Sadyrin EV, Pirozhkova TS, Novikova AA, Belikova MA (2019) Mechanical properties and size effects of self-organized film. J Tribol 141(5):051601
25. Puttichaem C, Souza GP, Ruthe KC, Chainok K (2021) Characterization of ultra-thin diamond-like carbon films by SEM/EDX. Coatings 11(6):729. https://doi.org/10.3390/coatings11060729
26. Kumar A, Das AK (2021) Evolution of microstructure and mechanical properties of Co–SiC tungsten inert gas cladded coating on 304 stainless steel. Eng Sci Technol, Int J 24(3):591–604
27. Akhtar MA, Mariotti CE, Conti B, Boccaccini AR (2021) Electrophoretic deposition of ferulic acid loaded bioactive glass/chitosan as antibacterial and bioactive composite coatings. Surf Coat Technol 405:126657
28. Scarano A, Piattelli M, Vrespa G, Caputi S, Piattelli A (2003) Bacterial adhesion on titanium nitride-coated and uncoated implants: an in vivo human study. J Oral Implant 29(2):80–85
29. Pushkariov VI, Nikolaev AL, Kaidashev EM (2014, October) Synthesis and characterization of ZnO nanorods obtained by catalyst-free thermal technique. J Phys: Conf Ser 541(1):012031. IOP Publishing
30. Raimondi MT, Pietrabissa R (2000) The in-vivo wear performance of prosthetic femoral heads with titanium nitride coating. Biomaterials 21(9):907–913
31. Ferrari M, Tricarico MG, Cagidiaco MC, Vichi A, Gherlone EF, Zarone F, Sorrentino R (2016) 3-year randomized controlled prospective clinical trial on different CAD-CAM implant abutments. Clin Implant Dent Relat Res 18(6):1134–1141

32. Brito RF, De Carvalho SR, Ferreira JR (2009) Thermal analysis in coated cutting tools. Int Commun Heat Mass Transfer 36(4):314–321
33. Ren Y, Gao X, Zhang C, Liu X, Sun S (2018) The electronic and elastic properties of Si atom doping in TiN: a first-principles calculation. Coatings 1–10
34. Chou W-J, Yu G-P, Huang J-H (2002) Mechanical properties of TiN thin film coatings on 304 stainless steel substrates. Surf Coat Technol 149:7–13
35. Sadyrin E, Karotkiyan R, Sushentsov N, Stepanov S, Zabiyaka I, Kislyakov E, Litvinenko A (2020) Mechanical properties derived by spherical indentation of TiN coating deposited by a method combining magnetron sputtering and arc evaporation. In: Advanced materials. Springer, Cham, pp 85–95
36. Mayrhofer PH, Kunc F, Musil J, Mitterer C (2002) A comparative study on reactive and non-reactive unbalanced magnetron sputter deposition of TiN coatings. Thin Solid Films 415(1–2):151–159
37. Lang F, Yu Z (2001) The corrosion resistance and wear resistance of thick TiN coatings deposited by arc ion plating. Surf Coat Technol 145(1–3):80–87
38. Bull SJ, Rickerby DS (1990) New developments in the modelling of the hardness and scratch adhesion of thin films. Surf Coat Technol 42(2):149–164
39. Field JS, Swain MV, Dukino RD (2003) Determination of fracture toughness from the extra penetration produced by indentation-induced pop-in. J Mater Res 18(6):1412–1419
40. Oliver WC, Pharr GM (1992) An improved technique for determining hardness and elastic modulus using load and displacement sensing indentation experiments. J Mater Res 7(6):1564–1583
41. Pharr GM, Oliver WC (1992) Measurement of thin film mechanical properties using nanoindentation. MRS Bull 17(7):28–33
42. Maxwell AS, Owen-Jones S, Jennett NM (2004) Measurement of Young's modulus and Poisson's ratio of thin coatings using impact excitation and depth-sensing indentation. Rev Sci Instrum 75(4):970–975
43. Tran MD, Poublan J, Dautzenberg JH (1997) A practical method for the determination of the Young's modulus and residual stresses of PVD thin films. Thin Solid Films 308:310–314
44. Sue JA (1992) X-ray elastic constants and residual stress of textured titanium nitride coating. Surf Coat Technol 54:154–159
45. ISO 14577-1:2002. Metallic materials—instrumented indentation test for hardness and materials parameters—part 1: test method
46. Argatov II (2011) Depth-sensing indentation of a transversely isotropic elastic layer: second-order asymptotic models for canonical indenters. Int J Solids Struct 48(25–26):3444–3452
47. Argatov II, Sabina FJ (2016) Small-scale indentation of an elastic coated half-space: the effect of compliant substrate. Int J Eng Sci 104:87–96
48. Bagault C, Nelias D, Baietto MC, Ovaert TC (2013) Contact analyses for anisotropic half-space coated with an anisotropic layer: effect of the anisotropy on the pressure distribution and contact area. Int J Solids Struct 50(5):743–754
49. Borodich FM (2014) The Hertz-type and adhesive contact problems for depth-sensing indentation. In: Advances in applied mechanics, vol 47. Elsevier, pp 225–366
50. Krenev LI, Volkov SS, Sadyrin EV, Zubar TI, Chizhik SA (2018) Mechanical material tests by the nanoindentation method at various indenter and specimen temperatures. J Eng Phys Thermophys 91(3):594–600
51. Vasiliev AS, Volkov SS, Sadyrin EV, Aizikovich SM (2020) Simplified analytical solution of the contact problem on indentation of a coated half-space by a conical punch. Mathematics 8(6):983
52. Duan P, Xia Y, Bull S, Chen J (2018) Finite element modeling of nanoindentation response of elastic fiber-matrix composites. J Mater Res 33(17):2494–2503
53. Kang JJ, Becker AA, Sun W (2012) Determining elastic–plastic properties from indentation data obtained from finite element simulations and experimental results. Int J Mech Sci 62(1):34–46

54. Gan L, Ben-Nissan B, Ben-David A (1996) Modelling and finite element analysis of ultra-microhardness indentation of thin films. Thin Solid Films 290:362–366
55. Zhang W, Subhash G (2001) An elastic–plastic-cracking model for finite element analysis of indentation cracking in brittle materials. Int J Solids Struct 38(34–35):5893–5913
56. Vasiliev AS, Sadyrin EV, Mitrin BI, Aizikovich SM, Nikolaev AL (2018) Nanoindentation of ZrN coatings on silicon and copper substrates. Russ Eng Res 38(9):735–737
57. Sadyrin E, Vasiliev A, Volkov S (2020) Mathematical modeling of experiment on Berkovich nanoindentation of ZrN coating on steel substrate. Acta Polytech CTU Proc 27:18–21
58. Sadyrin EV, Mitrin BI, Krenev LI, Nikolaev AL, Aizikovich SM (2017, October) Evaluation of mechanical properties of the two-layer coating using nanoindentation and mathematical modeling. In: International conference on physics and mechanics of new materials and their applications. Springer, Cham, pp 495–502

The manufacturer's authorised representative in the EU is Springer Nature Customer Service Centre GmbH, Europaplatz 3, 69115 Heidelberg, Germany. If you have any concerns regarding our products, please contact ProductSafety@springernature.com

Printed and bound by CPI Group (UK) Ltd, Croydon, CR0 4YY
28/09/2026
02224075-0007